Design and Mounting of Prisms and Small Mirrors in Optical Instruments

Tutorial Texts Series

- *Design and Mounting of Prisms and Small Mirrors in Optical Instruments*, Paul R. Yoder, Jr., Vol. TT32
- *Basic Electro-Optics for Electrical Engineers*, Glenn D. Boreman, Vol. TT31
- *Optical Engineering Fundamentals*, Bruce H. Walker, Vol. TT30
- *Introduction to Radiometry*, William L. Wolfe, Vol. TT29
- *An Introduction to Interpretation of Graphic Images*, Sergey Ablameyko, Vol. TT27
- *Thermal Infrared Characterization of Ground Targets and Backgrounds*, Pieter A. Jacobs, Vol. TT26
- *Introduction to Imaging Spectrometers*, William L. Wolfe, Vol. TT25
- *Introduction to Infrared System Design*, William L. Wolfe, Vol. TT24
- *Introduction to Computer-based Imaging Systems*, Divyendu Sinha and Edward R. Dougherty, Vol. TT23
- *Optical Communication Receiver Design*, Stephen B. Alexander, Vol. TT22
- *Mounting Lenses in Optical Instruments*, Paul R. Yoder, Jr., Vol. TT21
- *Optical Design Fundamentals for Infrared Systems*, Max J. Riedl, Vol. TT20
- *An Introduction to Real-Time Imaging*, Edward R. Dougherty and Phillip A. Laplante, Vol. TT19
- *Introduction to Wavefront Sensors*, Joseph M. Geary, Vol. TT18
- *Integration of Lasers and Fiber Optics into Robotic Systems,* Janusz A. Marszalec and Elzbieta A. Marszalec, Vol. TT17
- *An Introduction to Nonlinear Image Processing,* Edward R. Dougherty and Jaakko Astola, Vol. TT16
- *Introduction to Optical Testing,* Joseph M. Geary, Vol. TT15
- *Sensor and Data Fusion Concepts and Applications*, Lawrence A. Klein, Vol. TT14
- *Practical Applications of Infrared Thermal Sensing and Imaging Equipment,* Herbert Kaplan, Vol. TT13
- *Image Formation in Low-Voltage Scanning Electron Microscopy*, L. Reimer, Vol. TT12
- *Diazonaphthoquinone-based Resists*, Ralph Dammel, Vol. TT11
- *Infrared Window and Dome Materials*, Daniel C. Harris, Vol. TT10
- *An Introduction to Morphological Image Processing*, Edward R. Dougherty, Vol. TT9
- *An Introduction to Optics in Computers*, Henri H. Arsenault and Yunlong Sheng, Vol. TT8
- *Digital Image Compression Techniques*, Majid Rabbani and Paul W. Jones, Vol. TT7
- *Aberration Theory Made Simple*, Virendra N. Mahajan, Vol. TT6
- *Single-Frequency Semiconductor Lasers*, Jens Buus, Vol. TT5
- *An Introduction to Biological and Artificial Neural Networks for Pattern Recognition*, Steven K. Rogers and Matthew Kabrisky, Vol. TT4
- *Laser Beam Propagation in the Atmosphere*, Hugo Weichel, Vol. TT3
- *Infrared Fiber Optics*, Paul Klocek and George H. Sigel, Jr., Vol. TT2
- *Spectrally Selective Surfaces for Heating and Cooling Applications*, C. G. Granqvist, Vol. TT1

Design and Mounting of Prisms and Small Mirrors in Optical Instruments

Paul R. Yoder, Jr.
Consultant in Optical Engineering

Tutorial Texts in Optical Engineering
Volume TT32

Donald C. O'Shea, Series Editor
Georgia Institute of Technology

SPIE Optical Engineering Press
A Publication of SPIE—The International Society for Optical Engineering
Bellingham, Washington USA

Library of Congress Cataloging-in-Publication Data

Yoder, Paul R.
Design and mounting of prisms and small mirrors in optical instruments / Paul R. Yoder, Jr.
p. cm. – (Tutorial texts in optical engineering; v. TT32)
Includes bibliographical references and index.
ISBN 0-8194-2940-6 (softcover)
1. Catadioptric systems—Design and construction. 2. Lens mounts. 3. Mirrors.
4. Prisms. I. Title. II. Series.
TS517.5.C38Y63 1998
681'.42—dc21 98-19947
CIP

Published by

SPIE—The International Society for Optical Engineering
P.O. Box 10
Bellingham, Washington 98227-0010
Phone: 360/676-3290
Fax: 360/647-1445
Email: spie@spie.org
WWW: http://www.spie.org/

Printed in the United States of America.

Dedication

To Betty, who sat alone all those evenings
and weekends while I wrote my books.

SERIES INTRODUCTION

The Tutorial Texts series was begun in response to requests for copies of SPIE short course notes by those who were not able to attend a course. By policy the notes are the property of the instructors and are not available for sale. Since short course notes are intended only to guide the discussion, supplement the presentation, and relieve the lecturer of generating complicated graphics on the spot, they cannot substitute for a text. As one who has evaluated many sets of course notes for possible use in this series, I have found that material unsupported by the lecture is not very useful. The notes provide more frustration than illumination.

What the Tutorial Texts series does is to fill in the gaps, establish the continuity, and clarify the arguments that can only be glimpsed in the notes. When topics are evaluated for this series, the paramount concern in determining whether to proceed with the project is whether it effectively addresses the basic concepts of the topic. Each manuscript is reviewed at the initial state when the material is in the form of notes and then later at the final draft. Always, the text is evaluated to ensure that it presents sufficient theory to build a basic understanding and then uses this understanding to give the reader a practical working knowledge of the topic. References are included as an essential part of each text for the reader requiring more in-depth study.

One advantage of the Tutorial Texts series is our ability to cover new fields as they are developing. In fields such as sensor fusion, morphological image processing, and digital compression techniques, the textbooks on these topics were limited or unavailable. Since 1989 the Tutorial Texts have provided an introduction to those seeking to understand these and other equally exciting technologies. We have expanded the series beyond topics covered by the short course program to encompass contributions from experts in their field who can write with authority and clarity at an introductory level. The emphasis is always on the tutorial nature of the text. It is my hope that over the next few years there will be as many additional titles with the quality and breadth of the first seven years.

Donald C. O'Shea
Georgia Institute of Technology

CONTENTS

PREFACE

This tutorial text is intended to provide practitioners in the fields of optical engineering and optomechanical design with a comprehensive understanding of several different ways in which prisms and small mirrors typically are designed and mounted in optical instruments, the advantages and disadvantages of these various mounting arrangements, and some analytical tools that can be used to evaluate and compare different designs. The presentation does not include the theoretical background for these tools, but does cite the sources for the equations listed. Each section contains an illustrated discussion of the technology involved and, wherever feasible, one or more worked-out practical examples.

The text is based, in part, on short courses on *Precision Optical Component Mounting Techniques* offered by SPIE–The International Society for Optical Engineering, that I have had the privilege of teaching over a period of years. Techniques for mounting lenses which also are covered in those courses are discussed in another tutorial text *Mounting Lenses in Optical Instruments* published as TT21 in 1995 by the SPIE Press.[1] Because of page limitations, the mirrors considered here have major dimensions no larger than about 24 in. (61 cm). Much of the material included here is applicable to any sized mirrors, but the treatment of those topics in the context of larger sized optics is not comprehensive. In depth tutorial treatment of the design and mounting of larger mirrors would make a useful future contribution by some ambitious author.

The designs discussed here are drawn from the literature, my own experiences in optical instrument design and development, and the works of colleagues. I acknowledge the contributions of others with my deepest thanks and sincerely hope that I have accurately recorded and explained the information given to me. I further acknowledge and thank Robert Ginsberg, Alson Hatheway, and Donald O'Shea who reviewed this book in its preliminary form and offered many valuable suggestions for improvements.

The mounting stress theories discussed in Chapters 5 and 8 have not previously been covered explicitly to this depth in the literature and can only be considered as approximations. These theories would benefit from further investigation and refinement based on more precise computational methods such as finite element analysis. Comments, corrections, and suggestions for improvements in the presentations of these topics or in any other portion of this book would be welcomed.

I wish for the readers of this book a deepening understanding of the technologies discussed and success in the application of the concepts, designs, and analytical techniques presented here.

Paul R. Yoder, Jr.
Norwalk, Connecticut
June 1998

Design and Mounting of Prisms and Small Mirrors in Optical Instruments

CHAPTER 1

INTRODUCTION

This chapter addresses general issues that typically must be considered by the designer or engineer during the evolution of an optical instrument design. Subsequent chapters delve more deeply into specific design issues involving mounting prisms and small mirrors. In many cases, we include illustrative examples and descriptions of actual designs.

Wherever possible, numerical values given in this book are expressed in both the metric or International System of units (SI) and the U.S. Customary System of units (US) (formerly called the English System). Examples taken directly from the literature are expressed only in the system used by the original author. Changes of units from one system to the other can easily be accomplished through use of the conversion factors tabulated in Appendix A.

Effective engineering design of optical instruments requires advance knowledge of the adverse environments under which the product is expected to operate successfully as well as those it must survive. In this chapter, we summarize ways in which temperature, pressure, vibration, shock, humidity, corrosion, contamination, fungus, abrasion, erosion, and high-energy radiation can affect an instrument's performance and/or useful life. We also offer some general suggestions for designing the apparatus to withstand these adverse conditions. More specific design guidelines are included in later sections as parts of descriptions of successful instrument designs.

Because careful selection of materials is vital for maximizing environmental resistance and ensuring the proper operation of the product, we review the attributes of some of the most frequently used optical and mechanical materials. Tables of key optomechanical properties of selected materials may be found in Appendix C.

1.1 Applications of prisms and mirrors

The principal applications for most prisms and flat mirrors, i.e., components that do not contribute optical power and hence cannot form images, are as follows:

- To deviate or bend the system axis
- To displace the system axis laterally
- To fold an optical system into a given shape or package size
- To provide proper image orientation
- To adjust optical path length
- To divide or combine beams by intensity sharing or aperture sharing (at a pupil)
- To divide or combine images at an image plane
- To allow for dynamic scanning of a beam
- To disperse light spectrally (prisms)
- To modify the aberration balance of the optical system (prisms).

Mirrors with convex or concave surfaces do contribute optical power so can enter into the system's image formation process directly.

The number of reflections provided in a system including mirrors and/or prisms s important, especially in visual applications. An odd number produces a "left-handed" (reversed or reverted) image that is not directly readable while an even number of reflections gives a normal, "right-handed" image. The latter is readable even if inverted. See Fig. 1.1. Vector techniques, summarized well by Walles and Hopkins, are particularly powerful in determining how a particular combination of reflecting surfaces will affect location and orientation of images.[2]

(a) (b)

Fig. 1.1 (a) Left- and (b) right-handed images

In this book, we consider various design configurations for prisms and mirrors that accomplish their intended functions as well as typical ways in which these components can be mounted.

1.2 Environmental considerations

It is essential in a serious discussion of any instrument design to identify the environmental conditions under which the end item is expected to perform in accordance with given specifications as well as the extreme conditions that it must survive without permanent damage. The most important conditions to be considered are temperature, pressure, vibration, and shock. These external influences exert static and/or dynamic forces on hardware members that may cause deflections or dimensional changes which, in turn, may result in misalignment, buildup of adverse internal stresses, birefringence, or even component breakage. In some applications, "crash safety," wherein the instrument or portions thereof must not pose a hazard to personnel in an otherwise survivable violent-shock event, is also specified. Other important considerations include humidity, corrosion, contamination, fungus, abrasion, erosion, and high-energy radiation that can affect performance or lead to progressive deterioration of the instrument. It is important for intended users and system engineers to define the expected exposures of the instrument to adverse environments under operating, storage, and transportation conditions as early in the design process as possible so appropriate provisions can be made to minimize their effects on a timely basis.

1.2.1 Temperature

Key temperature effects to be considered include high and low extremes, thermal shock, and gradients. Military equipment is usually designed to withstand extreme temperatures of -62°C (-80°F) to 71°C (160°F) during storage or shipment. It usually must operate adequately at temperatures of -54°C (-65°F) to 52°C (125°F). Commercial equipment generally is designed for smaller ranges of temperatures. Special-purpose equipment such as a space payload may experience temperatures approaching absolute zero, while sensors intended to function in the interior of a furnace may have to operate at temperatures of several hundred degrees Celsius. See Table B1 in Appendix B for typical extremes.

Rapid changes in temperature can occur when an electro-optical sensor on a spacecraft leaves or reenters the atmosphere or when a camera is taken directly from a warm room to a frigid winter location. These "thermal shocks" can significantly affect performance or even cause damage to optics. Thermal diffusivity is a characteristic of all materials that can affect how quickly parts of an optical instrument respond to temperature changes. Many optical materials have low thermal conductivities so heat is not transferred rapidly through them. Slower changes of temperature affect performance mainly by introducing temperature gradients, changing part dimensions, or causing instrument parts to move relative to each other. Materials having inhomogeneous thermal expansion coefficients, such as are sometimes used in mirror substrates, can change dimensions under temperature changes differently at various locations within a component and hence modify critical dimensions or surface figure. Common effects of misalignments are loss of system resolution due to focus errors or asymmetry of images, loss of calibration in measuring devices, and pointing errors. Gradients can degrade the uniformity of refractive index in transmitting materials such as glass and cause similar effects. Focus compensation techniques, such as those summarized in Sec. 4.9 of Ref 1 can help reduce some of these adverse thermal effects. They generally cannot correct problems due to gradients.

1.2.2 Pressure

Pressure manifests itself in optical instrument design chiefly because: the refractive index of air changes with altitude; gas, moisture, or other contaminants "pump" through small leaks in instrument housings; or deflecting (and perhaps damaging) optical components exposed to pressure differentials. See Table B1 in Appendix B for typical extremes. Optical components exposed to aerodynamic or hydrodynamic overpressures in rapidly moving vehicles can also deflect excessively if not properly designed and mounted for these conditions. Everyday changes in barometric pressure can degrade the performance of very high resolution optical devices such as optical lithography projectors used to make microcircuits because the refractive index of the air adjacent to optical surfaces changes sufficiently to cause errors in focus or in magnification. The optics of some large optical instruments, such as sealed submarine periscopes or missile-tracking telescopes, also may suffer from air pressure variations when the temperature changes.

Other pressure-related effects include outgassing of certain materials (plastics, elastomers, lubricants, brazed joints, surface platings, etc.) and permeation of water vapor through seals and seemingly impervious walls. Very careful material selection and control of manufacturing processes help significantly here.

1.2.3 Vibration and shock

Vibration and shock environments both involve application of mechanical forces to the instrument. These forces cause the entire instrument or portions thereof to displace from their normal equilibrium positions. The displaced member tends to return to equilibrium under the action of restoring forces that may include internal elastic forces, as in the case of a mass attached to a spring, or gravitational forces, as in the case of a pendulous mass. If the driving force is periodic, the member may oscillate about the equilibrium position under a condition of sustained vibration; if that force is sudden and temporary, the member is said to be experiencing a shock impulse.

Since any physical structure has characteristic natural frequencies at which its members oscillate mechanically in various vibrational modes, application to that structure of a driving force at or near one of these particular frequencies (or a harmonic thereof) can cause a condition of resonance in which the amplitude of member oscillation increases until limited by internal or external damping. In most designs, resonances should be avoided. Usually, the amplitude, frequency, and direction of the forces externally applied to an instrument cannot be controlled by the designer, so the only corrective actions possible are to make the instrument's structure stiff enough so that its natural frequency is higher than those of the driving forces and/or to introduce means for damping the resultant vibrations. Table B2 in Appendix B lists vibrational power spectral densities and frequency ranges of typical military and aerospace environments.[3]

It is important to recognize that microscopic dimensional changes can occur in an optical component due to applied loading, either self-induced by gravity or by acceleration, vibration, or shock. These changes may be temporary under low loading or permanent under higher loading. The magnitude of either may need to be kept one or two orders of magnitude lower in optical instrumentation than in conventionally engineered mechanical hardware.[4]

Design techniques that can be used to increase resistance of an optical instrument to shock include ensuring adequate support for fragile members (such as lenses, windows, prisms, and mirrors), providing adequate strength of all structural members so as to minimize the risk of distortion beyond their elastic limit, and reducing supported masses.

Key to achievement of a successful design under specified vibration and shock conditions is knowledge of how the instrument will react to the imposed forces. Analytical (i.e., software) tools of ever increasing capability using finite element analysis (FEA) methods are available for modeling the design and predicting its behavior under imposed time-varying loading.[5-7] Some of these tools interface with optical design software so that optical performance degradation under specific adverse conditions can be evaluated directly. The effects of temperature changes, thermal gradients, and pressure changes also can be evaluated with these same analytical tools.

1.2.4 Other environmental conditions

In order to maximize an optical instrument's resistance to humidity, corrosion, and contamination, it is usually important to use compatible materials in the design (i.e., ones that do not naturally form a galvanic couple when in contact with each other), to assemble the instrument in a clean, dry environment, and to seal all paths where leakage could otherwise occur between the instrument's interior and the outside world. Techniques for sealing optical instruments are discussed briefly in Sec. 2.3 of Ref. 1. Once sealed, the interior cavities of the instrument may be purged with dry gas (such as nitrogen or helium) through valves or removable seal screws to remove traces of moisture that could condense on optical surfaces or other sensitive component surfaces at low temperatures. In some cases, the pressure of the residual gas within the instrument will intentionally be somewhat above ambient external pressure to help prevent entry of moisture and other contaminants when the temperature (and hence internal pressure) changes or when the instrument is exposed to driving rain or submerged in water. In special cases, external pressure sources may be attached to the instrument to maintain positive pressure over extended time periods. In other instruments, the optics are sealed to the housing, but a leakage path is intentionally provided so pressure differentials do not build up. In such cases, the leakage path is through a desiccator and a filter that prevents moisture, dust, and other particulate matter from entering.

The potential for damage to optics and coatings by fungus is greatest when the instrument is exposed simultaneously to high humidity and high temperature; conditions that primarily exist in tropical climates. The use of organic materials such as cork and leather, in optical instruments is specifically forbidden by military specifications; this is generally good practice for non-military applications unless the environment will be well controlled. Inorganic materials also may support growth of fungus due to residual contaminants on supposedly clean surfaces. Even glass may support fungus growth under certain conditions. This growth may then stain the glass or coatings and affect transmission and image quality.

Abrasion and erosion problems occur most frequently in devices with optical surfaces exposed to wind-driven sand or other abrasive particles or to raindrops or ice and snow particles moving at high relative velocity. Usually, the former damage occurs on land vehicles or helicopters, while the latter occurs on aircraft traveling at high speed [>200 m/s, (447 mph)]. Softer optical materials, such as infrared-transmitting crystals, are most affected by these conditions.[8] A limited degree of protection can be afforded by thin coatings of harder materials. In the space environment, exposure to micrometeorites and debris may cause damage to optics such as unprotected telescope mirrors. Retractable or disposable covers are used in some cases to provide temporary protection.

Limited protection can be provided for optics exposed to high-energy radiation in the form of gamma and x-rays, neutrons, protons, and electrons by shielding with materials that absorb these radiation types and/or by using optical materials, such as fused silica, that are relatively insensitive to such radiation or by using radiation-protected optical glasses. The latter suffer from slight reduction in blue-end visible light transmission before exposure to radiation but maintain their transmission characteristics over a broad spectral range after such exposure much better than unprotected glasses. Several types of radiation-protected glass incorporating cerium oxide are available from such suppliers as Schott Glass Technologies, Inc. These materials differ only very slightly in regard to their optical and mechanical properties as compared to the equivalent standard glasses.[9]

1.3 Materials properties

Key terms and mechanical properties of materials in the context of optical instrument design and the symbols and units used to represent them in this book are as follows:

Force (F) is an influence applied to a body that tends to cause that body to accelerate or to deform. The force is expressed in newtons (N) or pounds (lb).

Stress (S) is force imposed per unit area. It may be internal or external to a body and is expressed in pascals (Pa) [equivalent to newtons per square meter (N/m^2)] or pounds per square inch ($lb/in.^2$).

Strain ($\delta L/L$) is an induced dimensional change per unit length. It is dimensionless, but is commonly expressed in micrometers per meter (μm/m) or microinches per inch (μin./in.).

Young's modulus (E) is the rate of change of tensile stress with respect to linear strain. It is expressed in newtons per square meter (N/m^2) or pounds per square inch ($lb/in.^2$).

Yield strength is the stress at which a material exhibits a specified deviation from elastic behavior (proportional stress vs strain). It usually is taken to be 2×10^{-3} or 0.2% offset.

Microyield strength (or precision elastic limit) is the stress that causes one part per million of permanent strain in a short time.

Thermal expansion coefficient (CTE or α) is change in length per unit length per degree temperature change. It is commonly expressed in millimeters per millimeter per degree Celsius (mm/mm-°C) or inches per inch per degree Fahrenheit (in./in.-°F). It may be expressed as parts per million (ppm) per degree.

Thermal conductivity (k) is the quantity of heat transmitted per unit time through a unit area per unit of temperature gradient. It is commonly expressed in Watts per meter Kelvin (W/m K) or British thermal units per hour foot degree Fahrenheit (Btu/hr ft °F).

Specific heat (C_p) is the ratio of the quantity of heat required to raise the temperature of a body by one degree to that required to raise the temperature of an equal mass of water by one degree. It commonly is expressed in Joules per kilogram Kelvin (J/kg K) or British thermal units per pound degree Fahrenheit (Btu/lb °F).

Thermal diffusivity (D) quantifies the rate of heat diffusion within a body. It is a derived property expressed as thermal conductivity divided by density times specific heat ($k/\rho C_p$).

Poisson's ratio (υ) is the dimensionless ratio of lateral strain to longitudinal (or axial) strain in a bar under uniform longitudinal tension or compression.

Density (ρ) is mass per unit volume and is expressed in grams per cubic centimeter (g/cm^3) or pounds per cubic inch ($lb/in.^3$).

Materials of greatest importance to us here are optical glasses, plastics, crystals, and mirror substrate materials; metals and composites used for cells, retainers, spacers, lens barrels, mirror and prism mounts, and structures; and adhesives and sealants. A few general comments regarding each of these types of materials follow:

1.3.1 Optical glasses

Several hundred varieties of optical-quality glass are available from manufacturers worldwide. The "glass map" shown in Fig. 1.2 includes the majority of the glasses produced by Schott Glass Technologies, Inc., in the USA and/or Germany.[10] Other manufacturers produce essentially the same glasses. These are typically plotted by refractive index n_d (ordinate) and Abbe number v_d (abcissa) for yellow (helium) light as shown in the figure. They fall into 22 groups based upon chemical constituency. Figure 1.3 shows a page from the Schott catalog for a representative glass type (BK7) that illustrates the types of technical information available about each glass type for optomechanical design purposes.[10] Most mechanical properties of interest are listed in the figure under "Other Properties". The $\alpha_{-30/+70}$ item is the material's CTE for the temperature range of concern in instrument design, λ (here k) is thermal conductivity, ρ is density, E is Young's modulus, and μ (here υ) is Poisson's ratio. Resistance to humidity is indicated on a scale of 1 (high) to 4 (low) by the parameter CR. The rates of change of relative and absolute refractive index with temperature, listed in the figure, are of interest in temperature-compensated systems.

Table C1, of Appendix C, lists selected mechanical properties of 68 optical glasses designated by Walker[11] as the types he considered most useful to lens designers and that "span the most common range of index and dispersion and have the most desirable characteristics in terms of price, bubble content, staining characteristics and resistance to adverse environmental conditions." These glasses are listed in order of increasing 6-digit "glass code" (n_d-1 followed by 10 times v_d). The maximum and minimum values for each parameter are indicated by the symbol "a". It is interesting to note that these extreme values typically differ only by a factor of about two. So, as a rule of thumb, drastic mechanical differences do not occur when glass type is changed.

Table C2 compares 11 lightweight Schott glasses[12] (not included in Fig. 1.2 or Table C1) to the nearest optically equivalent standard-weight types. These glasses might well be chosen over the corresponding standard glasses for use in prisms and second-surface mirrors if weight is a prime concern. Mechanical properties of the lightweight varieties (other than density) are not greatly different from those of the standard varieties. Properties of radiation-resistant glass types are available from Schott.[13] These glasses also are not included in Fig. 1.2 or Table C1.

1.3.2 Optical plastics

A few types of commercially available plastics are suitable for use as optical components (lenses, windows, prisms, and mirrors) in some applications. Key types are identified and selected mechanical properties are listed in Table C3.

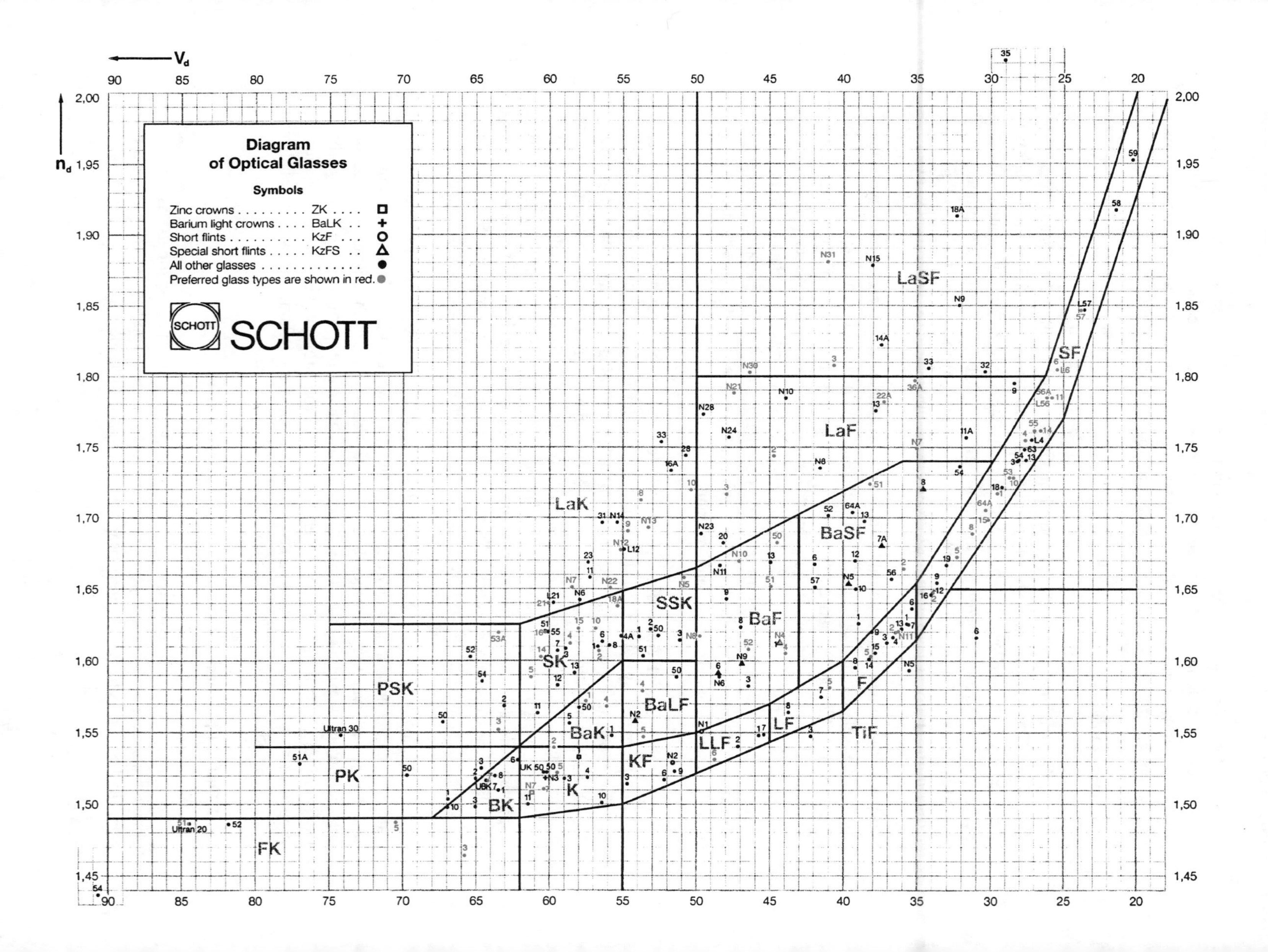
Diagram
of Optical Glasses
Symbols
Zinc crowns . . . . ZK . . . .
Barium light crowns . . . . BaLK . .
Short flints . . . . KzF . . .
Special short flints . . . . . KzFS . .
All other glasses . . . .
Preferred glass types are shown in red.
SCHOTT
SCHOTT
ν_d
n_d
LaSF
SF
LaF
BaSF
LaK
SSK
BaF
SK
PSK
BaLF
BaK
LLF
LF
F
TiF
KF
PK
K
BK
FK

BK7 517642	n_d = 1.51680	ν_d = 64.17	$n_F - n_C$ = 0.008054
	n_e = 1.51872	ν_e = 63.96	$n_{F'} - n_{C'}$ = 0.008110

Refractive Indices

	λ [nm]	
$n_{2325.4}$	2325.4	1.48921
$n_{1970.1}$	1970.1	1.49495
$n_{1529.6}$	1529.6	1.50091
$n_{1060.0}$	1060.0	1.50669
n_t	1014.0	1.50731
n_s	852.1	1.50980
n_r	706.5	1.51289
n_C	656.3	1.51432
$n_{C'}$	643.8	1.51472
$n_{632.8}$	632.8	1.51509
n_D	589.3	1.51673
n_d	587.6	1.51680
n_e	546.1	1.51872
n_F	486.1	1.52238
$n_{F'}$	480.0	1.52283
n_g	435.8	1.52668
n_h	404.7	1.53024
n_i	365.0	1.53627
$n_{334.1}$	334.1	1.54272
$n_{312.6}$	312.6	1.5486_2
$n_{296.7}$	296.7	
$n_{280.4}$	280.4	
$n_{248.3}$	248.3	

Constants of Dispersion Formula

B_1	1.03961212
B_2	$2.31792344 \cdot 10^{-1}$
B_3	1.01046945
C_1	$6.00069867 \cdot 10^{-3}$
C_2	$2.00179144 \cdot 10^{-2}$
C_3	$1.03560653 \cdot 10^{2}$

Constants of Formula for dn/dT

D_0	$1.86 \cdot 10^{-6}$
D_1	$1.31 \cdot 10^{-8}$
D_2	$-1.37 \cdot 10^{-11}$
E_0	$4.34 \cdot 10^{-7}$
E_1	$6.27 \cdot 10^{-10}$
λ_{TK} [μm]	0.170

Internal Transmittance τ_i

λ [nm]	τ_i (5 mm)	τ_i (25 mm)
2500.0		
2325.4	0.89	0.57
1970.1	0.968	0.85
1529.6	0.997	0.985
1060.0	0.999	0.998
700	0.999	0.998
660	0.999	0.997
620	0.999	0.997
580	0.999	0.996
546.1	0.999	0.996
500	0.999	0.996
460	0.999	0.994
435.8	0.999	0.994
420	0.998	0.993
404.7	0.998	0.993
400	0.998	0.991
390	0.998	0.989
380	0.996	0.980
370	0.995	0.974
365.0	0.994	0.969
350	0.986	0.93
334.1	0.950	0.77
320	0.81	0.35
310	0.59	0.07
300	0.26	
290		
280		
270		
260		
250		

Color Code

λ_{80}/λ_5	33/30

Remarks

Relative Partial Dispersion

$P_{s,t}$	0.3098
$P_{C,s}$	0.5612
$P_{d,C}$	0.3076
$P_{e,d}$	0.2386
$P_{g,F}$	0.5349
$P_{i,h}$	0.7483
$P'_{s,t}$	0.3076
$P'_{C',s}$	0.6062
$P'_{d,C'}$	0.2566
$P'_{e,d}$	0.2370
$P'_{g,F'}$	0.4754
$P'_{i,h}$	0.7432

Deviation of Relative Partial Dispersions ΔP from the "Normal Line"

$\Delta P_{C,t}$	0.0216
$\Delta P_{C,s}$	0.0087
$\Delta P_{F,e}$	−0.0009
$\Delta P_{g,F}$	−0.0009
$\Delta P_{i,g}$	0.0036

Other Properties

$\alpha_{-30/+70°C}$ [10^{-6}/K]	7.1
$\alpha_{20/300°C}$ [10^{-6}/K]	8.3
Tg [°C]	557
$T_{10^{13.0}}$ [°C]	557
$T_{10^{7.6}}$ [°C]	719
c_p [J/(g·K)]	0.858
λ [W/(m·K)]	1.114
ρ [g/cm³]	2.51
E [10^3 N/mm²]	82
μ	0.206
K [10^{-6} mm²/N]	2.77
$HK_{0.1/20}$	610
B	0
CR	2
FR	0
SR	1
AR	2.0
PR	2.3

Temperature Coefficients of Refractive Index

	$\Delta n_{rel}/\Delta T$ [10^{-6}/K]			$\Delta n_{abs}/\Delta T$ [10^{-6}/K]		
[°C]	1060.0	e	g	1060.0	e	g
−40/−20	2.4	2.9	3.3	0.3	0.8	1.2
+20/+40	2.4	3.0	3.5	1.1	1.6	2.1
+60/+80	2.5	3.1	3.7	1.5	2.1	2.7

SCHOTT Optical Glass Nr. 10 000 9/92

Fig. 1.3 A page from an optical glass catalog showing optical and mechanical properties of a typical glass (BK7). (Courtesy of Schott Glass Technologies, Inc., Duryea, PA)

In general, optical plastics are softer than glasses, so they tend to scratch easily and are hard to polish to a precise surface figure. Their CTEs are larger than those of glasses and of most crystals. Most plastics tend to absorb water from the atmosphere. The biggest advantages of using plastic optical components are their low densities and ease of manufacture in large quantities by low-cost molding techniques. It is relatively easy to mold integral mechanical mounting features into plastic components such as lenses, windows, prisms, or mirrors during manufacture.

1.3.3 Optical crystals

Natural and synthetic crystalline materials are used in optical components when transmission in the infrared or ultraviolet spectral regions is required. They also are used to provide special optical characteristics such as increased dispersion for some specific wavelengths. These materials fall generally into four groups: alkali and alkaline earth halides; infrared-transmitting glasses and other oxides; semiconductors; and chalcogenides. Mechanical properties of interest here are given in Tables C4 through C7 for the crystals most commonly used as optics.

1.3.4 Mirrors

Mirrors consist of a reflecting surface (usually a thin film coating) attached to or integral with a supporting structure or substrate. Their sizes can range from a few millimeters to many meters; we concentrate in this text on ones with maximum dimensions no larger than 24 in. (61 cm). The substrates can be made of glasses, low-expansion ceramics, metals, composites, or (rarely) plastics. Tables C8a and C8b list some mechanical properties of the most common mirror materials. Table C9 quantifies structural figures of merit for most of the same materials.[14] The figures of merit allow direct comparisons between candidate materials for a given application. For example, a commonly used figure of merit in mirror design is specific stiffness, E/ρ, which helps us determine which material would have the least mass or self-weight deflection for a given mirror geometry and size. The various figures of merit listed in Table C9 relate to the comparisons listed in the headings. Choice of which figure of merit to apply in a particular case depends upon the design requirements and constraints. Tables C10a through C10d list characteristics of aluminum alloys, aluminum matrix composites, several grades of beryllium, and major silicon carbide matrix types used in mirrors.

The method used in manufacturing a given mirror depends largely upon the type of material involved. Table C11 correlates common machining, surface finishing, and coating methods for various common materials.[4] It is important that the manufacturing processes, including plating or coating, do not introduce excessive internal or surface stress into the mirror substrate or coating.

1.3.5 Materials for mechanical components

The materials typically used for mechanical components of optical instruments such as instrument housings, lens barrels, cells, spacers, retainers, prism and mirror mounts, etc., are metals (typically aluminum alloys, beryllium, brass, Invar, stainless steel, and titanium). Composites (metal matrices, silicon carbide, and filled plastics) may be used in some structural applications. Some of these materials also are used as mirror substrates. Mechanical properties of selected versions of the metals and one metal matrix may be found in Table C12. The general qualifications of the metals in the context of optical component mounting applications are as follows:

> Aluminum alloys: Alloy 1100 has low strength, is easily formed by spinning or deep-drawing, and can be machined and welded or brazed. Alloy 2024 has high strength and good machinability but is hard to weld. Alloy 6061 is the general-purpose structural aluminum alloy with moderate strength, good dimensional stability, and good machinability; is easily welded and brazed. Alloy 7075 has high strength, machines well, but is not suited for welding. Alloy 356 is used for moderate- to high-strength structural castings; and it machines and welds easily. Most aluminum alloys are heat-treated to differing degrees of hardness depending on the application. Surfaces oxidize quickly, but can be protected by chemical films or anodic coatings. The latter may produce significant dimensional buildup. A black anodized finish reduces light reflections, so this type of finish is frequently used on aluminum parts for optical instruments. CTE match of aluminum alloys to glasses, ceramics, and most crystals is not close. Table C12 compares characteristics of several aluminum alloys used for mirror substrates.
>
> Beryllium is light in weight, has high stiffness, conducts heat well, resists corrosion and radiation effects, and is fairly stable dimensionally. It is relatively brittle, so machining is not easy. Telescope barrels of 12 in. (30.5 cm) diameters have been machined to typical dimensional tolerances of $\pm$0.0002 in. (5.1 μm). Beryllium parts are frequently formed by powder metallurgy techniques such as hot isostatic pressing (HIP). Precision grade beryllium is relatively expensive to purchase and to process so it is used primarily in optical instruments intended for sophisticated applications such as structures and mirror or grating substrates for use at cryogenic temperatures. It also is the material of choice in some space applications where radiation resistance or weight savings are vital and monetary costs are of lesser importance. Table C13 compares characteristics of several common beryllium grades. Paquin has countered claims as to the extreme hazards of working beryllium by pointing out that simple exhaust systems with suitable filters for particulate material and conventional means for collection and disposal of loose abrasive grinding/polishing slurry are very effective as safety precautions.[14]
>
> Brass is used where high corrosion resistance, good thermal conductivity, and/or ease of machining are required, but weight is not very critical. It is popular for screw-machined parts and marine applications. Brass can be blackened chemically.
>
> Invar, an iron-nickel alloy, is used most frequently in high-performance instruments for space and/or cryogenic applications to take advantage of its low CTE. It is quite heavy, and machining sometimes affects its thermal stability. Annealing is advised. A version called Super-Invar has even lower CTE over a

limited temperature range. It is not recommended for use at temperatures below -50°C (-58°F). To prevent oxidation, Invar frequently is chrome plated.

Stainless steels [sometimes called corrosion resistant steels (CRES)] are used in optical mounts primarily for their strength and their fairly close CTE match to some glasses. They are relatively dense, so a weight penalty must be paid to achieve these advantages. A chromium oxide layer that forms on exposed surfaces makes these steels resistant to corrosion. In general, these steels are harder to machine than aluminum alloys. Type 416 is the most easily machined and can be blackened chemically or with black chrome plating. Type 17-4PH has good dimensional stability. Stainless steels can be welded to like materials or brazed to many different metals.

Titanium is the material of choice in many high-performance systems where close CTE match to glass is essential. Flexures are sometimes made of titanium. It is about 60% heavier than aluminum. Titanium is somewhat expensive to machine and can be cast. Brazing is easy, but welding is more difficult; electron beam or laser welding techniques work best. Parts can also be made by powder metallurgy methods. Corrosion resistance is high.

Some plastics, particularly glass-reinforced epoxies and polycarbonates, are used in structural parts such as housings, spacers, prism and mirror mounts, and lens barrels for cameras, binoculars, office machines, and other commercial optical instruments. They are relatively lightweight, and most can be machined conventionally or by single-point diamond turning (SPDT), while some can be cast. Generally, plastics feature low cost. Unfortunately, they are not as stable dimensionally as metals and tend to absorb water from the atmosphere and to outgas in a vacuum. The CTEs of filled varieties can be customized to some extent.

1.3.6 Adhesives and sealants

Optical cements used to hold refracting surfaces of lenses or prisms together as, for example, to form cemented doublets, triplets, or beamsplitters must be transparent in the spectral region of interest, have good adhesion characteristics, have acceptable shrinkage, and (preferably) be able to withstand exposure to moisture and other adverse environmental conditions. The most popular optical cements are thermosetting and photosetting (ultraviolet light curing) types. Some mechanical properties of interest are given in Table C13 for a generic type of optical cement.

Structural adhesives most frequently used to hold optics to mounts and to bond mechanical parts together are two-part epoxies and urethanes. They cure best at elevated temperatures and suffer some (up to 6%) shrinkage during cure. Some adhesives emit volatile ingredients during cure or if exposed to vacuum or elevated temperatures. The emitted material may then condense as contaminating films on nearby cooler surfaces such as lenses. Typical properties of representative types are summarized in Table C14.

Sealants used today are usually room-temperature-vulcanizing (RTV) elastomers that cure into flexible, form-fitting masses with reasonably good adherence properties. They are typically poured or injected into gaps between lenses and mounts or between mount components to seal leaks and/or to help hold the optics in place under vibration, shock, and temperature change. Some outgas or emit effluents during cure or in vacuum more than others. Use of primers prior to application of the sealants is recommended by the manufacturers for many of these products. Typical physical characteristics and mechanical properties of a few representative sealants are given in Tables C15a and C15b. The cure times, colors, and certain physical properties can be modified significantly through use of additives and/or catalysts.

CHAPTER 2
ATTRIBUTES OF THE SUCCESSFUL OPTIC-TO-MOUNT INTERFACE

The prime purpose of the optical component-to-mount interface is to hold the component (here a prism or small mirror) with minimal self-weight distortion in its proper position and orientation within the optical instrument. This implies the application of mechanical constraints, i.e., external forces that limit component motions in the normal gravity environment, when the temperature changes, or when external mechanical disturbances occur. How these constraints should be applied is the first topic considered in this chapter. The advantages of semi-kinematic mounting techniques are explained. We then explore how the applied forces may cause stresses and birefringence to develop within the component. The generalizations on mounting stresses given here are the subjects of more detailed, quantitative discussions in Chaps. 5 and 8. The chapter closes with brief considerations of cost and manufacturability of the components and their mounts.

2.1 Mechanical constraints

Under all operating conditions, it is important that each optical component be constrained so it remains within decentration, tilt, and axial spacing budgets and that induced stresses, surface deformations, and birefringence are tolerable. Both lateral and axial constraints are needed for each component. Contact with all components should occur outside the optically used apertures, if possible. Ideally, the mechanical interfaces should be such that all six degrees of freedom (three mutually perpendicular translations and three rotations about the translation axes) of the optic are independently constrained without redundancy. If these contacts are infinitesimal in area, i.e., points, no bending moments can be transferred to the optic. Six such point contacts are required to constrain the component. Such a mounting would be called kinematic; it cannot be achieved in the real world because stresses developed at the contacts would be too large. A semi-kinematic mount is one with six constaints, each of which has a finite but small area to distribute force and minimize contact stress.

Semi-kinematic mounting may not be economically feasible with optical components having circular symmetry such as lenses, some windows, and some mirrors. In these cases, mounting designs are usually created so that contact areas at a few locations are as large as practical or large numbers of smaller contact areas are provided. The latter approach distributes forces so they are not concentrated. A rigid body with more than six constraints is overconstrained and the mounting interface is non-kinematic. Location of an overconstrained optic may be uncertain and the optic may be deformed by the imposed forces. Careful design and manufacture of the interfacing surfaces will help reduce these problems. Multiple-contact supports frequently are used as axial and radial supports for mirrors. Some prism designs are more amenable than others to semi-kinematic mounting. The use of flexures in a semi-kinematic mounting interface allows deterministic mounting of an optic without introducing undesirable moments.

Figure 2.1, adapted from Smith,[15] indicates how the required six constraints might be applied to a rectangular parallelepiped-type body such as a simple cube prism. In View (a), we see six identical balls attached to three mutually perpendicular flat surfaces. The dashed construction lines indicate how the balls might be located symmetrically. If the cube prism is held in contact with all six balls, it will be uniquely constrained. The bottom of the prism rests on the three contacts parallel to the X-Z plane so it cannot translate in the Y direction nor tilt about the X or Z axes. Two contacts parallel to the Y-Z plane prevent translation along the X axis and rotation about the Y axis. The single contact at the X-Y plane controls translation along the Z axis. A single force applied to

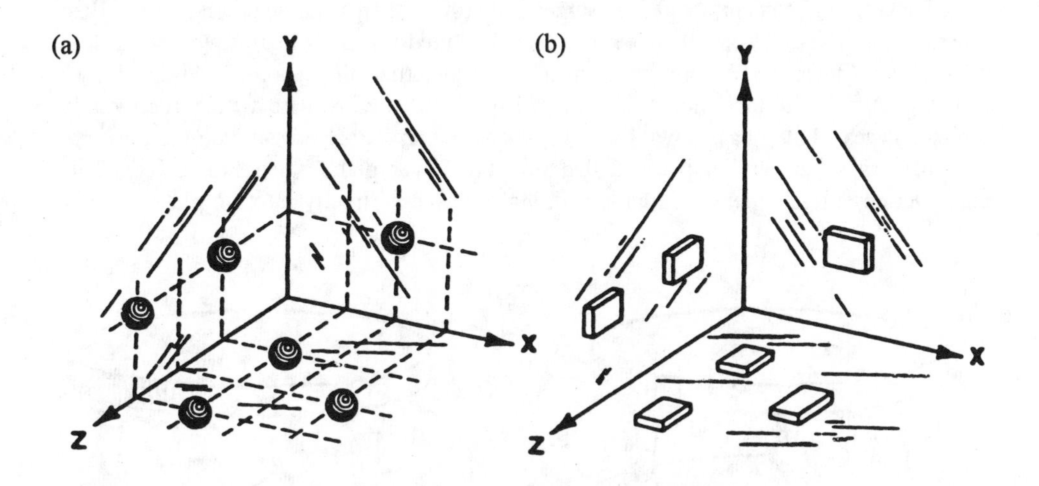

Fig. 2.1 (a) Kinematic and (b) semi-kinematic position-defining registration surfaces intended for interfacing with a cube-shaped prism (not shown). (Adapted from Smith[15])

the near corner of the prism and aimed toward the origin would hold the prism in place. This force would pass through the center of gravity of the cube prism. Three forces, each applied normal to one of the exposed prism surfaces and directed toward a contact point or a point midway between adjacent contact points, also would hold the prism. Unfortunately, this particular multiple-force condition is not very practical for an optical application since all six faces of the prism would be at least partially obscured. By increasing the separations of some of the balls it may be possible to clear the apertures without destroying symmetry or the kinematic condition.

Figure 2.1(b) shows conceptually how the point contacts on balls could be replaced by small areas on raised pads to distribute the mechanical preloads on the prism surfaces. The design now is called semi-kinematic. If the pads are machined or lapped coplanar and mutually perpendicular, introduction of stress at the contacts is minimized. If not accurately aligned, one or more area contacts may degenerate into line or point contacts thereby causing contact stress to increase. Forcing the prism against non-coplanar areas also would tend to introduce moments that could distort the optical surfaces.

Frequently it is necessary for subassemblies of prisms or mirrors in their mounts to be removed from the optical instrument and replaced in the identical location and orientation. Kinematic interfaces allow this to be accomplished with high accuracy. Figure 2.2(a) shows (schematically) one such interface. It comprises lower and upper plates; the prism or mirror is attached to the upper plate (thereby forming the removable subassembly) and the lower plate is permanently attached to the instrument structure. Attached to the bottom of the upper plate are three balls, symmetrically located on a given "bolt circle" diameter. The lower plate has three "sockets": a "vee," a trihedral-shaped hole, and a flat surface. The balls fit repeatably into the sockets establishing six positional constraints whenever the plates are clamped together. The three balls and the three sockets can be manufactured or purchased with posts that can be pressed into drilled holes in the plates. A conical socket can be substituted for the trihedral with slight loss of accuracy (because contact occurs on a line rather than three points). Figure 2.2(b) shows a similar interface mechanism in which the three balls mate with three radially directed "vees." Kittel has shown how "vees" can be formed by pressing dowel pins into six holes as three parallel pairs drilled into the lower plate.[16] See Fig. 2.2(c). This construction is less expensive than machining the "vees" directly into the plate.

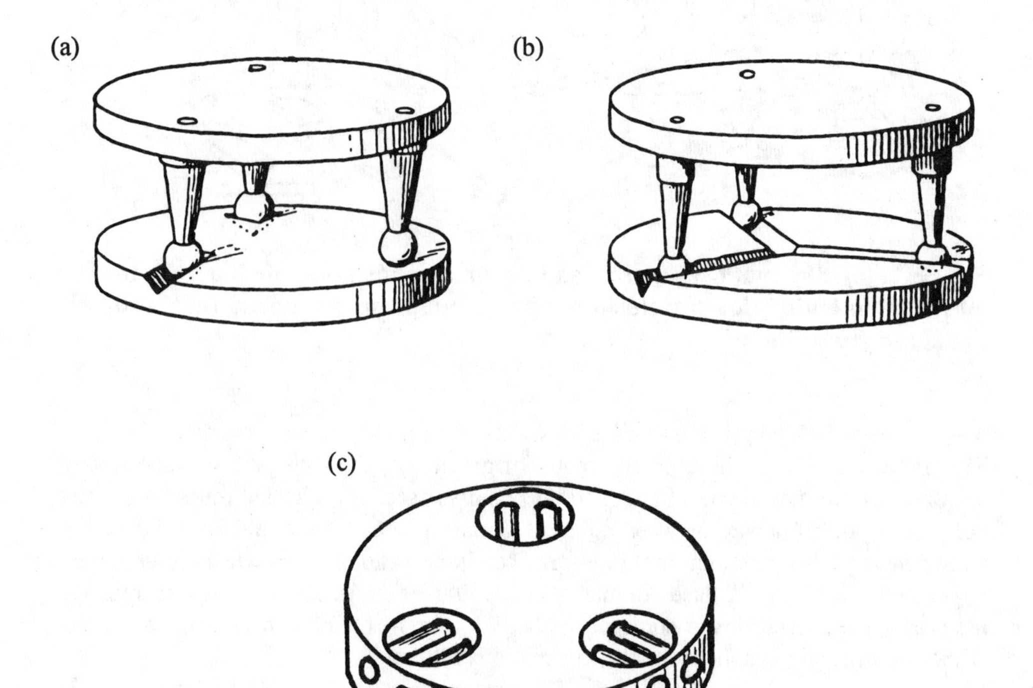

Fig. 2.2 (a) and (b) concepts for zero degree of freedom separable interfaces. (From Strong[17]), (c) concept for providing "vees" with parallel rods (from Kittel[16])

A non-kinematic technique frequently used for mounting prisms involves glass-to-metal bonds with thin layers of adhesives. These designs generally result in reduced interface complexity and compact packaging while providing mechanical strength adequate for withstanding the severe shock, vibration, and temperature changes characteristic of military and aerospace applications. This method of mounting also is used in some less rigorous applications because of its inherent simplicity and reliability.[8]

2.2 Consequences of mounting forces

Mounts with glass-to-metal contact are sometimes called "hard mounts." The mechanical preload imposed by such a contact at a given point on the optical component's surface at a given temperature is here designated by the symbol P_i and is measured in pounds or newtons. Usually, when the temperature changes, this preload also changes due to differential expansion between the component material and the mount material. Equations for estimating the value for P_i in typical mounting configurations are given in discussions of design examples in Chaps. 5 and 8.

Mounting preloads tend to compress (or strain) any optical component and produce corresponding elastic stresses within the component. In Chaps. 5 and 8, we show how to estimate the magnitudes of these stresses and to determine if they appear to be tolerable. Forces concentrated on point areas cause localized stresses of high magnitude; these are particularly undesirable, since they can lead to excessive distortion of malleable materials such as plastics or breakage of brittle materials such as glass, ceramics, or crystals. Metal optics (mirrors) are less susceptable to this type damage, but even they can be distorted by applied forces. As noted earlier, distributing forces over areas reduces the stress. Reaction forces are also exerted on the mount by mechanical preloads on optics. These can distort the mount temporarily or permanently or, in extreme cases, cause the mount to fail.

Lower levels of applied forces can introduce birefringence (induced spatial inhomogeneity of refractive index) into normally isotropic refracting optical materials. This affects the propagation speeds of the perpendicular and parallel components of polarized light passing through the material, so these components become out of phase. The magnitude of birefringence occuring per unit length in a particular sample of material under a given level of stress depends on the stress optic coefficient of the material. Birefringence is most important in optical systems using polarized light such as polarimeters, many interferometers, most laser systems, and some high-performance cameras.

Even lower levels of applied forces can cause optical surfaces to deform, especially if the forces are not applied symmetrically. Minute surface deformations (measured in fractions of a wavelength of light) may affect system performance. The significance of a given deformation depends, in part, on the location of the surface within the optical system, the specific design of the prism- or mirror-to-mount interface, and the performance requirements of the system containing the optical surface in question. Larger departures from perfection can be tolerated on surfaces near an image than near a system pupil. Mirror deformations are more significant than refracting surface deformations at the same location in the system because reflected wavefront errors are twice the surface error while refracting surface errors are (n - 1) times the surface error. Because of these system- and application-dependent factors, no general methods of estimating surface deformations nor guidelines in regard to how much surface deformation can be tolerated are given here.

2.3 Cost and manufacturability

Key factors in determining cost and manufacturability of the optical and the mechanical components of an optical instrument are the physical dimensions of the optic because extremely large or extremely small parts are hard to make and test. Since the majority of prisms and mirrors used in common optical instruments are of modest size, we here limit our considerations to prism apertures of 0.4 to 6 in. (1 to 15.2 cm) and to mirror apertures no larger than about 24 in. (61 cm).

Another key factor is the set of dimensional tolerances assigned during the design process. Smith[18] and other authors[19,20] have discussed philosophies for assigning tolerances, while Willey and Durham[21] and Willey and Parks[22] have described techniques and algorithms useful in evaluating assigned tolerances in terms of their impact on the cost of optics and interfacing mechanical parts. These latter techniques allow designers to predict whether a given design can be built to specified tolerances and to estimate the manufacturing costs of the optical components used. The above-mentioned authors also point out that early knowledge that some tolerances are excessively tight in terms of available fabrication technology would, in some cases, lead to timely redesign rather than discovering the need for redesign after failing in the attempt to make parts that meet these requirements. Review of the design by individuals skilled in the applicable manufacturing procedures well before releasing drawings for manufacture also tends to prevent unfortunate surprises during fabrication, inspection, assembly, or testing.

Cost and manufacturability are also driven, usually to a lesser extent, by material choices. Very soft or very hard optical materials are hard to grind and polish. Many crystals fall into the former category while fused silica, sapphire, and diamond fall into the latter. Similarly, aluminum and brass are easy to machine while titanium and Invar are more difficult. Some materials are easily machined to very high accuracy by single point diamond turning (SPDT) so this is a popular technique for making high performance optical and mechanical components even though it may be somewhat more expensive than using conventional machines and tool bits.

CHAPTER 3
PRISM DESIGN

Many types of prisms have been designed for use in various optical instrument applications. Most have unique shapes as demanded by the geometry of the ray paths, reflection and refraction requirements, compatability with manufacture, weight reduction considerations, and provisions for mounting. Before we consider how to mount these prisms we should understand how they are designed. Our first topics in this chapter are refractive effects, total internal reflection, and the construction and use of tunnel diagrams. We then see how to determine aperture requirements and reference analytical means for calculating third order aberration contributions from prisms. The chapter closes with design information for 27 types of individual prisms and prism combinations frequently encountered in optical instrument design.

3.1 Geometric considerations

3.1.1 Refraction and reflection

The laws of refraction and reflection of light govern the passage of rays through prisms and mirrors. In Fig. 3.1, we see a comparison of ray paths from an object passing

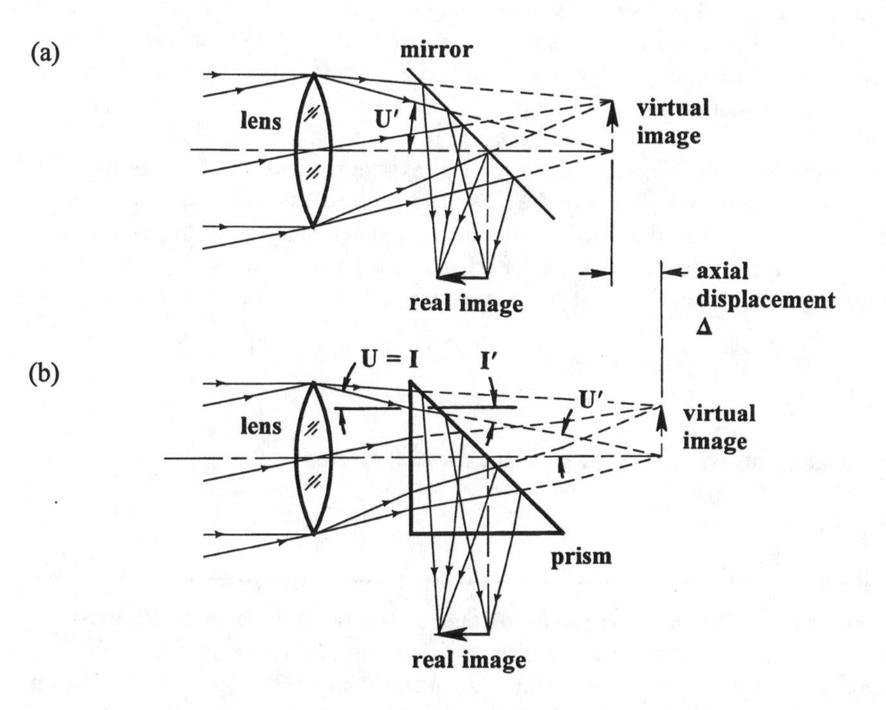

Fig. 3.1 Illustrations of 90° deviations by reflection of rays (a) at a 45° mirror and (b) in a right-angle prism. In (b), angles U, U′, I, and I′ pertain to the first surface of the prism

through a lens and a reflector enroute to the image. In View (a), the reflector is a flat mirror while in View (b) it is a right-angle prism in which reflection occurs at an internal surface. The most significant differences are the ray deviations that occur at the prism's refracting surfaces and the axial displacement of the image due to replacement of air by glass in part of the path in View (b). Refraction, of course, follows Snell's law which may be written as

$$(n_i)(\sin I_i) = (n'_i)(\sin I'_i), \tag{3.1}$$

where n_i and n'_i are refractive indices in object and image spaces of surface "i" and I_i and I'_i are the ray angles of incidence and refraction, respectively at that surface. Reflection follows the familiar relationship

$$I_i' = I_i, \tag{3.2}$$

where I_i and I'_i are the values of the ray angles of incidence and reflection at surface "i". The angles in these equations are measured with respect to the surface normal at the point of incidence of the ray on the surface or the axis. Algebraic signs of the angles are not shown in either of these equations.

The entrance and exit faces of most prisms are oriented perpendicular to the optical axis of the optical system since this promotes symmetry and reduces aberrations for non-collimated beams passing through the prism. Notable exceptions are the Dove prism, the double-Dove prism, wedge prisms, and prisms used to disperse light as required for use in monochromators and spectrographs.

A prism with faces normal to the optical axis refracts rays exactly as would a plane-parallel plate oriented normal to the axis. The geometrical path length, t, through the prism measured along the axis is the same as the thickness of the plate. Any reflections occurring inside the prism do not affect this behavior. The axial displacement, Δ (see Fig. 3.1), of an image formed by rays passing through the prism is given by:

$$\Delta = t\left(1 - \frac{\tan U'}{\tan U}\right) = \frac{t}{n}\left(n - \frac{\cos U}{\cos U'}\right) \tag{3.3}$$

For small angles, this equation reduces to the paraxial version

$$\Delta = (n-1)t/n \tag{3.4}$$

Numerical Example No. 1: Image axial displacement due to insertion of a prism. Assume that an aplanatic lens images a distant object with a f/4 beam. How much does the image move axially when a FN11 right-angle prism with t = 38.1 mm is inserted into the beam?

The marginal ray angle for this f/4 beam is sin U′ = 0.5/(f-no.) = 0.5/4 = 0.12500; hence, U′ = 7.1808°. From Table C1, the refractive index for FN11 glass is 1.621. Since the entrance face of the prism is normal to the axis, I = U′ so, by Eq. (3.1),

$$\sin I' = \sin 7.1808° / 1.621 = 0.07711 \quad \text{and } I' = 4.4227°$$

By Eq. (3.4), the image moves by

$$\Delta = (38.1/1.624)(1.621 - (\cos 7.1808° / \cos 4.4227°)) = 14.711 \text{ mm } (0.579 \text{ in.})$$

By Eq. (3.5), the paraxial approximation of this displacement is

$$\Delta = (1.621 - 1)(38.1) / 1.621 = 14.596 \text{ mm } (0.575 \text{ in.})$$

Reflection within a prism folds the light path. In Fig. 3.1(b), the object (an arrow, not shown) is imaged by the lens through the prism as the indicated virtual image. After reflection, the real image is located as shown. If the page were to be folded along the line representing the reflecting surface, the real image and the solid-line rays would coincide exactly with the virtual image and the dashed-line rays. A diagram showing both the original prism (ABC) and the folded counterpart (ABC′) is called a "tunnel diagram." See Fig. 3.2. The rays a-a′ and b-b′ represent actual reflected paths while rays a-a″ and b-b″ appear to pass directly through the folded prism with proper refraction, but without the reflection. Multiple reflections are handled by successive folds of the page. This type diagram, which can be drawn for any prism, is particularly helpful when designing an optical instrument using prisms since it simplifes the estimation of required apertures and, hence, the size of those prisms.

To illustrate the use of a tunnel diagram, let us consider the telescope optical system of Fig. 3.3. This could be a spotting telescope or one side of a binocular. The pair of Porro prisms serve to erect the image as indicated by the "arrow crossed with a drumstick" symbols at various locations in the figure. Figure 3.4(a) shows the front

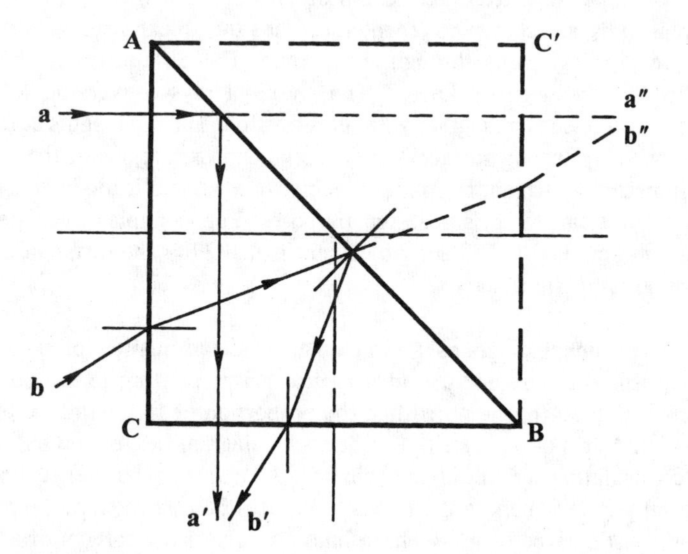

Fig. 3.2 Illustration of a tunnel diagram for a right angle prism

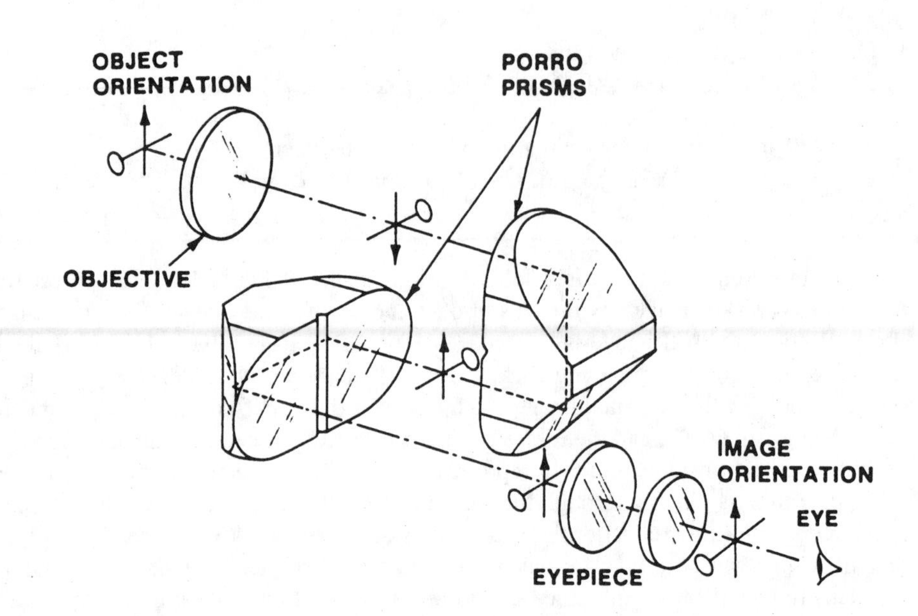

Fig. 3.3 Optical system of a typical telescope with a Porro prism erecting system. (From Yoder[8] by courtesy of Marcel Dekker, Inc.)

portion of the same system with the Porro prisms represented by tunnel diagrams. Folds in the light path are indicated by the diagonal lines. We designate all the apertures of the prisms as dimension "A"; the axial path length of each prism is then 2A. In Fig. 3.4(b) the prism path lengths are shown as 2A/n; these are the thicknesses of air optically equivalent to the physical paths through the prisms. The air-equivalent thickness is sometimes called the "reduced thickness." The marginal rays converging to the axial image point are drawn in such a diagram without refraction. The ray heights at each prism surface (including the reflecting surfaces) are paraxial approximations of the true values that would be indicated by trigonometric ray tracing. In most applications, this degree of approximation is adequate for prism design purposes. For example, an angle of 7° is 0.12217 radians; its sine is 0.12187 and its tangent is 0.12278. The differences between these values are not very significant.

Smith used tunnel diagrams to illustrate the determination of the minimum required Porro prism apertures for use in a typical prism erecting telescope.[15] With a diagram similar to Fig. 3.4(b), he noted that the proportion of face width A_i to reduced thickness was A_i: $(2A_i/n_i)$ or $n_i/2$. He then redrew the diagram in the form shown in Fig. 3.5 to facilitate calculating the minimum value for A_1 and A_2. The dashed lines drawn from the top front prism corners to the opposite vertices both have slopes, m, equalling one-half the ratio just derived or $n_i/4$. These lines are loci of the corners of a family of prisms with the proper proportions. The intersections of these two dashed lines with the

(a)

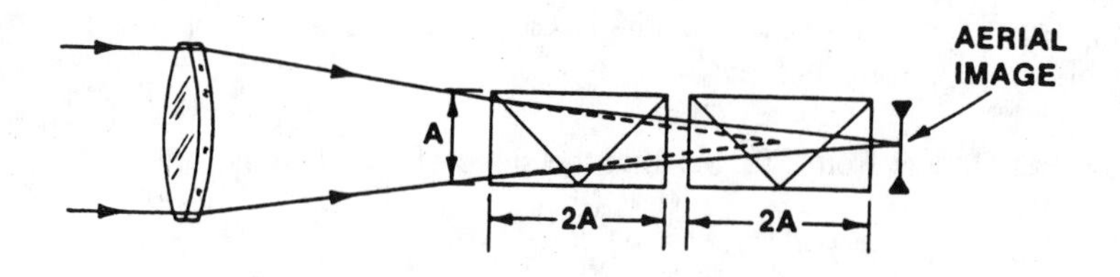

(b)

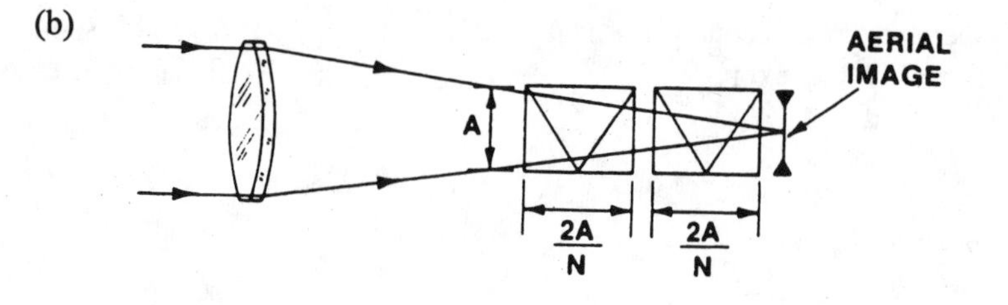

Fig. 3.4 Lens and Porro prisms from Fig. 3.3 with prisms shown (a) by conventional tunnel diagrams and (b) by tunnel diagrams with reduced thicknesses. (Adapted from Smith[15])

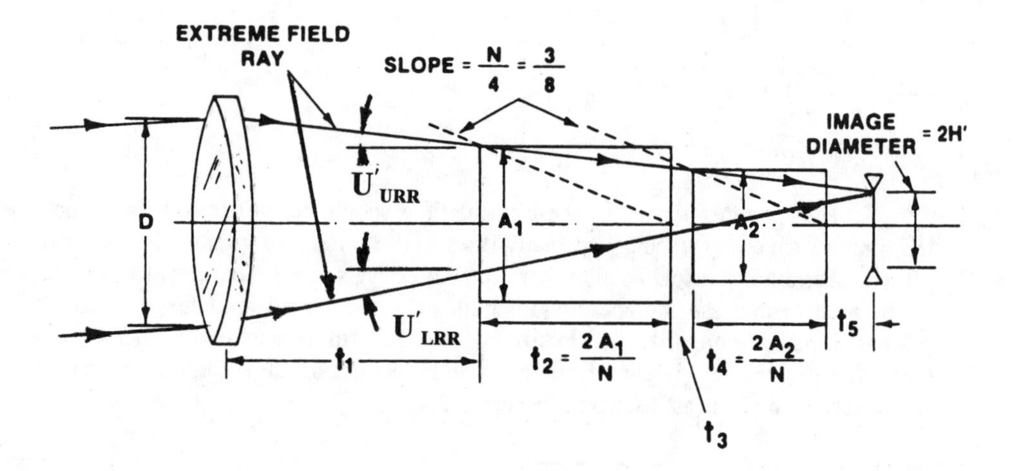

Fig. 3.5 Determination of minimum prism apertures from geometric proportions and the outermost unvignetted full field rays. (Adapted from Smith[15])

outermost full-field ray (frequently called the upper rim ray or URR) locate the corners of the two porro prisms. Note that the airspaces between optical components must be known for this procedure to succeed.

It is easy to see from Fig. 3.5 that the slope of the URR is

$$\tan U'_{URR} = \frac{((D/2) - H')}{EFL_{OBJ}} \tag{3.5}$$

and that the semi-aperture of the second prism is $A_2/2 = H' + (t_4 + t_5)(\tan U')$. This semi-aperture also is given by the expression $A_2/2 = (m)(t_4) = (n_i)(t_4)/4$. Equating these expressions for A_2 we find that

$$t_4 = \frac{(t_5)(\tan U'_{URR}) + H'}{(n_i/4) - \tan U'_{URR}} \tag{3.6}$$

and

$$A_2 = (n_i)(t_4)/2. \tag{3.7}$$

By similar logic, we can write expressions for the axial thickness and aperture of the first Porro prism:

$$t_2 = \frac{(t_3 + t_4 + t_5)(\tan U'_{URR}) + H'}{(n_i/4) - \tan U'_{URR}} \tag{3.8}$$

and

$$A_1 = (n_i)(t_2)/2. \tag{3.9}$$

The apertures derived by these calculations should be confirmed by more precise techniques such as ray tracing; especially if a specific amount of vignetting is needed for off-axis aberration control. To allow for protective bevels and dimensional tolerances, we might need to increase the apertures of both prisms by small amounts such as a few percent of their apertures. The prism size may also have to be increased further to prevent ghost (i.e., unwanted) images from internal reflections if they cannot be controlled adequately with slots (as discussed in Sect. 3.3.4).

Numerical Example No. 2: Prism size calculation.

Determine the required minimum apertures of both prisms in a system per Fig. 3.5 with the following characteristics: EFL_{OBJ} = 177.800 mm (7 in.), objective aperture = 50 mm

(1.968 in.), image diameter = 15.875 mm (0.625 in.), t_3 = 3.175 mm (0.125 in.), t_5 = 12.7 mm (0.500 in.), and prism index = 1.500.

By Eq. (3.5), $\tan U'_{URR} = ((50.000/2) - (15.875/2)) / 177.800 = 0.09506$
$U'_{URR} = 5.4813°$

By Eq. (3.6), $t_4 = ((12.700)(0.09596) + (15.875/2)) / ((1.5/4) - 0.09596)$
$= 32.813$ mm (1.292 in.)

By Eq. (3.7), $A_2 = (1.5)(32.813) / 2 = 24.610$ mm (0.969 in.)

By Eq. (3.8), $t_2 = ((3.175 + 32.813 + 12.700)(0.9596) + (15.875/2)) / ((1.5/4) - 0.09596)$
$= 45.189$ mm (1.779 in.)

By Eq. (3.9), $A_1 = (1.5)(45.189) / 2 = 33.892$ mm (1.334 in.)

The general geometric technique just described can be adapted to determine the required apertures of other types of prisms used in converging or diverging beams.

3.1.2 Total internal reflection

A special case of refraction can occur when a ray is incident upon an interface where n is greater than n′ as, for example, at the hypotenuse surface (surface 2) inside a right angle prism. In the last section we assumed that all rays would reflect; as indeed they would if the surface had a reflective coating such as silver or aluminum. If that surface is uncoated, however, Snell's law [Eq. (3.1)] says that, for small angles of incidence and low values of prism index, a ray can refract through the surface into the surrounding air. See ray a-a′ in Fig. 3.6. This ray is vignetted and does not contribute to the image formed below the prism. If we increase the ray angle I_2, the angle I'_2 also increases. For some value of I_2, I'_2 can reach 90°. Then, sin I'_2 is unity. Since this sine cannot exceed unity, we find that, for still larger values of I_2, the ray reflects internally just as if the surface were silvered. The particular value of I_2 corresponding to $I'_2 = 90°$ is called the critical angle, I_C, where

$$\sin I_C = n_2' / n_2 \qquad (3.10)$$

Usually, the medium beyond the surface 2 is air so n'_2 is unity and $\sin I_C = 1/n_2$.

We can take advantage of total internal reflection (TIR) in prisms by choosing the refractive index high enough that all rays that we want to reflect do exceed I_C at the surface in question. Then the reflections take place without photometric loss and reflective coatings are not needed on that surface. It is important to note that TIR occurs only at clean surfaces, so special care must be taken not to let the surface become contaminated with condensed water, fingerprints, or other foreign matter that can change the refractive index outside that surface.

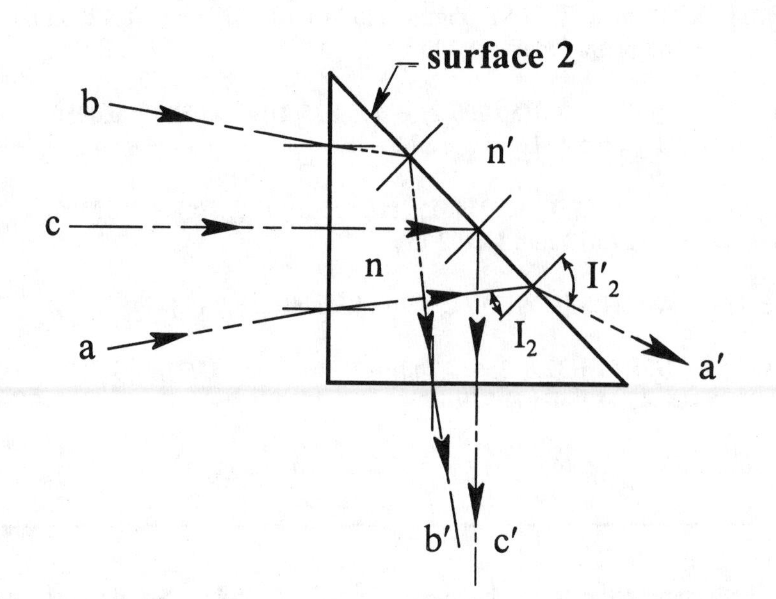

Fig. 3.6 Ray paths through an unsilvered right angle prism of low refractive index. Ray a-a′ is at an angle of incidence I_2 smaller than I_C so it "leaks" through the surface while I_2 for each of rays b-b′ and c-c′ exceeds I_C so they "totally reflect" internally.

Numerical Example No. 3: Unvignetted field of view for TIR in a prism-erecting telescope.

Assume that the prisms of Example No. 2 are not silvered and are made of F2 glass with index of 1.620. What field of view can the prisms transmit without vignetting due to loss of TIR?

From Eq. (3.10), $\sin I_C = 1 / 1.620 = 0.61728$, $I_C = 38.1181°$

From geometry, I′ at the entrance face of each prism is $(45° - I_C) = 6.8819°$

From Eq. (3.1), the ray angle in front of each prism is

$$\sin I = (1.620)(\sin 6.8819°) = 0.19411, \quad I = 11.1930°$$

This ray angle equals the angle of the lower rim ray (LRR) passing from the bottom of the lens aperture to the top of the image so $U'_{LRR} = 11.1930°$ and $\tan U'_{LRR} = 0.19788$

Modifying Eq. (3.5) to apply to the LRR, we get

$$\tan U'_{PR} = ((D/2) + H') / EFL_{OBJ} = ((50.000/2) + H') / 177.8 = 0.019788$$

Solving for H′ we get H′ = 10.18277 mm

Since $H' = (EFL_{OBJ})(\tan U'_{PR})$, we find that the unvignetted telescope field of view is $\pm \arctan(H' / EFL_{OBJ}) = \pm 3.2778°$

3.2 Prism aberration contributions

As mentioned earlier, prisms usually are designed so their entrance and exit faces are perpendicular to the optical axis of the transmitted beam. If that beam is collimated, no aberrations are introduced. Aberrations do result if the beam is not collimated. In a converging or diverging beam, a prism introduces longitudinal aberrations (spherical, chromatic, and astigmatism) as well as transverse aberrations (coma, distortion, and lateral chromatic). Smith provided exact and third-order equations for calculating the aberration contributions of a plane parallel plate or the equivalent prism.[15] Ray-tracing programs give aberration contributions of prisms in a given design on a surface-by-surface basis.

3.3 Typical prism configurations

Chapter 13 of *MIL-HDBK-141 Optical Design* (1962)[23] gives generic dimensions, the axial path length, and tunnel diagrams of many types of common prisms. Most of these designs were described earlier in *ORDM 2-1, Design of Fire Control Optics*, a two-volume treatise on telescope design written by Frankford Arsenal's long-time Chief Lens Designer, Otto K. Kaspereit, and published by the U.S. Army in 1953. Since copies of these books are hard to find and they, like later excerpts,[2,15,24] do not always include all information we need to design mounts for the prisms, we include here design data for 27 types of prisms, some of which were not included in any of the above references.

In the description of each prism type, we include orthographic projections, generic dimensions, axial path length, and, in some cases, isometric views, tunnel diagrams, approximate prism volume, and bonding area information (see Sect. 4.3). The following parameter definitions apply:

A	= aperture for collimated beam passage;
B, C, D, etc.,	= other linear dimensions;
a, b, c, etc.,	= typical bevels;
δ, θ, φ, etc.,	= angular dimensions;
t	= axial path length;
V	= prism volume (neglecting small bevels);
ρ	= glass density;
a_G	= acceleration factor measured as "times gravity,"
Q	= minimum bond area in cm^2 for adhesive with joint strength of 1.38×10^7 Pa and a safety factor of 2, and
Q_{MAX}	= maximum circular (C) or racetrack (RT) bond area in cm^2 achievable on prism mounting surface.

3.3.1 Right angle prism

Figure 3.1(b) shows the function of this prism in its most common role as a means for deviating a beam by 90° whereas Fig. 3.2 shows its tunnel diagram. Fig. 3.7 shows three views and design equations for this prism. A typical bonded interface to a cylindrical mounting pad is indicated. Variations of this prism are used as the Porro prism, the Dove prism, and the double Dove prism. Each of these designs is considered later.

3.3.2 Beamsplitter (beamcombiner) cube prism

Two right angle prisms cemented together at their hypotenuse surfaces with a partially reflecting coating at the interface form a cube-shaped beamsplitter or beam-combiner. This type prism is shown in Fig. 3.8. If this prism (or any multiple-component prism) is to be bonded to a mechanical mount, the adhesive joint should be constrained to one component; the bond would then not bridge the cemented joint. This is because the two glass surfaces may not be accurately co-planar and the strength of the bond may be degraded by differences in adhesive thickness. If the adjacent surfaces are re-ground after cementing, bonding across the joint may be acceptable.

Most of the design equations for the beamsplitter cube apply also to a monolithic cube such as might be used as a rotating prism in a high-speed camera. With a solid cube, the bond area, Q, can be as large as $Q_{MAX} = 0.78A^2$ cm^2.

3.3.3 Amici prism

The Amici prism (see Fig. 3.9) is essentially a right-angle prism with its hypotenuse configured as a 90° "roof" so a transmitted beam makes two reflections instead of just one. A right-handed image is produced. The prism can be designed so the transmitted beam is split by the dihedral edge between the roof surfaces or so the beam hits the roof surfaces in sequence. These possibilities are illustrated in Fig. 3.10(a) and (b), respectively. In the former case, the dihedral angle must be accurately 90° (i.e., within a few arc-seconds) in order not to produce a noticeable double image. This makes the smaller component's cost higher because of the added labor or fixturing required to correct the roof angle. The prisms of Fig. 3.10 are shown to be of equal size so the beam transmitted without splitting must be smaller than that with splitting.

3.3.4 Porro prism

A right angle prism arranged so the beam enters and exits the hypotenuse surface, as shown in Fig. 3.11(a), is called a Porro prism. Ray a-a′ travels nearly parallel to the axis while rays b-b′ and c-c′ enter at slightly different field angles. Note that rays a-a′ and b-b′ turn around and exit parallel to the entering rays; this shows that the prism is retrodirective in the plane of refraction. Path c-c′ represents a ray entering near the edge of the prism. It intercepts the hypotenuse A-C internally and hence has three reflections and produces an inverted image. Such a ray is called a "ghost" ray since it does not contribute useful information to the main image. It does add stray light so should be eliminated. The groove cut into the center of the hypotenuse does just that so it is a usual

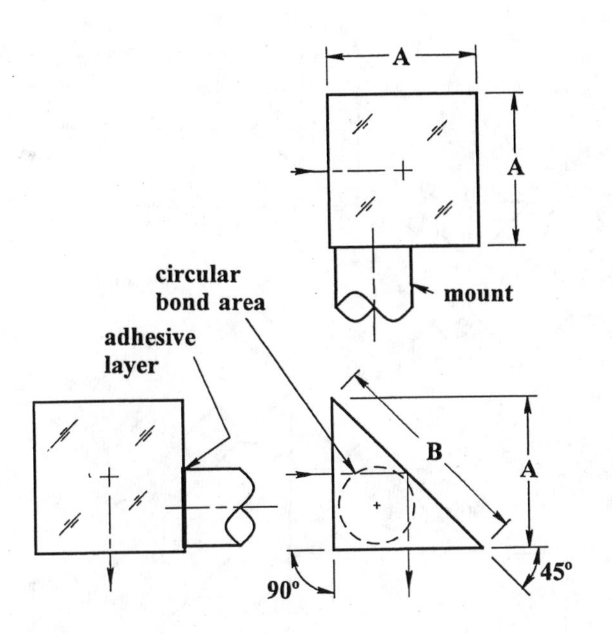

Fig. 3.7 Right angle prism
$t = A$; $B = 1.414A$; $V = 0.5A^3$
$Q = 7.10 \times 10^{-6} A^3 \rho\ a_G$; $Q_{MAX\ (C)} = 0.27A^2$

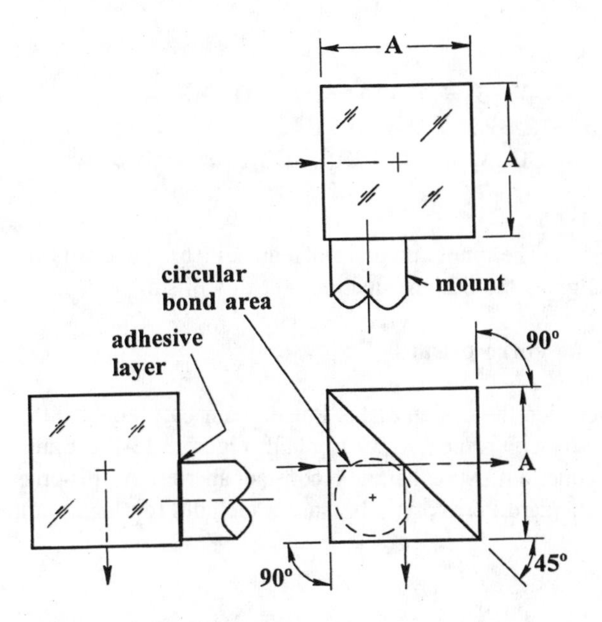

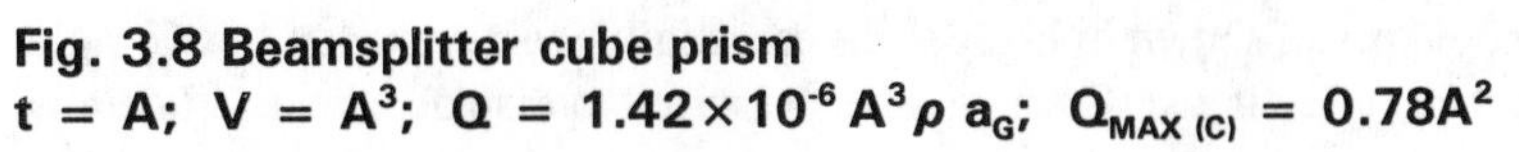
Fig. 3.8 Beamsplitter cube prism
$t = A$; $V = A^3$; $Q = 1.42 \times 10^{-6} A^3 \rho\ a_G$; $Q_{MAX\ (C)} = 0.78A^2$

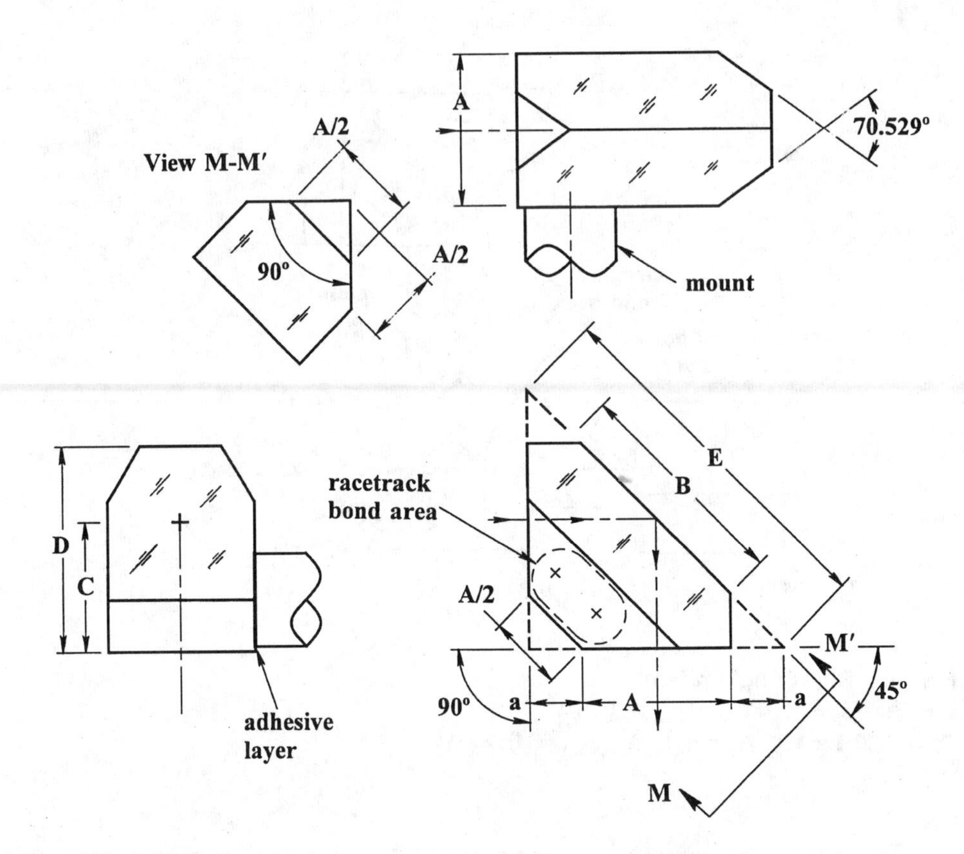

Fig. 3.9 Amici prism
t = 2.707A; a = 0.354A; B = 1.414A; C = 0.8536A
D = 1.354A; E = B + 2.828a; V = 0.78A³
$Q = 1.10 \times 10^{-5} A^3 \rho\, a_G$; $Q_{MAX\,(C)} = 0.16A^2$; $Q_{MAX\,(RT)} = 0.31A^2$

feature of the Porro design. The tunnel diagram of Fig. 3.11(b) shows all these rays and the groove. Figure 3.12 gives the design equations for this prism.

3.3.5 Abbe version of the Porro prism

Ernst Abbe modified the design of the Porro prism by rotating the bottom half about the optic axis by 90° with respect to the top half. Figure 3.13 illustrates this prism and provides its design equations. Note that, with constant aperture A, this prism is shown slightly larger than the standard Porro prism because it includes (optional) larger bevels.

3.3.6 Rhomboid prism

The rhomboid prism shown in Fig. 3.14 is essentially the integration of two right angle prisms with their reflecting surfaces parallel. It is used to displace the axis laterally

(a)

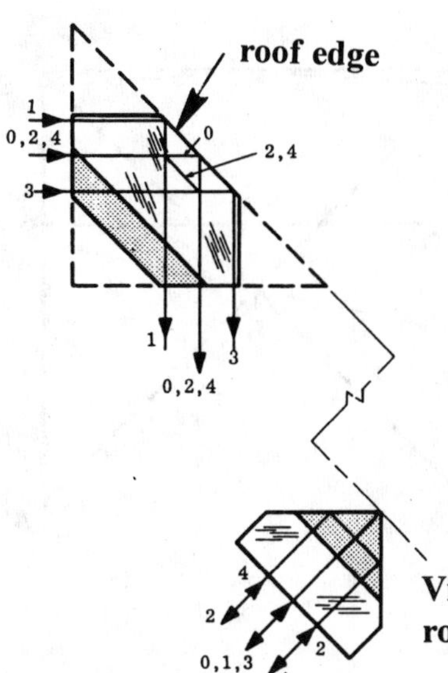

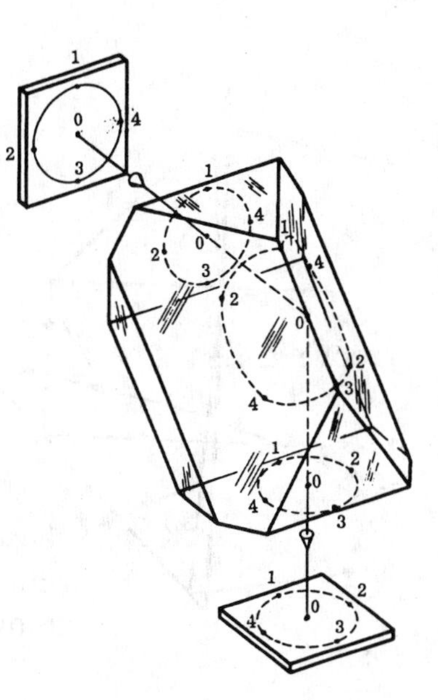

(b)

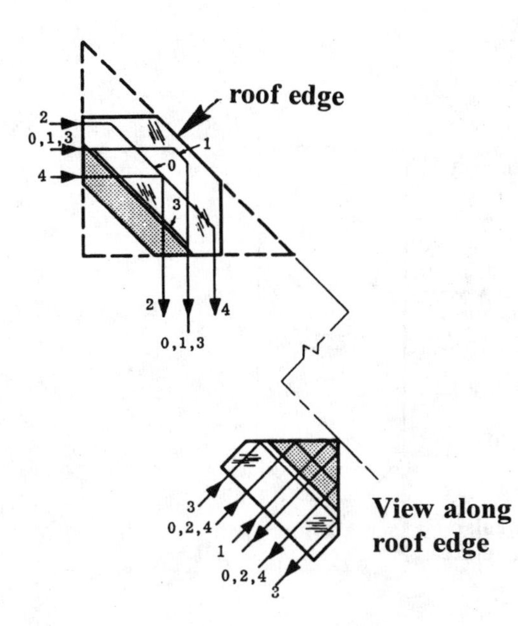

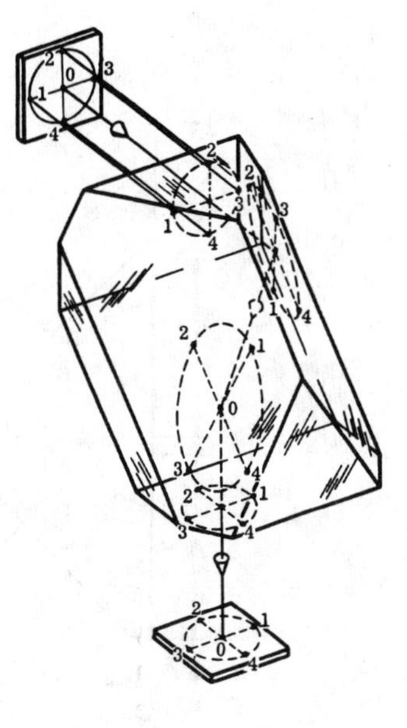

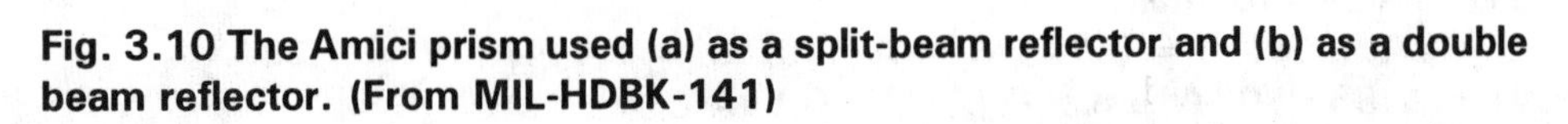

Fig. 3.10 The Amici prism used (a) as a split-beam reflector and (b) as a double beam reflector. (From MIL-HDBK-141)

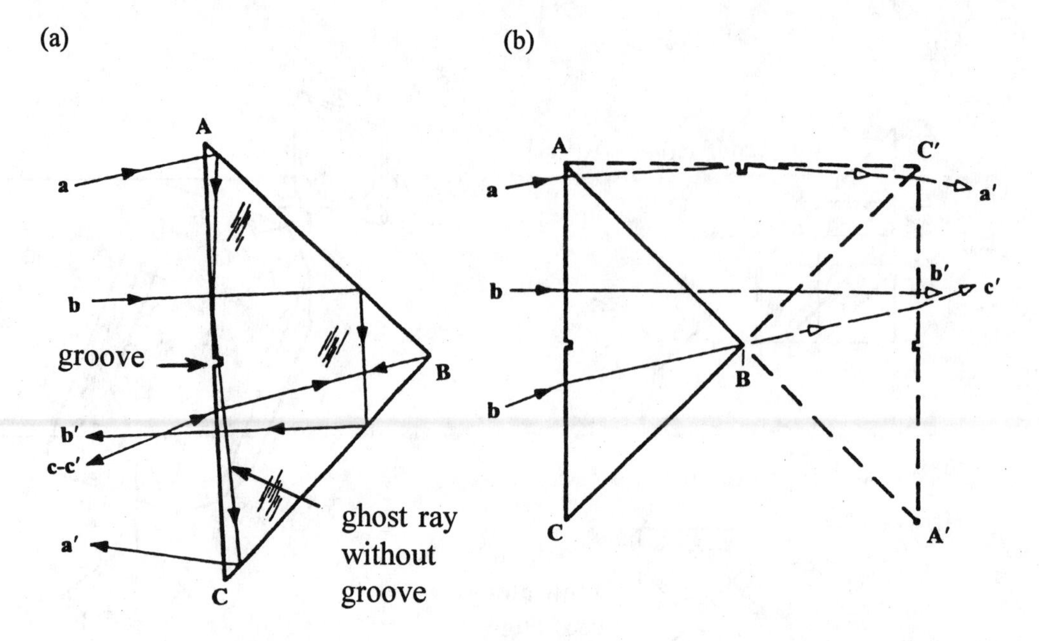

Fig. 3.11 (a) Typical ray paths through a Porro prism. (b) Its tunnel diagram

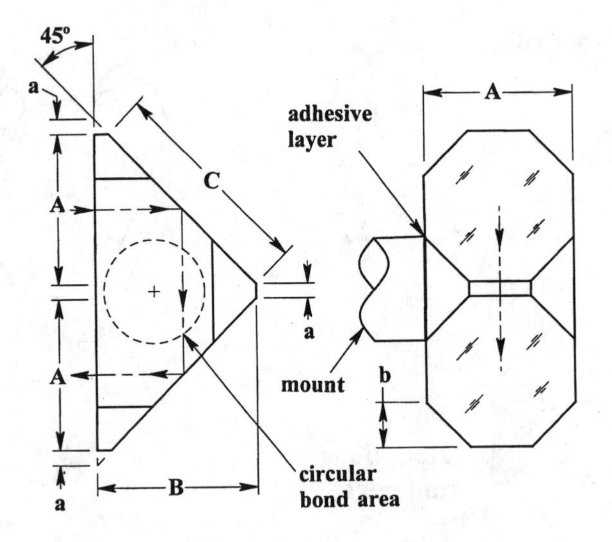

Fig. 3.12 Porro prism
$t = 2A + 3a$; $a = 0.1A$; $B = A + a$; $C = 1.414A$; $V = 1.29A^3$
$Q = 1.83 \times 10^{-5} A^3 \rho\, a_G$; $Q_{MAX\,(C)} = 0.51A^2$

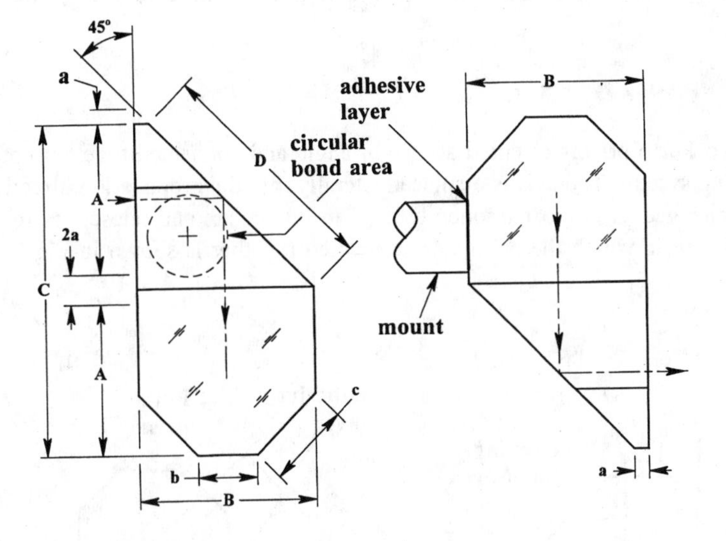

Fig. 3.13 Abbe version of the Porro prism
$t = 2A + 4a$; $a = 0.1A$; $b = 0.414A$; $B = A + 2a$; $C = 2.2A$
$D = 1.556A$; $V = 1.83A^3$
$Q = 2.62 \times 10^{-5} A^3 \rho\, a_G$; $Q_{MAX\,(C)} = 0.39A^2$

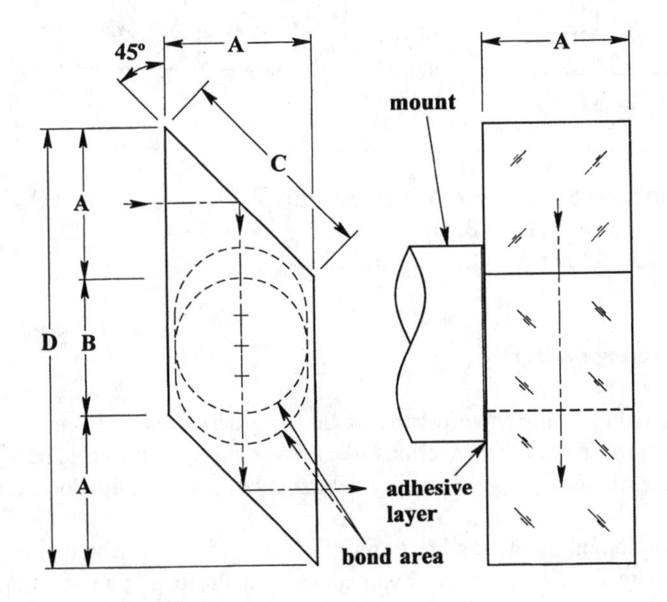

Fig. 3.14 Rhomboid prism
$t = 2A + B$; $B =$ variable; $V = A^3 (A + B)$
$Q = 1.4 \times 10^{-5} A^2 (A + B) \rho\, a_G$
For $B = 0$: $Q_{MAX\,(C)} = 0.39A^2$; $Q_{MAX\,(RT)} = 0.68A^2$
For $B > 0.414A$: $Q_{MAX\,(C)} = 0.78A^2$; $Q_{MAX\,(RT)} = 0.58A^2 + 0.5AB$

without changing the axis direction. The prism is insensitive to tilt in the plane of refraction.

3.3.7 Porro erecting system

Two Porro prisms oriented at a right angle and connected together constitute a Porro erecting system. The axis is displaced laterally, but the propagation direction of the beam is unchanged. It is most frequently used in binoculars and telescopes to erect the image. A design in which the prisms are cemented together is shown in Fig. 3.15.

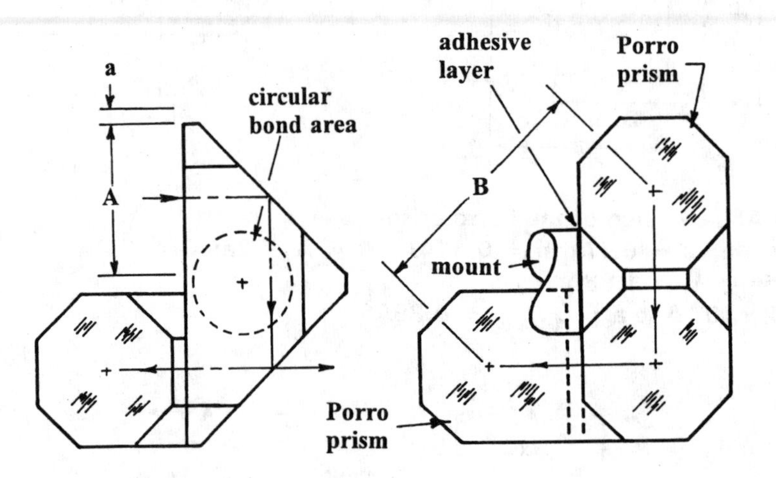

Fig. 3.15 Porro erecting system (cemented)
$t = 4A + 2a$; $a = 0.1A$; $B = 1.556A$; $V = 2.57A^3$
$Q = 3.67 \times 10^{-5} A^3 \rho a_G$; $Q_{MAX\ (C)} = 0.51A^2$

3.3.8 Abbe erecting system

The combination of two prisms of the type described in Sect. 3.3.5 makes an erecting prism subassembly that functions like a Porro erecting system. For a given prism aperture, the lateral offset is approximately 77% smaller than with the Porro arrangement.

A variation on the design has two right-angle prisms cemented side-by-side, but facing in opposite directions on the hypotenuse of a Porro prism. See Fig. 3.16. The resulting subassembly can have a larger bond area than the two-prism variety.

3.3.9 Penta prism

The penta prism neither reverts nor inverts the image; it merely turns the axis by

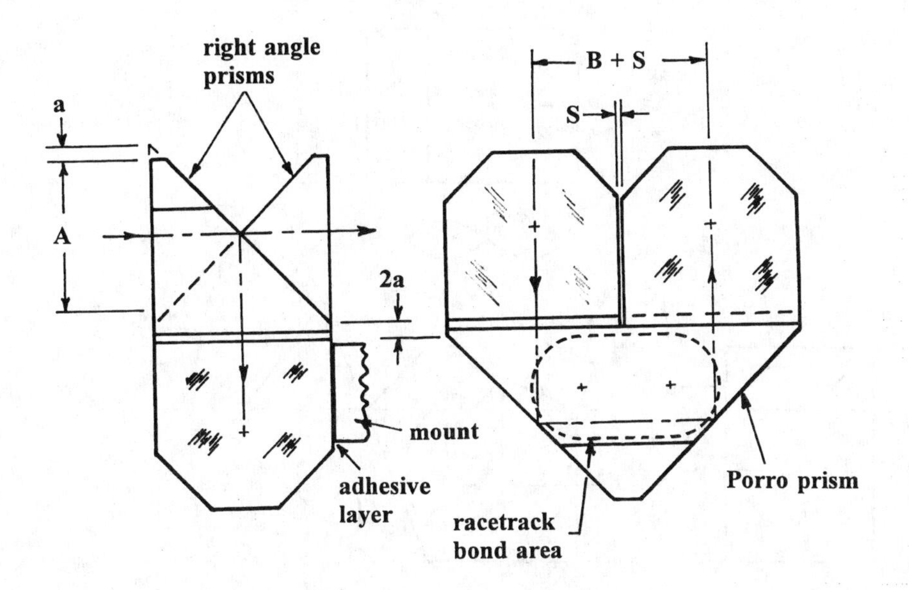

Fig. 3.16 Abbe erecting system (cemented)
$t = 4A + 6a + S$; $a = 0.1A$; $B = 1.150A$; $V = 3.69A^3$
$Q = 5.24 \times 10^{-5} A^3 \rho a_G$; $Q_{MAX(C)} = 0.50A^2$; $Q_{MAX(RT)} = 0.96A^2$

90°. The design is given in Fig. 3.17. A useful characteristic of this prism is that it is insensitive to tilt as a rigid body in the plane of refraction. For this reason, it is used in applications such as optical rangefinders and surveying equipment.

3.3.10 Roof penta prism

If we convert one reflecting surface of the penta prism into a 90° roof, the component inverts the image in the direction normal to the plane of refraction. For a given aperture and material, the roof penta is about 17% larger and 19% heavier than the standard penta. Adding the roof does not change the penta's constant deviation characteristic. The design for the roof penta is shown in Fig. 3.18.

3.3.11 Amici/penta erecting system

A combination of the Amici prism with a penta prism provides two reflections in each direction perpendicular to the axis so can be used as an erecting system. Frequently the prisms are cemented together as illustrated in Fig. 3.19(a). This design has been used in some binoculars. A functionally similar erecting system can be obtained by combining a right angle prism with a roof penta prism. See Fig. 3.19(b). For a particular

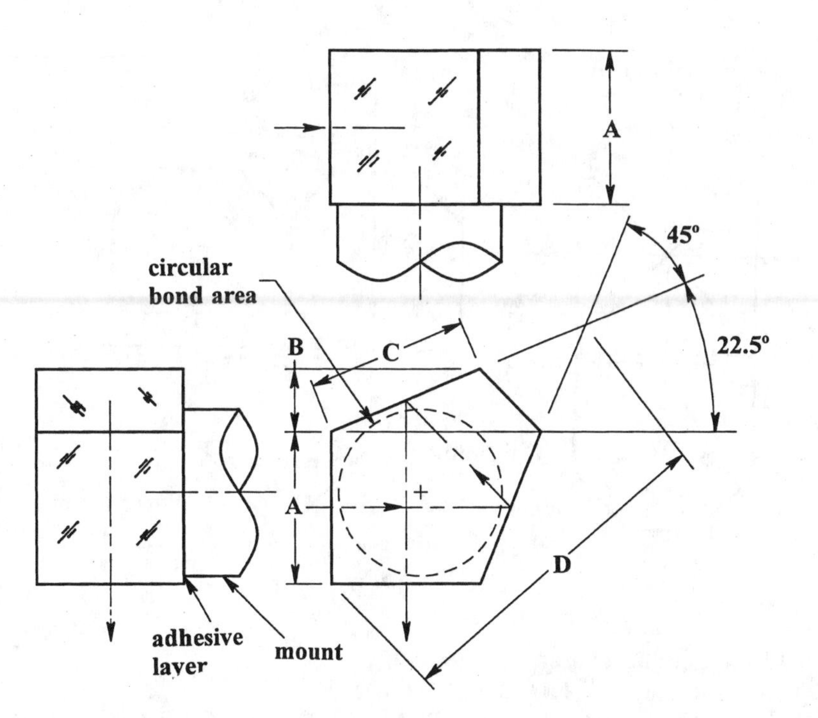

Fig. 3.17 Penta prism
$t = 2A + B + 0.707(A + B) = 3.414A$; $B = 0.414A$; $C = 1.082A$
$D = 2.414A$; $V = 1.50A^3$; $Q = 2.13\times10^{-5} A^3 \rho\, a_G$; $Q_{MAX\,(C)} = 1.13A^2$

aperture A, the indicated height dimensions differ by about 4%. This system has primarily been used in military periscopes. A variation of this design was used in an experimental compact military binocular.[25] This prism is illustrated in Fig. 3.20.

3.3.12 Dove prism

The Dove prism is a right angle prism with the top section removed and the optical axis parallel to the hypotenuse face as shown in Fig. 3.21. This single-reflection prism inverts the image in the plane of refraction. It is most commonly used to rotate the image by turning the prism about its axis; the image rotates at twice the speed of the prism. Because of the oblique incidence of the axis at the entrance and exit faces, the prism can be used only in a collimated beam. The prism dimensions depend upon the refractive index because of the refraction at the tilted faces. A specific design using BK7 glass is defined in the figure. Alternate versions can have entrance and exit faces tilted at angles other than 45°.

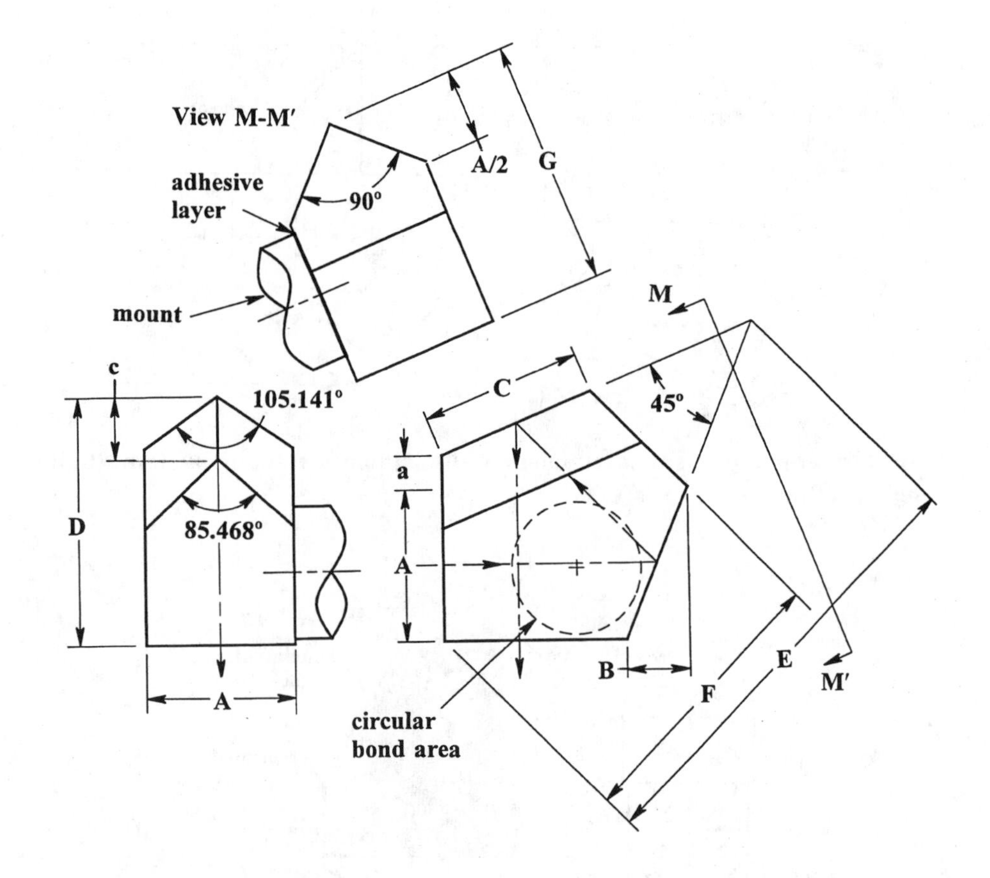

Fig. 3.18 Roof penta prism
$t = 4.223A$; $a = 0.237A$; $c = 0.383A$; $B = 0.414A$; $C = 1.082A$
$D = 1.651A$; $E = 2.986A$; $F = 1.874A$; $G = 1.621A$; $V = 1.79A^3$
$Q = 2.55 \times 10^{-5} A^3 \rho a_G$; $Q_{MAX(C)} = 0.82A^2$

3.3.13 Double Dove prism

This prism comprises two Dove prisms, each of aperture A/2 by A, attached together at their hypotenuse faces. Figure 3.22 shows the configuration. It is commonly used as an image rotator. The prisms can be airspaced by a small distance and mechanically held. TIR then occurs. They also can be cemented together. In this case, a reflecting coating is placed on one prism face before cementing to keep the light from passing through the interface. For a given aperture, A, the double Dove prism is only one-half the length of the corresponding standard Dove prism. To minimize light loss, the leading and trailing edges of both prisms are given only minimal protective bevels. These bevels are ignored in the design equations.

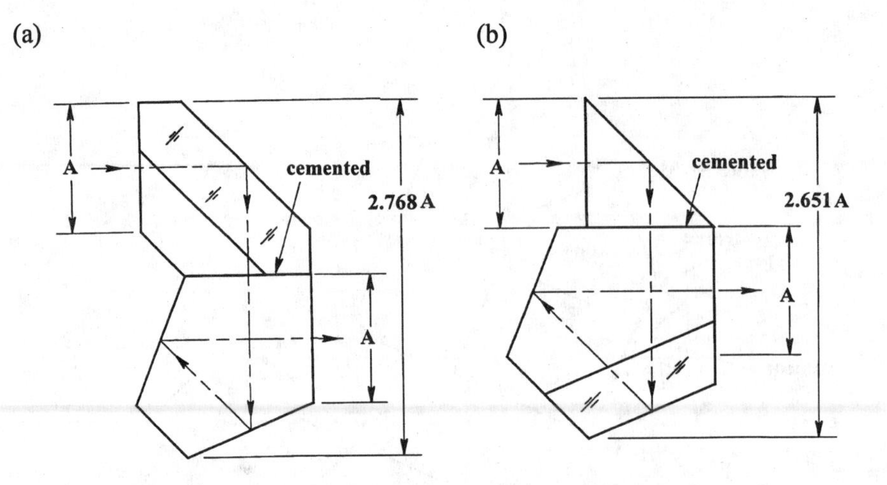

Fig. 3.19 Erecting prism assemblies: (a) Amici/penta erecting system; (b) Right angle/roof penta

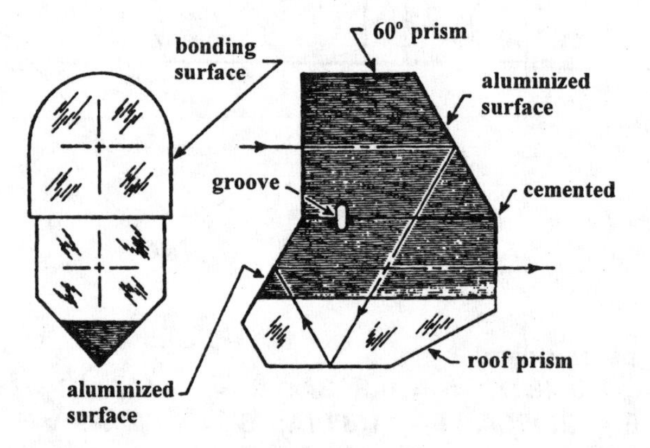

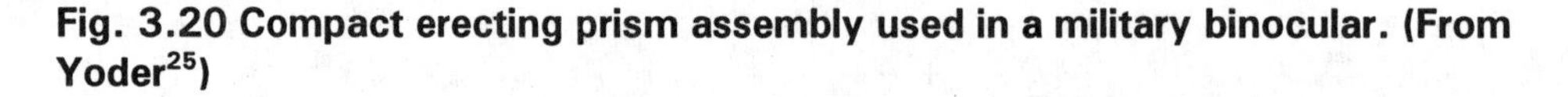

Fig. 3.20 Compact erecting prism assembly used in a military binocular. (From Yoder[25])

As shown in the end view of Fig. 3.22, the shape of a circular beam entering a double Dove prism is changed into a pair of "Dee-shaped" beams with curved edges adjacent. If vignetting is to be avoided, the apertures of subsequent optics must be large enough to accept the divided beam of diameter $\sqrt{2}A$. The MTF of the optical system in which the prism is used is somewhat degraded by the divided aperture. The 45° angles of the prisms must be quite accurate in order to minimize image doubling.

A cemented "cube-shaped" version of the double Dove prism is sometimes used as a line-of-sight scanning means. The prism is then rotated about an axis normal to the plane of refraction and passing through the prism's center. When located in front of a

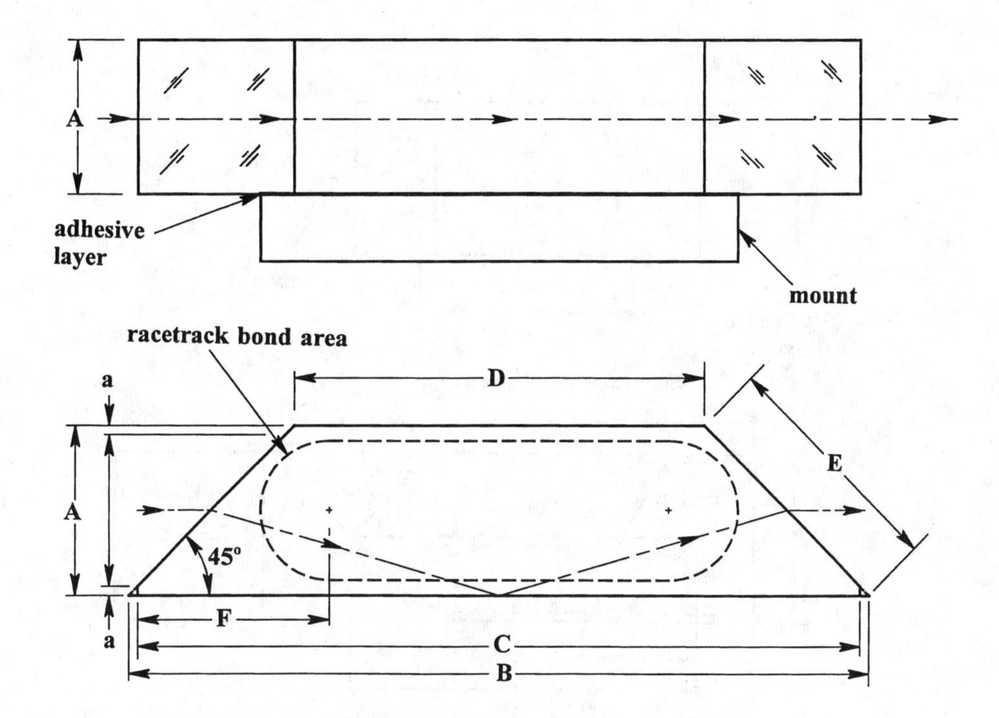

Fig. 3.21 Dove prism

$\theta = 45°$

$t = (n)(A + 2a) / (\sin\theta)((n^2 - \sin^2\theta)^{1/2} - \sin\theta)$

$a = 0.05A$

$B = (A + 2a)[(((n^2 - \sin^2\theta)^{1/2} + \sin\theta) / ((n^2 - \sin^2\theta)^{1/2} - \sin\theta))) + 1]$

$C = B - 2a$; $D = B - 2(A + 2a)$; $E = (A + a) / \cos\theta = 1.485A$

$F = (A + 2a) / (2\tan(\theta / 2) = 1.328A$

$V = (A)(A + 2a)(B) - (A)(A + 2a)^2 - a^2A = 3.905A^3$

$Q = 1.42 \times 10^{-5} V\rho a_G$; $Q_{MAX\ (C)} = \pi((A + a) / 2)^2 = 0.95A^2$

$Q_{MAX\ (RT)} = Q_{MAX\ (C)} + (A + a)(B - (2)(0.55A / \tan(\theta / 2)))$

For BK7 glass, $n = 1.517$

$t = 3.716A$; $B = 4.650A$; $C = 4.550A$; $D = 2.450A$;

$E = 1.485A$; $F = 1.328A$; $V = 3.905A^3$;

$Q = 1.39 \times 10^{-4} A^3G$; $Q_{MAX\ (C)} = 0.95A^2$; $Q_{MAX\ (RT)} = 3.14A^2$

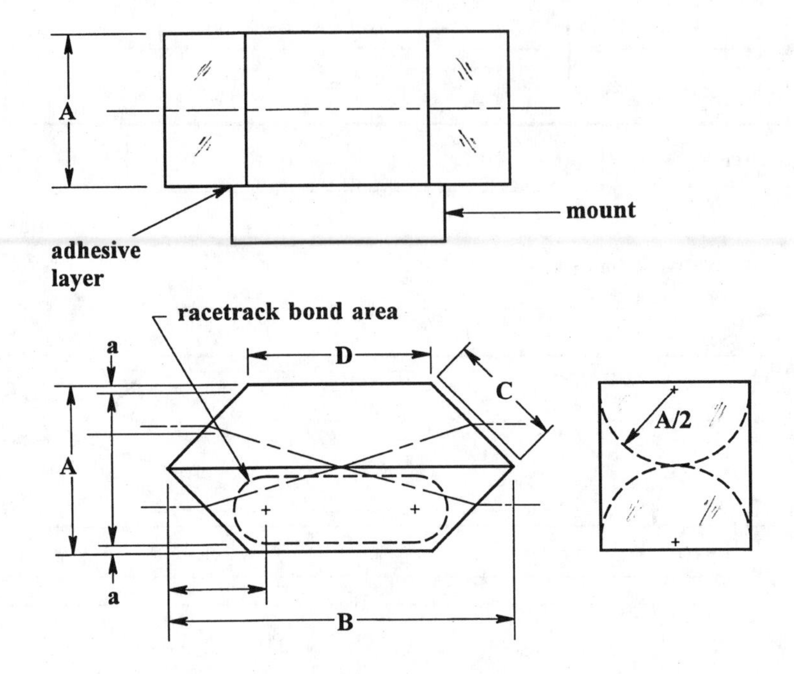

Fig. 3.22 Double Dove prism

$\theta = 45°$
$t = (n)((A / 2) + a) / (\sin \theta)((n^2 - \sin^2 \theta)^{1/2} - \sin \theta)$
$a = 0.05A$
$B = ((A / 2) + a)[(((n^2 - \sin^2 \theta)^{1/2} + \sin \theta) / ((n^2 - \sin^2 \theta)^{1/2} - \sin \theta)) + 1]$
$C = A + 2a = 1.1A$; $D = B - 2((A / 2) + a)$; $E = ((A / 2) + a) / \cos \theta$
$F = ((A / 2) + a) / (2)(\tan (\theta / 2)) = 0.664A$
$V = (A)(B)(A + 2a) - (2)(A)((A / 2) + a)^2 = 1.952A^3$
$Q = 2.44 \times 10^{-5} A^3 \rho a_G$; $Q_{MAX\ (C)} = (\pi)((A / 2) + a)^2 = 0.95A^2$
$Q_{MAX\ (RT)} = Q_{MAX\ (C)} + (A + a)(B - (2)(0.55A / \tan (\theta / 2))) = 1.5A^2$

For BK7 glass, $n = 1.517$
$t = 3.716A$; $B = 4.650A$; $C = 4.550A$; $D = 2.450A$
$E = 1.485A$; $F = 1.328A$; $V = 3.905A^3$
$Q = 1.39 \times 10^{-4} A^3 a_G$; $Q_{MAX\ (C)} = 0.95A^2$; $Q_{MAX\ (RT)} = 1.50A^2$

camera, periscope, or other optical instrument, such a prism can scan the system axis well over 180° in object space.

3.3.14 Reversion prism

This two-component (cemented) prism is shown in Fig. 3.23. It functions as an image rotator just like a Dove prism, but it can be used in converging or diverging beams. The central reflecting face must have a reflecting coating to prevent refraction through it. This surface usually is then covered by a protective coating such as electroplated copper and paint like a "back surface" mirror. Another version of this prism has the central reflecting surface replaced by a 90° roof to invert the image in the direction perpendicular to the plane of refraction.

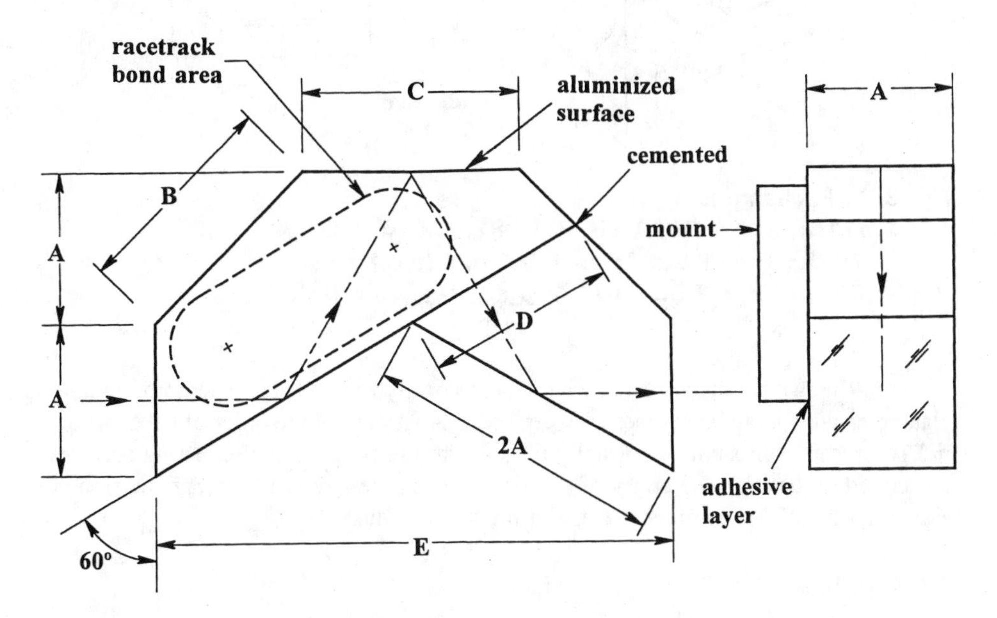

Fig. 3.23 Reversion prism
t = 5.196A; B = 1.414A; C = 1.464A; D = 1.2679A
E = 3.464A; V = 4.20A^3
Q = 5.97×10^{-5} A^3 ρ a_G; $Q_{MAX\,(C)}$ = 1.09A^2; $Q_{MAX\,(RT)}$ = 1.99A^2

3.3.15 Pechan prism

Because it has an odd number (5) of reflections, the Pechan prism is frequently used as a compact image rotator in place of the Dove or double Dove prisms because it can be used in convergent or divergent beams. The design is given in Fig. 3.24. The optical axis of the nominal design is displaced very slightly due to the central air space, but it is not deviated. The two outer reflecting surfaces must have reflecting coatings and protective overcoat and/or paint while the internal reflections occur by TIR so those surfaces are not coated.

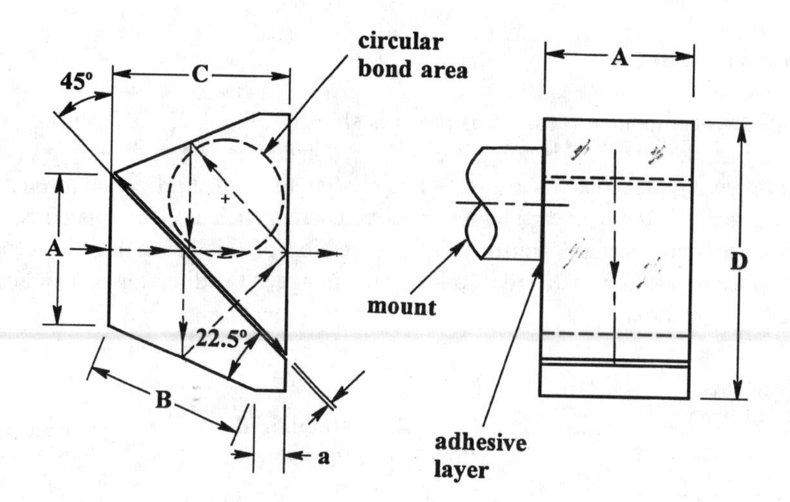

Fig. 3.24 Pechan prism
t = 4.621A; a = 0.207A; B = 1.082A; C = 1.207A
D = 1.707A; E = 1.828A; b = 0.1 mm (typ.)
$V = 1.80A^3$, $Q = 2.56 \times 10^{-5}\ A^3 \rho\ a_G$, $Q_{MAX\ (C)} = 0.60A^2$

The two prisms are usually held mechanically or bonded to a common mounting plate to create a narrow air space between them. A spacing of the order of 0.1 mm (0.004 in.) is typical. Thin shims are usually placed near the edges of the reflecting surfaces in a clamped mounting. The edges of the air space can be covered by a narrow ribbon of sealant such as RTV to prevent entry of moisture or dust.

3.3.16 Delta prism

Figure 3.25 shows the path of an axial ray through this triangular prism. TIR occurs in sequence at the exit and entrance faces. The intermediate face must be silvered to make it reflect. With the proper choice of index of refraction, apex angle, and prism height, the internal path can be made symmetrical about the vertical axis of the prism; the exiting axial ray then is co-linear with the entering axial ray.

With an odd number of reflections (3), the delta prism can be used as an image rotator. Since it has tilted entrance and exit faces, it can be used only in a collimated beam. For a given aperture, the overall size of the delta prism rotator is smaller than the Dove prism.[26] It has fewer lossy reflections and a shorter glass path than the Pechan prism so it should have better light transmission than the latter.

Design of this prism starts with choice of index of refraction. A value for the angle of incidence, I_1, at the first surface is then assumed. This angle equals the apex angle, θ. We vary I_1 until the same value for I_1' is obtained by Eqs. 3.11 and 3.12.

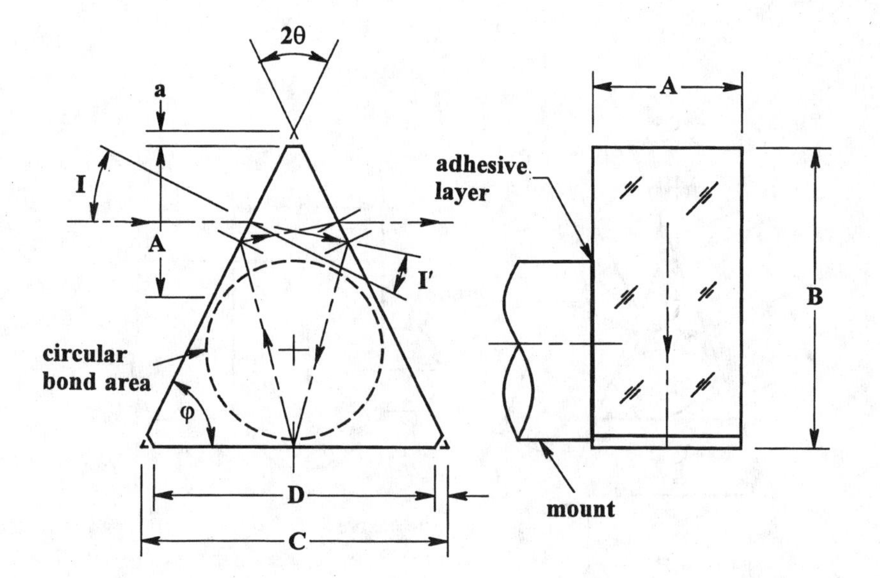

Fig. 3.25 Delta prism
Given n and 2θ calculated as explained in text.
$a = 0.1A;\ B = ((A + 2a)(\sin(180° - 4\theta)) / (2)(\cos\theta)(\sin\theta)) - a$
$C = 2B\tan\theta;\ D = (A\cos\theta) / (\cos\theta\cos(45° - (\theta/2)))$
$b = (C - D) / 2;\ \varphi = 90° - \theta$
$t_1 = ((A / 2) + a)(\sin 2\theta) / (\cos\theta\sin(90° - 2\theta + I'))$
$t_2 = (B - A - a - t_1\sin\theta) / \cos\theta;\quad t = 2(t_1 + t_2)$
$V = A((B + a)C - a^2\tan\theta - 2b^2\cos(\varphi / 2)\sin(\varphi / 2))$
$Q = 1.42\times 10^{-5} V \rho a_G;\ Q_{MAX} = \pi((C^2 / 4)\tan^2(\varphi / 2))$

$$I_1' = \theta = \arcsin(\sin I_1 / n) \tag{3.11}$$

$$I_1' = 4\theta - 90° \tag{3.12}$$

We then calculate $I_2 = (90° - I_2') / 2$ and check to see if TIR occurs at the exit surface, i.e., $I_2 > I_C$, for the chosen glass using Eq. (3.10). If not, a new glass with higher index must be chosen. Once these conditions are satisfied, we apply the equations of Fig. 3.25.

3.3.17 Schmidt prism

This roof prism will invert and revert the image so is usually used as an erecting system in telescopes. It also deviates the axis by 45° which allows an eyepiece axis orientation with respect to the objective axis that is convenient for some applications. The entrance and exit faces are normal to the axis. Figure 3.26 applies. If a roof is added to

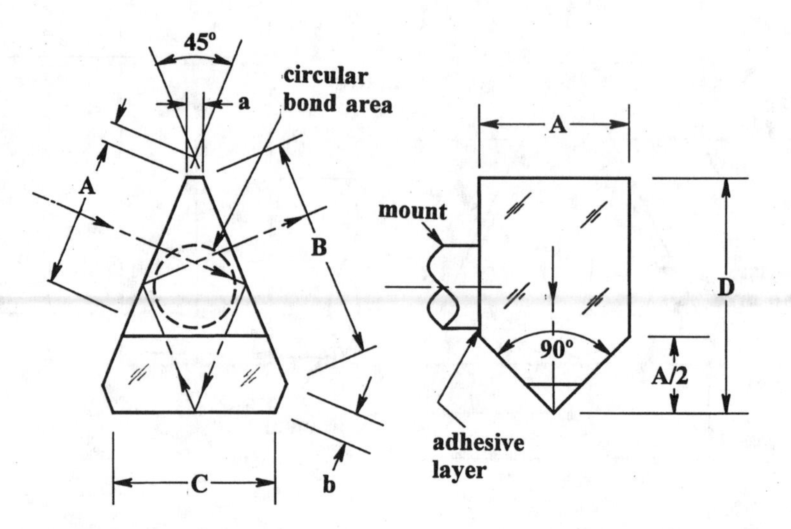

Fig. 3.26 Schmidt prism
a = 0.1A; b = 0.185A; c = 0.121A; d = 0.131A
B = 1.468A; C = 1.082A; D = 1.527A; E = 0.7071A
t = 3.045A; V = 0.863A^3; Q = 1.227 × 10^{-5}A$^3\rho a_G$; Q$_{MAX\ (C)}$ = 0.318A^3

the above-described delta prism, an erecting system with co-axial input and output optical axes results. This prism would resemble the Schmidt prism, but the entrance and exit faces would be tilted with respect to the axis so the modified prism must be used in a collimated beam.

3.3.18 45° Bauernfeind prism

This prism provides a 45° deviation of the axis using two internal reflections. The first reflection is by TIR while the second takes place at a coated reflecting surface. The smaller element of the Pechan prism is of this type. Figure 3.27 shows the design. A 60° deviation version of this prism also has been used in many applications.

The combination of a Schmidt prism with a 45° Bauernfeind prism forms a popular erecting system for binoculars because of its compact design. It sometimes is called the Schmidt-Pechan roof prism. An example of a mounting for such a prism is shown in Section 4.3.3.

3.3.19 Internally reflecting axicon prism

With conical surfaces as their active optical surfaces, axicons are frequently used to change a small circular laser beam into a larger outside diameter (OD) annular beam.

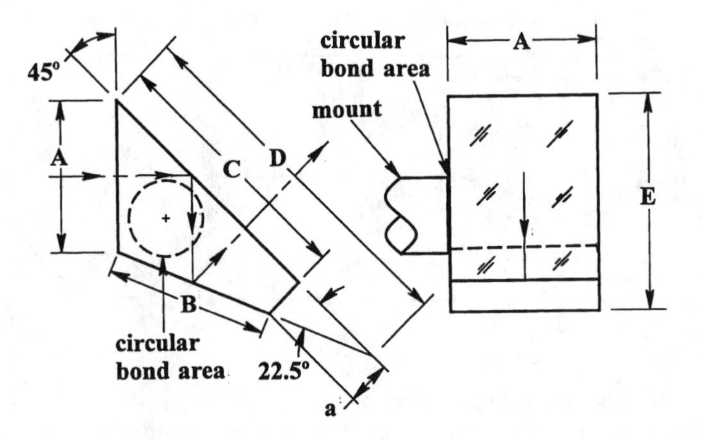

Fig. 3.27 45° Bauernfeind prism
a = 0.293A; B = 1.082A; C = 1.707A; D = 2.414A; E = 1.414A
t = 1.707A; V = 0.75A³; Q = 1.066 × 10⁻⁵ A³ ρ a_G; Q_MAX (c) = 0.331A²

The version shown in Fig. 3.28 has a coated reflecting surface to return the beam to and through the conical surface. Because of its rotational symmetry, this axicon is made with a circular cross-section. The apex is sharp or carries a very small protective bevel. A centrally perforated flat mirror at 45° provides a convenient way to separate the coaxial beams if located in front of this prism.

A refracting version of this axicon has been used to accomplish the same function. It is twice as long and is more expensive because it has two conical surfaces.

3.3.20 Cube corner prism

A corner cut symmetrically and diagonally from a solid glass cube creates a prism in the geometrical form of a tetrahedron (4-sided polyhedron). It has been referred to as a cube corner, corner cube, or tetrahedral prism. Light entering the diagonal face reflects internally from the other three faces and exits through the diagonal face. TIR usually occurs at each internal surface for common refractive index values. The return beam comprises six segments; one from each of the pie-shaped areas within the circular aperture shown in Fig. 3.29 as a dashed line. If the three dihedral angles between the adjacent reflecting surfaces are exactly 90°, the prism is retrodirective, even if the prism is significantly tilted. If one or more of these dihedral angles differs from 90°, the deviation differs from 180° and the reflected beams diverge.[27] This feature is used to advantage in applications such as laser tracking in space or those involving widely separated transmitter and receiver optical systems used for large-baseline ranging by triangulation.

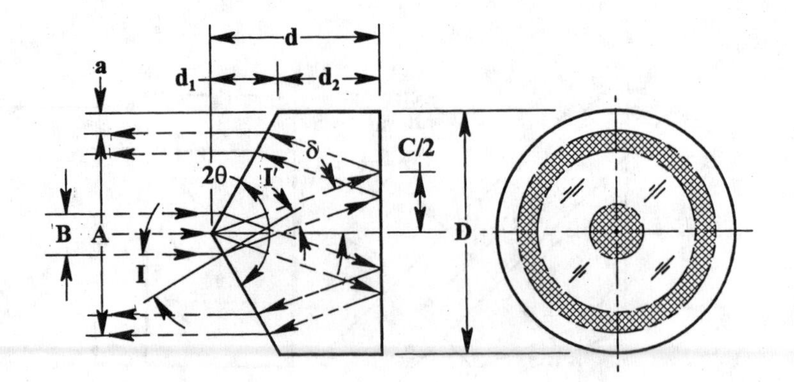

Fig. 3.28 Internally reflecting axicon prism
A = annulus OD; B = input beam OD; B/2 = annulus width; a = 0.1A
$I = 90° - \theta$; $I' = \arcsin(\sin I_1 / n)$; $\delta = I - I'$; $C = (2d)\tan\delta$; $D = A + 2a$
$d = (A / 4)((1 / \tan\theta) + (1 / \tan\delta))$; $d_1 = ((A/2) + a) / \tan\theta$; $d_2 = d - d_1$
$t = A / (2\sin\delta)$; $V = (0.785d_2 + 0.262d_1)A^2$

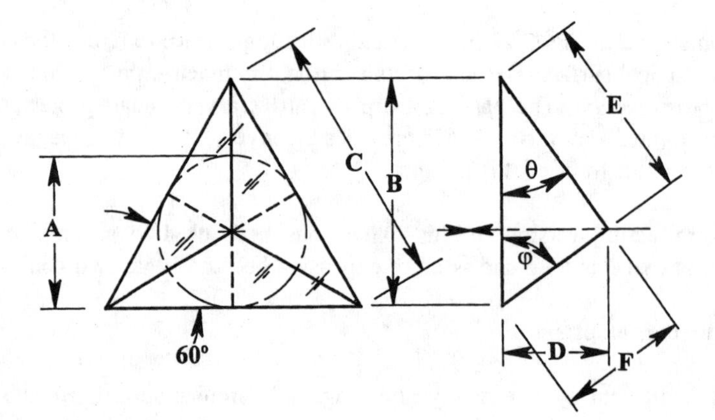

Fig. 3.29 Cube corner prism
A = aperture; $B = ((A / 2) / \sin 30°) + (A / 2) = 1.5A$
$C = (2B)\tan 30° = 1.732A$; $D = 0.707A$; $E = 1.225A$; $F = 0.866A$
$\varphi = 54.736°$; $\theta = 35.264°$, $t = 1.414A$

Figure 3.29 shows a cube corner prism with sharp corners. Usually, its rim is given a circular shape circumscribing the aperture (the dashed line). Figure 3.30 shows an actual circular-aperture prism. This is one of the 426 fused silica prisms used on the Laser Geodynamic Satellite (LAGEOS) launched by NASA in 1976 to provide scientists with extremely accurate measurements of movements of the earth's crust as a possible aid to understanding earthquakes, continental movement, and polar motion. The dihedral angles of the prisms were each 1.25 arc-seconds greater than 90°. A laser beam transmitted to the satellite was returned with sufficient divergence to reach a receiver telescope even though the satellite moved significantly during the beam's round trip transit time.

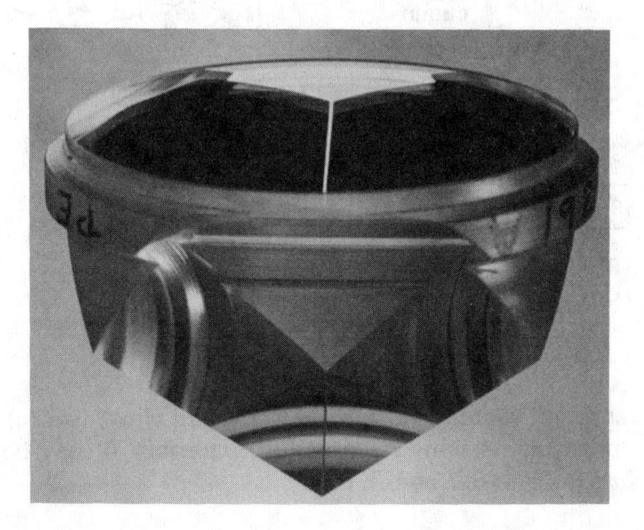

Fig. 3.30 Photograph of a typical cube corner prism with 3.8 cm (1.5 in.) circular aperture. (Courtesy of Raytheon Optical Systems Corp., Danbury, CT)

Another possible cube corner prism configuration has the rim cut to a hexagonal shape circumscribing the prism's circular clear aperture. This allows several of the prisms to be grouped tightly together so as to form a mosaic of closely-packed retrodirective prisms, thereby increasing the aperture of the group. Mirror versions of the cube corner prism frequently are used when operation outside the transmission range of normal refracting materials is needed. This, so called, hollow cube corner also has reduced weight for a given aperture.[28]

3.3.21 Biocular prism system

Attributed to Carl Zeiss, the prism system shown in Fig. 3.31 can be used in telescopes and microscopes when both eyes are to observe the same image presented by the objective. It does not provide stereoscopic vision, hence is called "biocular". From View (a), it can be seen to comprise four prisms: a right angle prism, P_1, cemented to a rhomboid prism, P_2, with a partially reflecting coating at the diagonal interface; optical path equalizing block, P_3; and a second rhomboid prism, P_4. The observer's interpupillary distance is designated as IPD. By rotating the prisms about the input axis, the IPD is changed to suit the individual using the instrument. Typically, the IPD is adjustable at least from 56 to 72 mm (2.20 to 2.83 in.). An external scale usually is provided to allow

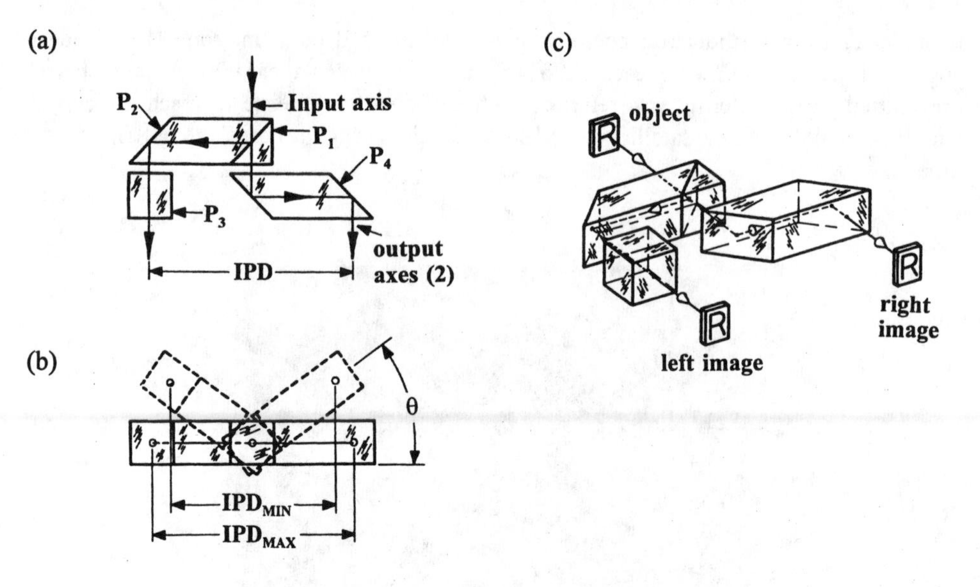

Fig. 3.31 Biocular prism system; (a) top view, (b) end view, and (c) isometric view. Individual prism dimensions for a uniform aperture A may be obtained from the equations in Figs. 3.7 and 3.14.

3.3.22 Dispersing prisms

Prisms are commonly used to disperse polychromatic light beams into their constituent colors in instruments such as spectrometers and monochromators. The index of refraction, n, of the optical material varies with wavelength so the deviation, measured with respect to the initial incident ray direction, of any ray transmitted at other than normal incidence to the prism's entrance and exit surfaces will depend upon n_λ, the angle of incidence at the entrance face, and the prism's apex angle, θ. Figure 3.32 illustrates two typical dispersing prisms. In each case, a single ray of "white" light is incident at I_1. Inside each prism, this ray splits into a spectrum of various colored rays. For clarity, the angles between rays are exaggerated in the figures. After refraction at the exit faces, rays of blue, yellow, and red wavelengths emerge with different deviation angles, δ_λ. The blue ray is deviated the most because $n_{BLUE} > n_{RED}$. If the emerging rays are imaged onto film or a screen by a lens, a multiplicity of images of different colors will be formed at slightly different lateral locations. While we refer here to colors as blue, yellow, and red, it should be understood that the phenomenon of dispersion applies to all wavelengths so we really mean the shorter, intermediate, and longer wavelength radiation under consideration in any given application. In the design shown in View (b), the deviation is unchanged for small rotations of the prism about an axis perpendicular to the plane of refraction; hence the name "constant deviation." The refractive index in this case usually is chosen to be large enough to cause TIR at the intermediate surface.

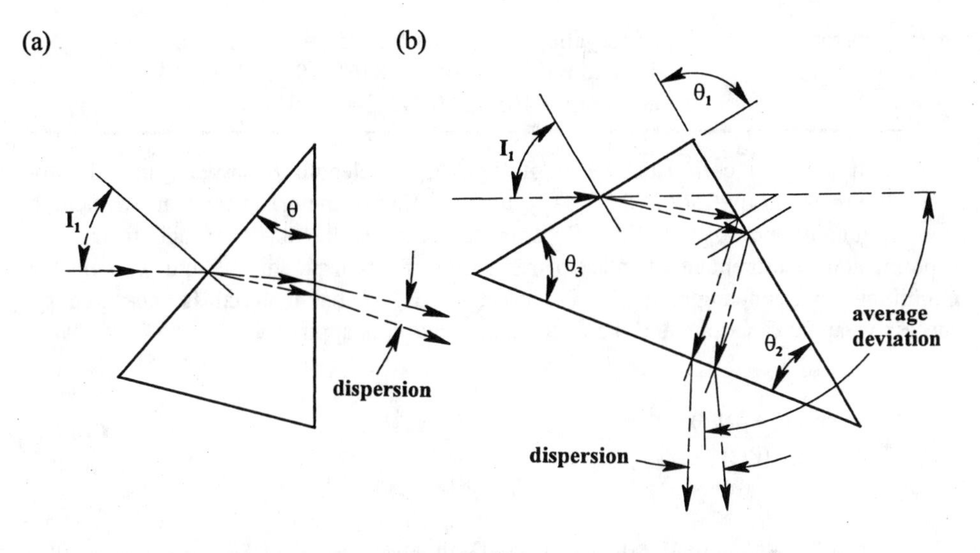

Fig. 3.32 Dispersion of a white light ray by (a) a simple prism and (b) by a constant deviation prism involving TIR

Numerical Example No. 4: Dispersion through a single prism.

A BK7 prism with apex angle, θ, of 30° disperses a white light collimated beam generally as shown in Fig. 3.32(a). Let the incident angle be $I_1 = 15°$. Applying Eq. (3.1) (Snell's law), what are the angular separations between the exiting blue, yellow, and red beams? If focused by a perfect 105 mm focal length lens onto a screen, what are the linear separations of the blue (F), yellow (d), and red (C) images at the screen?

wavelength (μm)	0.486 (F)	0.588 (d)	0.656 (C)
apex angle, θ (°)	30	30	30
I_1 (°)	15	15	15
sin I_1	0.25882	0.25882	0.25882
n_λ	1.52238	1.51680	1.51432
sin I_1'	0.17001	0.17063	0.17091
I_1' (°)	9.7884	9.8247	9.8410
$I_2 = I_1' - \theta$ (°)	-20.2116	-20.1753	-20.1590
sin I_2	-0.34549	-0.34489	-0.34463
sin I_2'	-0.52597	-0.52313	-0.52188
I_2' (°)	-31.7332	-31.5427	-31.4581
$\delta = I_1 - I_2' - \theta$ (°)	16.7332	16.5427	16.4581

The angles between the: blue and yellow beams = 16.7332° - 16.5427° = 0.1905°
yellow and red beams = 16.5427° - 16.4581° = 0.0846°
red and blue beams is 16.7332° - 16.4581° = 0.2751°

Image separation: blue to yellow = 105 tan 0.1905° = 0.3491 mm
yellow to red = 105 tan 0.0846° = 0.1550 mm
blue to red = 105 tan 0.2751° = 0.5041 mm

If a ray or collimated beam of light of wavelength λ passes symmetrically through a prism so $I_1 = - I_2'$ and $I_1' = - I_2$, the deviation of the prism for that wavelength is a minimum and $\delta_{MIN} = 2I_1 - \theta$. This condition is the basis of one means for experimental measurement of index of refraction of a transparent medium wherein the minimum deviation angle, δ_{MIN}, of a prism made of that material is measured by successive approximations and the following equation is applied:

$$n_{PRISM} = \frac{\sin((\theta + \delta)/2)}{\sin(\theta/2)}. \tag{3.12}$$

If we want any two of the various colored rays to emerge from the prism parallel to each other, we must use a combination of at least two prisms made of different glasses. Usually, these prisms are cemented together. Such a prism is called an achromatic prism. Figure 3.33 shows one configuration for an achromatic prism. All such prisms can be designed by choosing refractive indices and the first prism's apex angle then repeatedly applying Snell's law to find the appropriate incident angle and second prism apex angle that gives the desired deviation for a chosen wavelength and the desired dispersion for two other wavelengths that bracket the chosen one. The angle between the exiting rays with shortest and longest wavelengths is called the primary chromatic aberration; it here should be essentially zero. The angle between either of these extreme wavelength rays and that with intermediate wavelength is called the secondary chromatic aberration of the prism.

To illustrate a typical design procedure, in a two-element prism of the type shown in Fig. 3.33, we might specify that a yellow ray should enter the first prism at I_1 equal to the value for the minimum deviation condition if that prism were immersed in

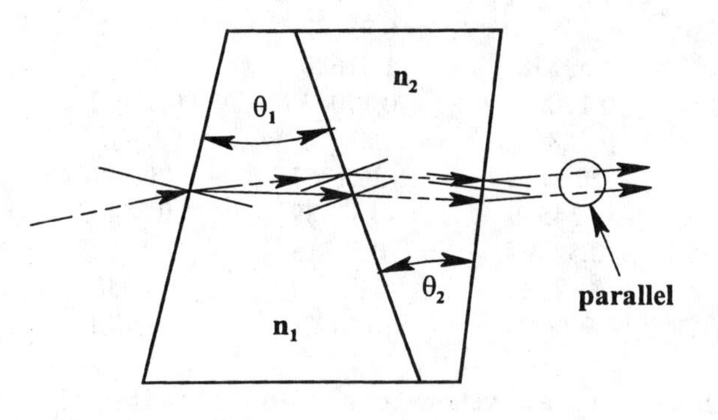

Fig. 3.33 A typical achromatic dispersing prism

air. The blue and red rays would then be dispersed. We would assume a value for θ_1 and calculate $I_1' = -I_2 = \theta_1/2$ and obtain I_1 from Snell's law. We would then add the second prism and redetermine I_2'. The following equation could then be used to find θ_2:

$$cotan\,\theta_2 = -\left(\frac{\Delta n_2}{2\,\Delta n_1 \sin(\theta_1/2)\cos I_2'}\right) + \tan I_2'. \tag{3.13}$$

Other than determining the prism glasses and angles requied to produce the desired chromatic effect, first-order design of a dispersing prism requires only calculation of the required apertures. Usually we assume a collimated input beam and make the apertures of the prism large enough not to vignet any of the dispersed beams. There are so many dispersing prism types that space here does not allow a comprehensive listing of the pertinent equations for computing these apertures. The techniques discussed above for the more often used prism types serve as guidelines for establishing these equations. This task is left to the ingenuity of the reader.

3.3.23 Thin wedge prisms

Prisms with small apex angles and (usually) axial thicknesses small compared to the component aperture are called optical wedges. Such a wedge is shown in Fig. 3.34.

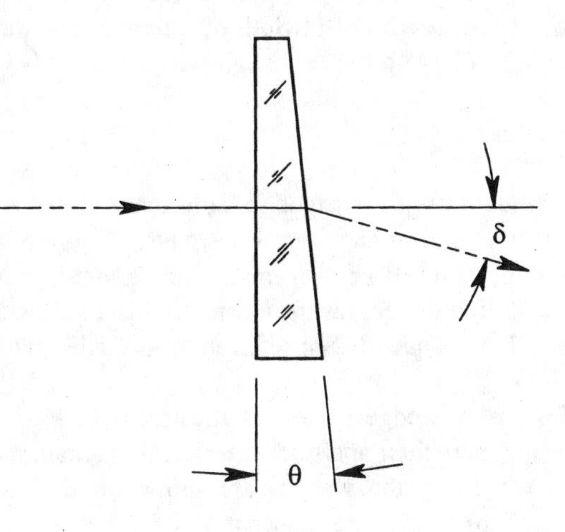

Fig. 3.34 A typical thin wedge

Since it's apex angle is small, we can assume that the angle expressed in radians equals its sine and, rewriting Eq. (3.12), we obtain the following simple equation for the wedge deviation:

$$\delta_\lambda = (n_\lambda - 1)\theta. \tag{3.14}$$

Differentiating this equation, we obtain the following expression for the dispersion, i.e., chromatic aberration, of the wedge:

$$d\delta_\lambda = dn_\lambda \theta. \tag{3.15}$$

Numerical Example No. 5: An optical wedge.

Assume a wedge has an apex angle of 1.9458°. What is its deviation if the glass index is 1.51680? What is its chromatic aberration for wavelengths corresponding to indices of 1.51432 and 1.52238?

By Eq. (3.14), $\delta = (1.51680 - 1)(1.9458) = 1.0056°$

By Eq. (3.15), $d\delta = (1.52238 - 1.51432)(1.9458) = 0.0157°$

A wedge so designed is one of minimum deviation. A common arrangement in optical instruments has the incident beam normal to the entrance face. Then $I_2 = \theta$, $I_2' = \arcsin(n \sin I_2)$, and $\delta = I_2' - \theta$. If not otherwise specified, we would assume n to apply to the center wavelength of the spectral bandwidth of interest. The deviation angle will differ from that given by Eq. (3.14) only very slightly.

3.3.24 Risley wedge system

Two identical optical wedges arranged in series and rotated equally in opposite directions about the optical axis form an adjustable wedge. They are used in collimated beams to provide variable pointing of laser beams, to align angularly one portion of an optical system to another portion thereof, as the means for distance measurement in some optical rangefinders, etc. They frequently are referred to as Risley wedges.

The action of a Risley wedge system is illustrated in Fig. 3.35. Usually the wedges are circular in shape; here their apertures are shown as small and large rectangles for clarity. In Views (a) and (c), the wedges are shown in their two positions for maximum deviation. The wedge bases are adjacent and $\delta_{SYSTEM} = \pm 2\delta$, where δ is the deviation of one wedge. If the wedges are turned from either maximum deviation position in opposite directions by β (see View (d)), the deviation becomes $\delta_{SYSTEM} = \pm 2\delta\cos\beta$ and the change in deviation from the maximum achievable value is $2\delta(1 - 2\cos\beta)$. If we continue to turn the wedges until $\beta = 90°$, we obtain the condition of View (b) where the bases are opposite, the system acts as a plane parallel plate, and the deviation is zero.

Since counter rotation of the wedges in a Risley wedge system provides variable

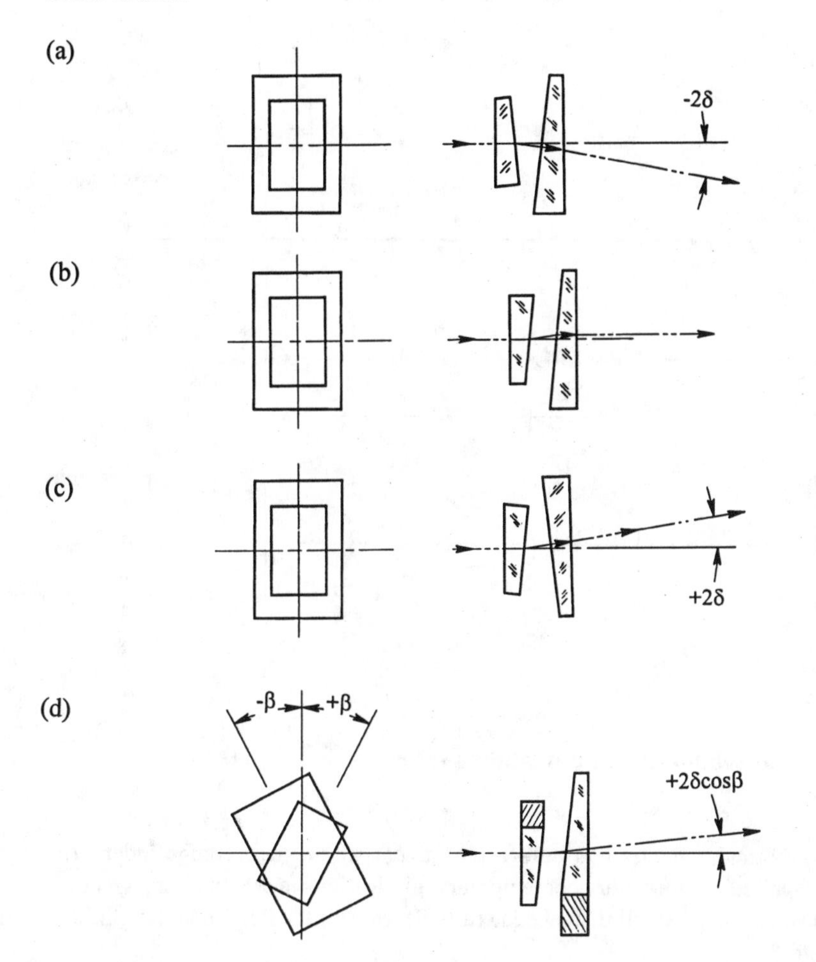

Fig. 3.35 Function of a Risley wedge prism system: (a) bases down, (b) bases opposed, (c) bases up, wedges counter rotated by β

deviation in one axis, a second such system, usually identical to the first, is sometimes added to provide independent variation in the orthogonal axis. The deviations from the two systems add vectorially in a rectangular coordinate system. Another arrangement has a single Risley wedge system mounted so both wedges can be rotated together about the optical axis as well as counterrotated. This provides variation of deviation in a polar coordinate system.

3.3.25 Sliding wedge

A wedge prism located in a converging beam will deviate the beam so the image is displaced laterally by an amount proportional to the wedge deviation and the distance from the wedge to the image plane. See Fig. 3.36 for a schematic of the device. If the prism is moved axially, the image displacement varies. This device most frequently was

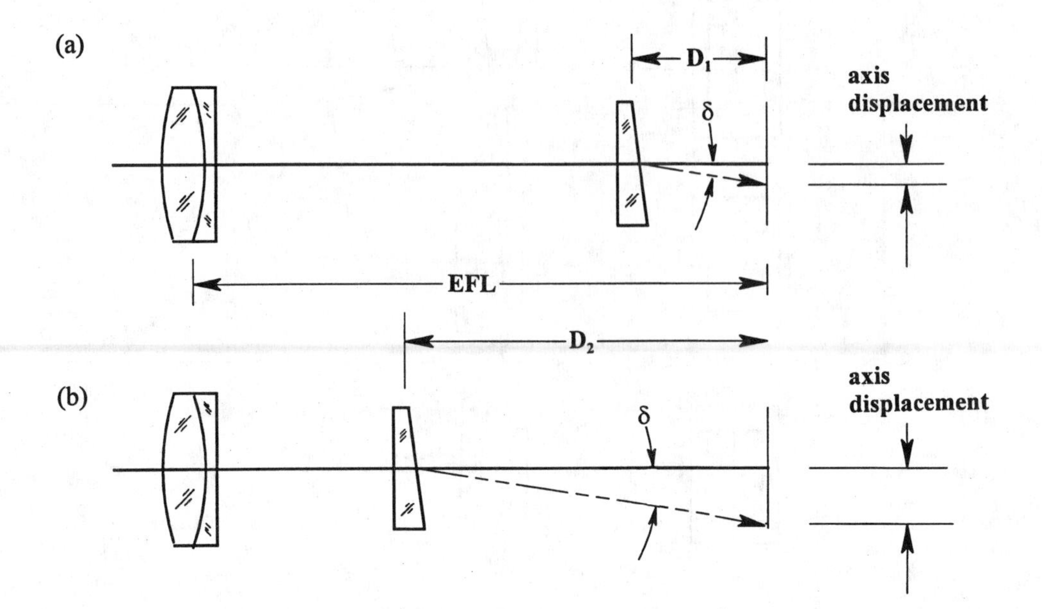

Fig. 3.36 A sliding wedge beam deviating system

used in military optical rangefinders before the advent of the laser rangefinder. The principle can be applied to other more contemporary applications in which an image needs to be variably displaced by a small distance laterally. If used with a long focal length lens, the wedge should be achromatic.

3.3.26 Focus-adjusting wedge system

Two identical optical wedges arranged with bases opposite and mechanized so each can be translated laterally by equal amounts relative to the optical axis provide a variable optical path through glass. Figure 3.37 shows the principle of operation of the device schematically. At all settings, the two wedges act as a plane parallel plate. If located in a convergent beam, this system allows the image distance to be varied and can be used to bring images of objects at different distances into focus at a fixed image plane. This type focus-adjusting system is sometimes used in large-aperture aerial cameras and telescopes such as those for tracking missiles or spacecraft launch vehicles where target range changes rapidly and the image-forming optics are large and heavy so cannot be moved rapidly and precisely by small distances. To first order, $t_i = t_0 \pm \Delta y_i \tan\theta$ and the focus variation is $\pm 2t_i((n - 1) / n)$. Here, t_0 is the axial thickness of one wedge at its center. Figure 3.38 shows the optical schematic for a typical application featuring a focus-adjusting wedge system. The changes in glass path as the wedges are moved may cause the aberration balance of the optical system to change. This would limit the focus adjustment range in high performance applications.

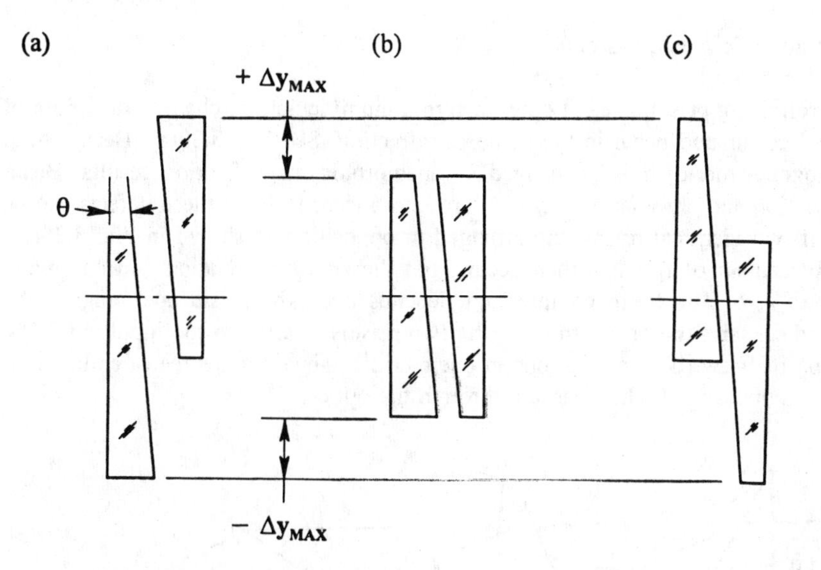

Fig. 3.37 A focus-adjusting wedge system: (a) minimum path, (b) nominal path, (c) maximum path

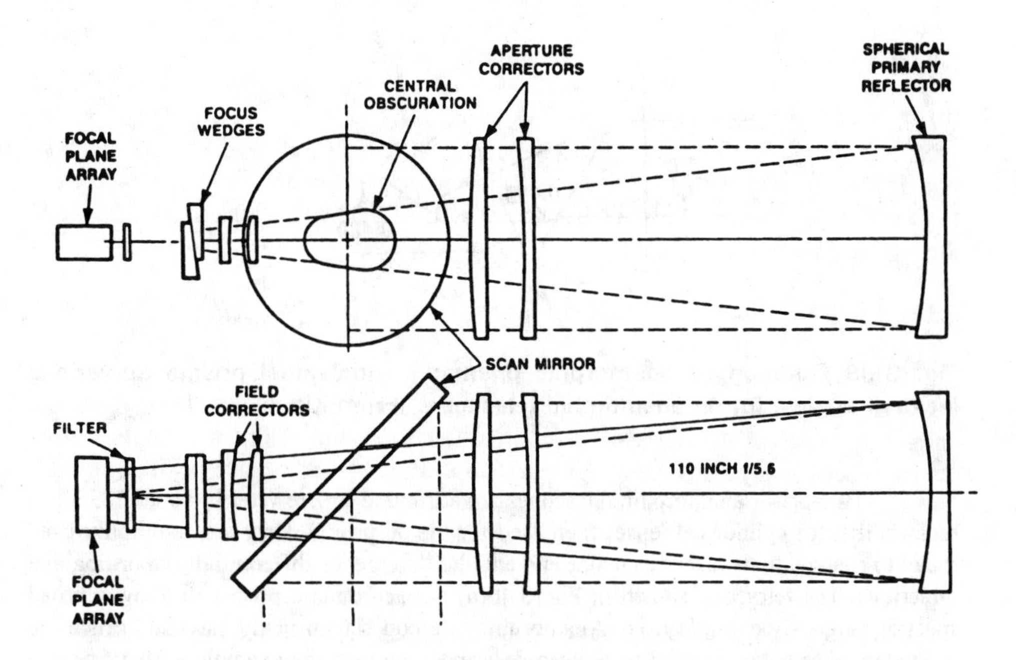

Fig. 3.38 Top and side views of the optical system for a 110-in. focal length, 19.6 in. aperture aerial reconnaissance lens featuring a focus-adjusting wedge system. (From Ulmes[29])

3.3.27 Anamorphic prism systems

A refracting prism, used at other than minimum deviation, changes the width of a transmitted collimated beam in the plane of refraction. See Fig. 3.39(a). Beam width in the orthogonal meridian is unchanged so anamorphic magnification results. Beam angular deviation and chromatic aberration are introduced. Both of these defects can be eliminated if two identical prisms are arranged in opposition as shown in Fig. 3.39(b). Lateral displacement of the axis then occurs, but the angular deviation and chromatic aberration are zero. The beam compression depends upon the prism apex angles, the refractive indices, and the orientations of the two prisms relative to the input axis. The configuration of View (b) is a telescope in one meridian since the degree of collimation of the beam is unchanged while passing through the optics.

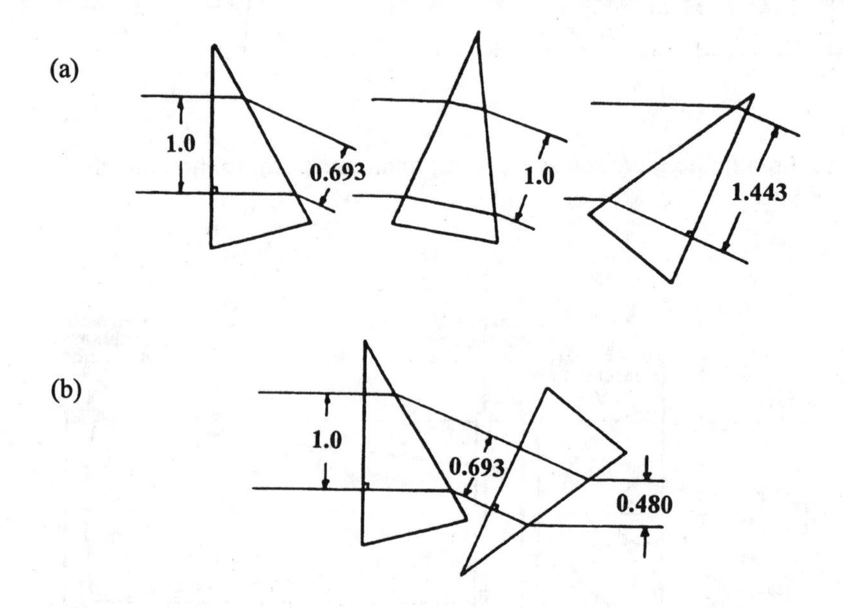

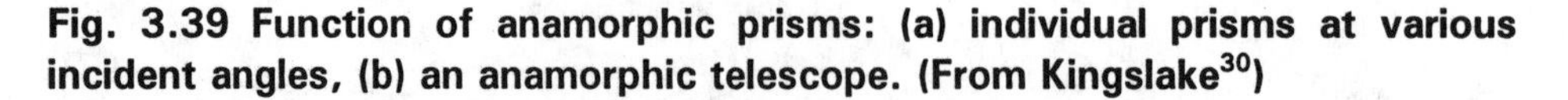
Fig. 3.39 Function of anamorphic prisms: (a) individual prisms at various incident angles, (b) an anamorphic telescope. (From Kingslake[30])

Two-prism anamorphic telescopes are attributed to Brewster in about 1835 as a replacement for cylindrical lenses then used for the purpose.[30] They are commonly used today to change diode laser beam size and angular divergence differentially in orthogonal directions. The telescope shown in Fig. 3.40(a) has achromatic prisms to allow a broad spectral range to be employed.[31] Anamorphic telescopes with many cascaded prisms to produce higher magnification have been described. An extreme example with 10 prisms is shown in View (b). This configuration is reported to be optimal for single-material achromatic expanders of moderate to large magnifications.[32] A telescope example comprising only one prism is shown in View (c).[33] It has three active faces, one of which functions by TIR. The entrance and exit faces can be oriented at Brewster's angle so the surface reflection (Fresnel) losses are eliminated for properly oriented polarized beams.

Anamorphic prism pairs have been used quite successfully to convert rectangular Excimer laser beams into more suitable square ones for materials processing and surgical applications.

(a)

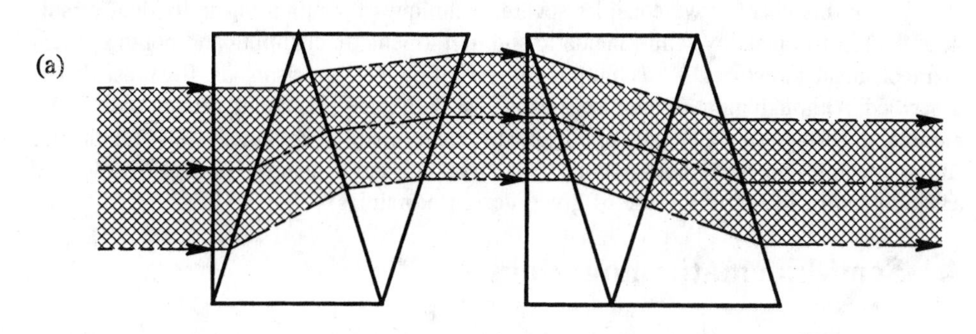

(b)

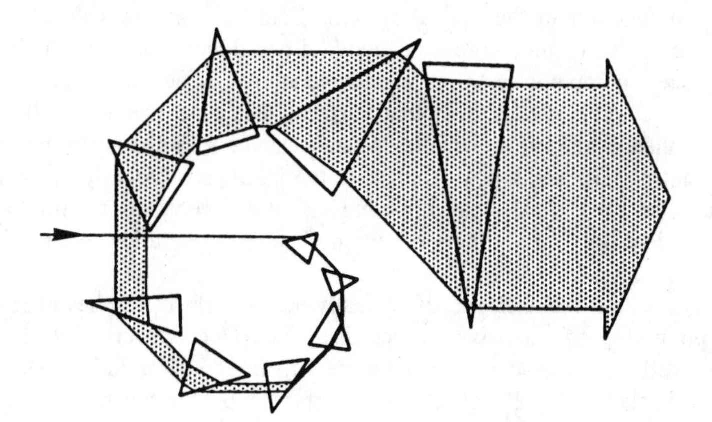

(c)

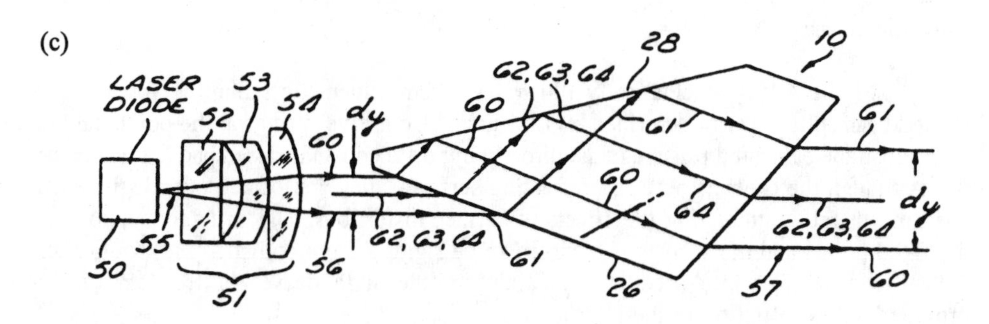

Fig. 3.40 Three anamorphic prism telescope designs: (a) achromatic prism assembly (from Lohmann and Stork[31]), (b) cascaded assembly (from Trebino[32]), and (c) single-prism telescope (from Forkner[33])

CHAPTER 4

PRISM MOUNTING TECHNIQUES

In this chapter, we consider several techniques for mounting individual prisms in optical instruments by semi-kinematic and non-kinematic clamping and bonding them in mechanical structures. A technique for mounting larger prisms on flexures also is described. Although most of the discussions deal with glass prisms interfacing with metal mounts, the designs are generally applicable to prisms made of optical crystals and plastics and to attaching prisms to non-metallic cells, brackets, and housings. Numerous examples are included to illustrate the use of given design principles.

4.1 Semi-kinematic mountings

The position (translations) and orientation (tilts) of a prism generally must be established during assembly and then carefully controlled to tolerances dependent upon its location and function in the optical system. Control is accomplished through the prism's interfaces with its mechanical surround. If possible, the optical material should always be placed in compression at all temperatures. Kinematic mounting avoids overconstraints that might distort the optical surfaces. The point contacts with potentially high stresses inherent in true kinematic mounts can be avoided by providing semi-kinematic mounting with small area contacts at the interfaces. Properly designed spring forces applied over these finite areas allow expansion and contraction with temperature changes while adequately constraining the prism against acceleration forces.

Contact within the apertures of optically active surfaces implies obscuration as well as the possibility of excessive surface distortion. Hence, such contact should be avoided. Since reflecting surfaces are much more sensitive to deformation than refracting ones they are especially critical. Note that TIR surfaces must not touch anything that will frustrate the index mismatch that causes the internal reflection to take place. If periodic cleaning of a TIR surface is required, the instrument design should provide the necessary convenient access.

Figure 4.1(a) schematically illustrates a semi-kinematic mounting for a cube-shaped beamsplitter prism described by Lipshutz.[34] Here, five springs at the points labled K_1 preload the cemented prism against directly opposite coplanar raised pads indicated as K_∞. Although the contacts occur on refracting surfaces, they are located outside the used aperture, thereby minimizing the effects of surface distortions. The structure supporting the fixed points and the springs is assumed to be rigid; a more realistic case would take the structure design into account. Constraint of the sixth degree of freedom (Z) is provided only by friction in the interfaces.

As shown in View (b), this beamsplitter is used to divide a beam converging towards an image plane; each beam then forming an image on a separate detector. In order for these images to maintain their proper alignment relative to each other, to the detectors, and to the structure of the optical instrument with temperature changes, the prism must not translate in the XY plane of the figure nor rotate about any of the three

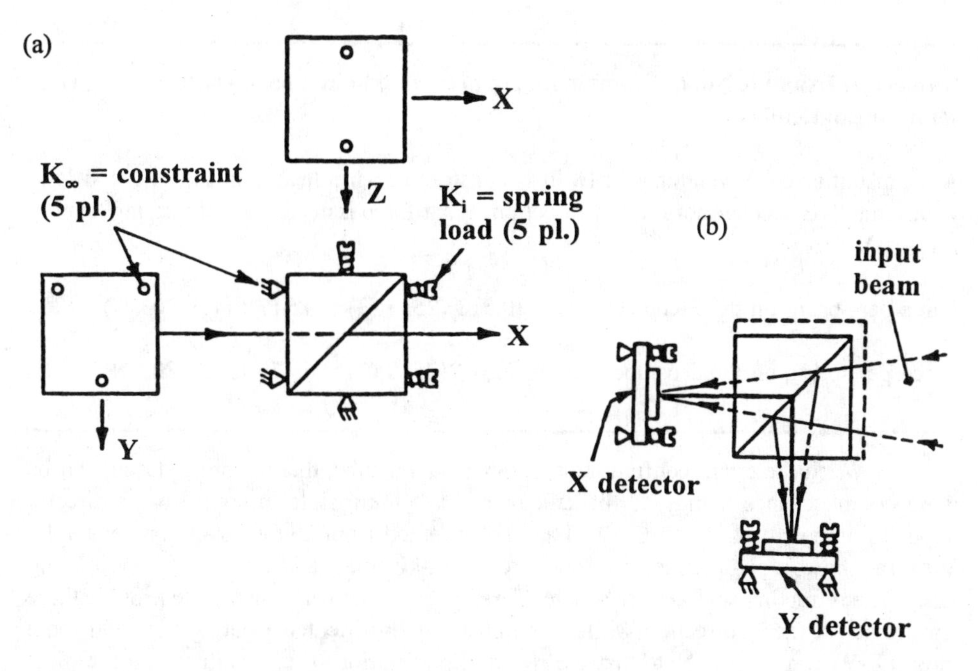

Fig. 4.1 (a) Three views of a semi-kinematic mount for a cube-shaped beamsplitter prism, (b) schematic of optical function. (Adapted from Lipshutz[34])

orthogonal axes. Translation in the Z direction has no optical effect. This motion is limited to a small value by mechanical stops (not shown). Once aligned, the springs ensure that the prism always presses against the five pads. The dashed outlines indicate how the prism will expand if the temperature increases. Registry of the prism surfaces against the pads does not change and the light paths to the detectors do not deviate. This is also true when the temperature decreases.

The preload force, P_i (in pounds), to be exerted by one spring onto the prism with the mounting just described may be calculated with the aid of Eq. (4.1).

$$P_i = \left(\frac{w}{N_i}\right)\sum a_i \tag{4.1}$$

where w is the weight of the prism, N_i is the number of springs active in the direction "i," and $\sum a_i$ is the vector sum, in the direction "i," of the maximum anticipated externally applied accelerations such as those due to constant acceleration, random vibration (3σ), amplified resonant vibration (sinusoidal), acoustic loading, and shock. These accelerations are expressed as multiples of gravity. Since all types of external accelerations do not generally occur simultaneously, the magnitude of the summation term does not need to be taken literally. Note that, if the prism weight is expressed in kilograms, Eq. (4.1) must include a multiplicative factor of 9.866 to convert units. The preload is then in newtons. Friction usually is ignored.

Numerical Example No. 6: Clamping force needed to hold a beamsplitter cube prism semi-kinematically.

A beamsplitter cube weighing 0.518 lb is constrained as indicated in Fig. 4.1 and is to withstand 25 G accelerations in any direction. What force is needed at each spring? Apply Eq. (4.1).

Force per spring on the 3-contact face = (0.518)(25) / (3) = 4.317 lb (19.203 N)

Force per spring on the 2-contact face = (0.518)(25) / (2) = 6.475 lb (28.802 N)

When the prism configuration is other than a cube, the mounting design can be more complex since it may be difficult or perhaps impossible to apply forces directly opposite support pads. Figures 4.2(a) and (b), adapted from Durie[35] show one example. View (a) shows a right angle prism mounted semi-kinematically on one of its refracting faces. The refracting surfaces are square. Three pads on the perforated plate provide three constraints in the Y direction while three horizontally oriented locating pins add three more (X-Y) constraints. Note that the required perforations (i.e., apertures) in the plate are not shown in Fig. 4.2(a) and (b). Ideally, all pads and pins contact the prism outside its optically active apertures. In View (b), the same prism is shown in side view. The forces F_2 and F_2 are oriented perpendicular to the hypotenuse face and touch the prism near the longer edges of the hypotenuse. F_1 is aimed symmetrically between the nearest pad ("b") and the nearest pin ("d") while F_2 is aimed symmetrically between pads "a" and "c" and pin "e." Horizontal force F_X holds the prism against pin "f" and horizontal and vertical components of F_1 and F_2 hold the prism against the three pads and remaining two locating pins. Although not optimum in terms of freedom from bending tendencies (i.e., moments), this arrangement is adequate since the prism is relatively stiff.

Figure 4.2(c) shows a right angle prism referenced to one triangular ground face. Again, the prism is pressed against three coplanar raised pads and three locating pins. The top plate presses the prism through a resilient (elastomeric) pad under clamping action of three screws. A leaf spring presses the prism against the pins. Other spring types could, of course, be used for the latter purpose. An attractive feature of this mounting is that it can be configured so the circular clear apertures of optically active surfaces are not obscured and those surfaces are not likely to be distorted by the imposed forces.

In Fig. 4.2(d), the hypotenuse face of a Porro prism is positioned against three coplanar raised pads on a perforated plate (perforations again not shown) while one edge touches two pins and one edge touches a third pin. The pins are perpendicular to the plate surface. A force, F_Z, directed parallel to and slightly above the plate holds the prism against two pins ("d" and "e") while force F_X, also just above the plate, holds it against the the third pin ("f"). A third force, F_Y, holds the prism against the three pads ("a," "b," and "c"). This force acts against the dihedral edge of the prism at its center. Once again, the prism is stiff enough that surface distortion is minimal.

Example No. 7: Clamping force needed to hold a Porro prism semi-kinematically.

A Porro prism is constrained as indicated in Fig. 4.2(d). It is made of SF8 glass and has an aperture A of 2.875 cm (1.132 in.). It is to withstand 10 G accelerations in any of the three axis directions. What should be P_X, P_Y, and P_Z? Ignore friction effects.

From Fig. 3.11, $V = 1.29A^3 = 23.764\ \text{cm}^3$
From Table C1, $\rho = 4.22\ \text{g/cm}^3$
Hence, $m = (23.764)(4.22) = 100.823\ \text{g} = 0.100\ \text{kg}$
Eq. (4.1) applies.

Then: $P_X = P_Y = P_Z = (9.866)(0.100)(10)/(1) = 9.87\ \text{N}\ (2.20\ \text{lb})$

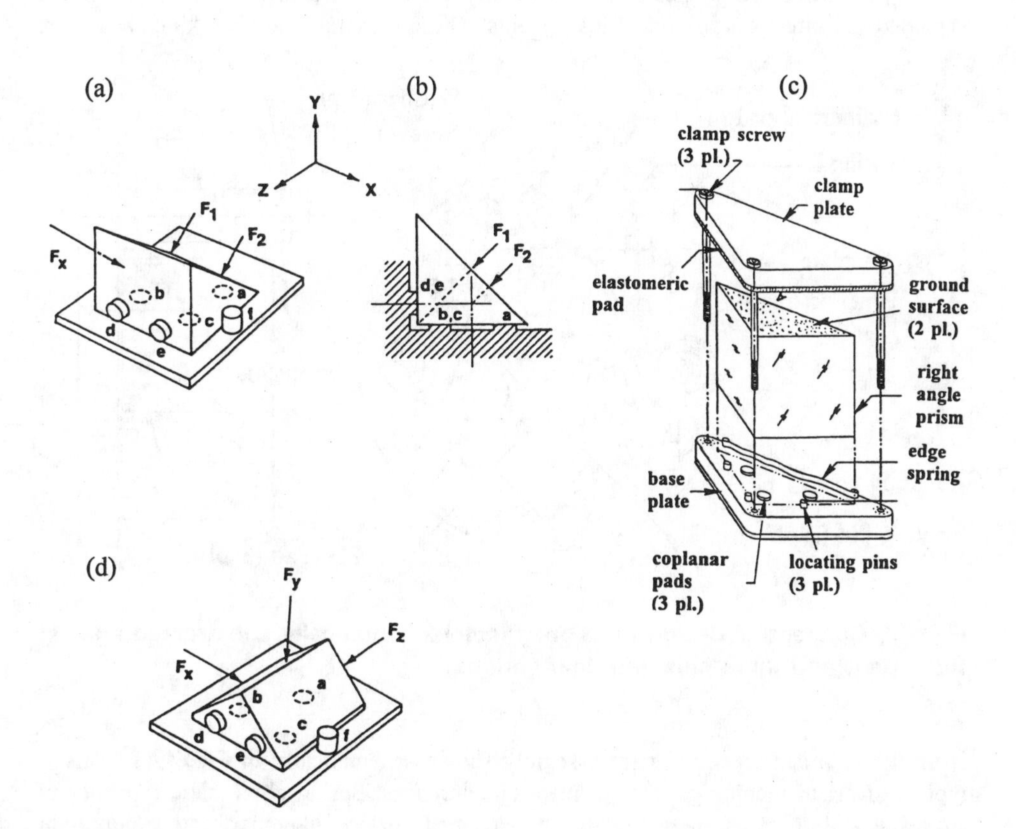

Fig. 4.2 Schematics of semi-kinematic mounts for (a) and (b) a right angle prism referenced to one refracting face, (c) a right angle prism referenced to one ground face, and (d) a Porro prism referenced to its hypotenuse face. ((a), (b), and (d) adapted from Durie,[35] (c) adapted from Vukobratovich[36])

Another semi-kinematic mounting is illustrated schematically in Fig. 4.3. Here, a penta prism is pressed against three coplanar pads on a baseplate in much the same manner as the Porro prism of Fig. 4.2(c). Three cantilevered spring clips provide the necessary preload directly through the prism against the pads. This provides three constraints: one translation and two tilts. The remaining two translations and one tilt are constrained by three locating pins with a leaf spring to provide preload against the pins. The equations needed to design the springs are given and their use illustrated in Sect. 5.1. A mounting design of this type is considered in detail in Sect. 9.1.

Applications of Eq. (4.1) are, of course, not limited to the prism mounting configurations just shown. We will have occasion to apply this equation to other spring-constrained optical component types later in this text.

4.2 Mechanically clamped non-kinematic mountings

Spring or strap means are frequently used to hold prisms in place against extended flat interfaces in optical instruments. These techniques are not kinematic. One

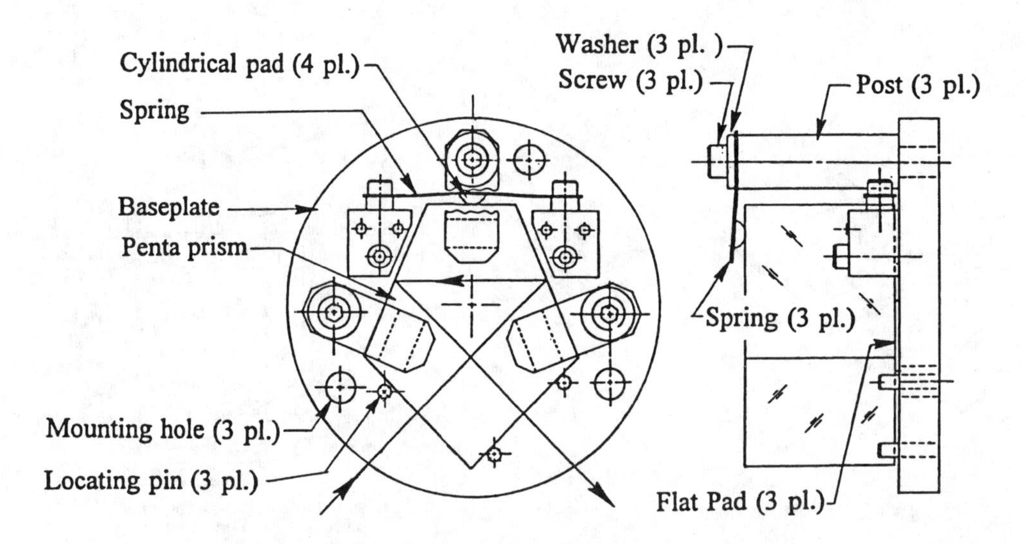

Fig. 4.3 Conceptual design for a prism semi-kinematically constrained against three pads and three pins with leaf springs

is the Porro prism erecting prism assembly shown schematically in Fig. 4.4.[37] This is typical of prism mountings in binoculars or telescopes. Spring clips made typically of spring steel hold each prism against a machined surface in a perforated aluminum mounting shelf that is, in turn, fastened with screws and locating pins to the instrument housing. Area contact occurs over large annular areas on the racetrack-shaped hypotenuse faces of the prisms while lateral constraints are provided by recessing those faces slightly into opposite sides of the shelf. Elastomeric-type adhesive is used to keep the prism from sliding in the necessary clearances around the recessed faces. In this design, the prisms

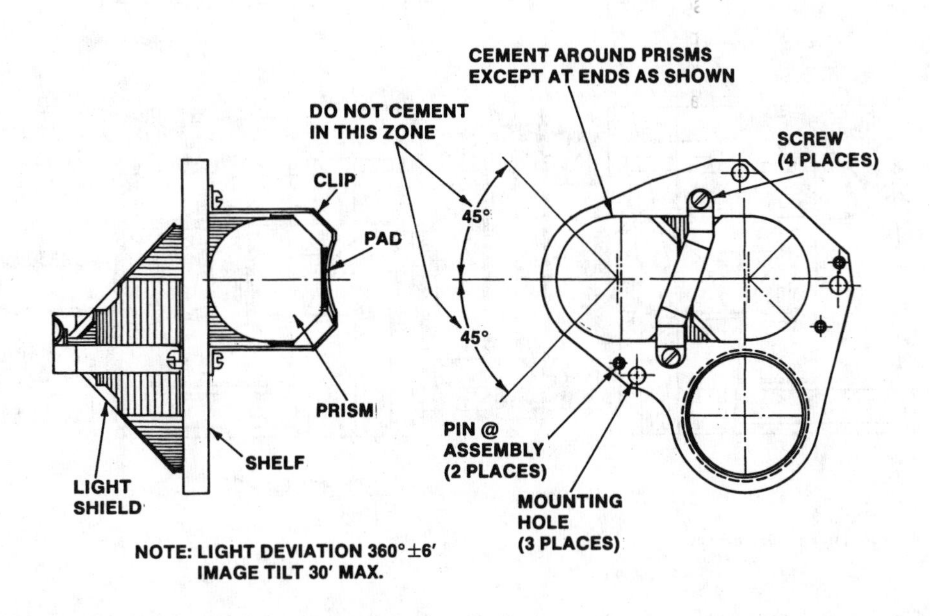

Fig. 4.4 Schematic of a typical strap mounting for a Porro prism erecting subassembly. (Adapted from Yoder[37])

are made of high-index (flint) glass so TIR occurs at the four reflecting surfaces. Thin aluminum light shields help prevent stray light from reaching the image plane. These shields have bent tabs that touch the edges of the prism faces to space the shields a short distance away from the reflecting surfaces.

Figure 4.5 is a portion of an assembly drawing showing a typical spring clip constraint for Porro prisms in a commercial binocular made by Swarovski Optik KG in Austria. The springs straddle the beveled apexes of the prisms. One end of each spring fits into a slot on the inside of the housing while the other end is held by a screw. The prism shelf and threaded holes for the screws are cast into the housing.

Another example of the many types of non-kinematic mounts for prisms is shown in Fig. 4.6. Here, an Amici prism is held by a flat spring clip against nominally flat, slightly resilient reference pads inside the triangular housing of a small 3-power military elbow telescope. The screw pressing against the center of the spring forces the ends of the spring against the prism. It is very important that the screw does not protrude far enough into the housing for the spring to touch the roof edge of the prism. Careful design would specify the correct screw length while explicit instructions in the manufacturing procedures to check the actual clearance would help avoid damage at assembly, or during exposure to shock, vibration, or extreme temperatures. Constraint perpendicular to the plane of the figure is provided by resilient pads attached to the inside surfaces of triangular-shaped covers that are attached with screws to the cast housing.

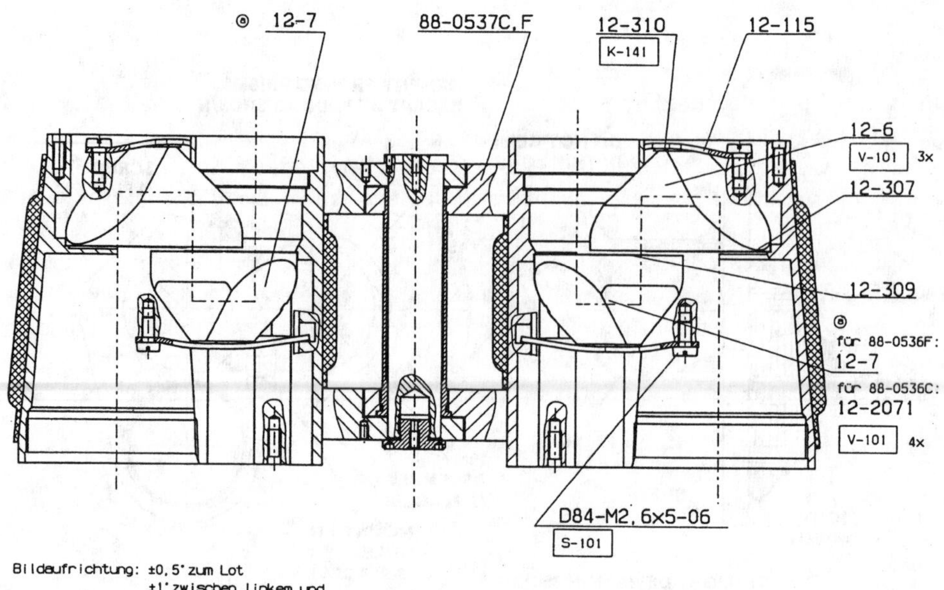

Fig. 4.5 Porro prism mounting in a contemporary commercial binocular. (Courtesy of Swarovski Optik KG, Absam/Tyrol, Austria)

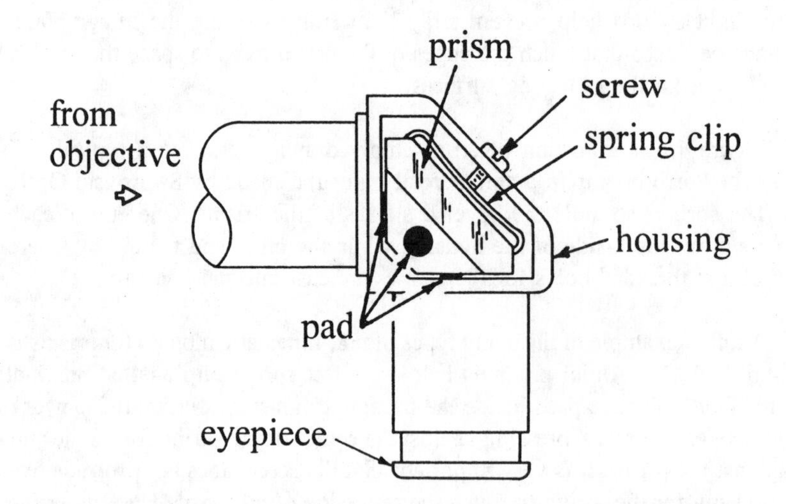

Fig. 4.6 Schematic of a small military elbow telescope with an Amici prism spring-loaded against resilient reference pads. (Adapted from Yoder[64])

Figure 4.7 shows the scanning head assembly from a military periscope. It uses a single prism that resembles a Dove prism with angles between faces of 35°, 35°, and 110°. The prism can be tilted about the horizontal transverse axis to scan the periscope's line of sight, in elevation, from the zenith to about 20° below the horizon. The prism is held in place in its cast aluminum mount by four spring clips each attached with two screws to the mount adjacent to the entrance and exit faces of the prism. The edges of the reflecting surface (hypotenuse) of the prism rest on narrow lands machined onto the casting. The lands do not extend into the optical aperture of that surface. The prism faces protrude slightly [about 0.5 mm (0.02 in.)] from the mount so that the proper preload is obtained when the clips bottom against the mount. Once centered, the prism cannot slide parallel to the long edge of its hypotenuse because of the convergence of the clamping forces. The vector sum of these forces is nominally perpendicular to the mounting surface and the tangential force components cancel each other. Motion laterally is constrained by friction and limited by a close fit into the mount.

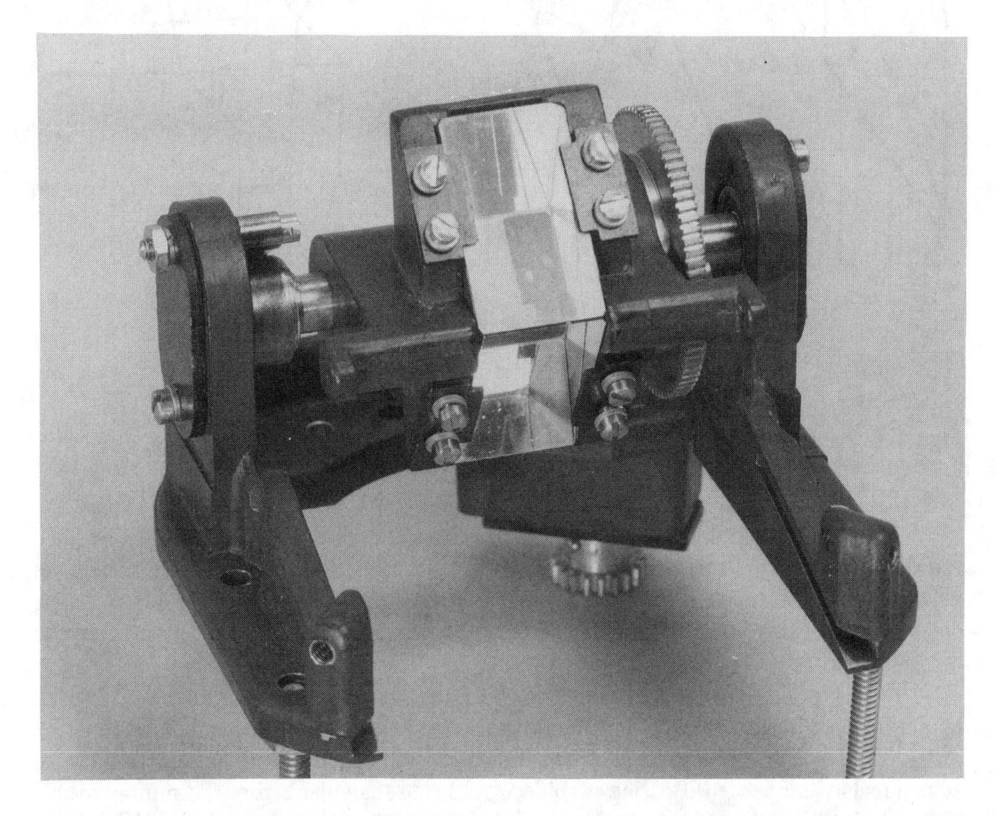

Fig. 4.7 A clamped Dove-type prism used in the elevation scanning head subassembly of a military periscope. (From Yoder[8] by courtesy of Marcel Dekker, Inc.)

Figure 4.8 illustrates the scanning function of the prism in the above periscope. This motion usually is limited optically by vignetting of the refracted beam at its top or bottom. Mechanical stops usually are built into the instrument to limit physical motion so the end-point vignetting is acceptable for the application.

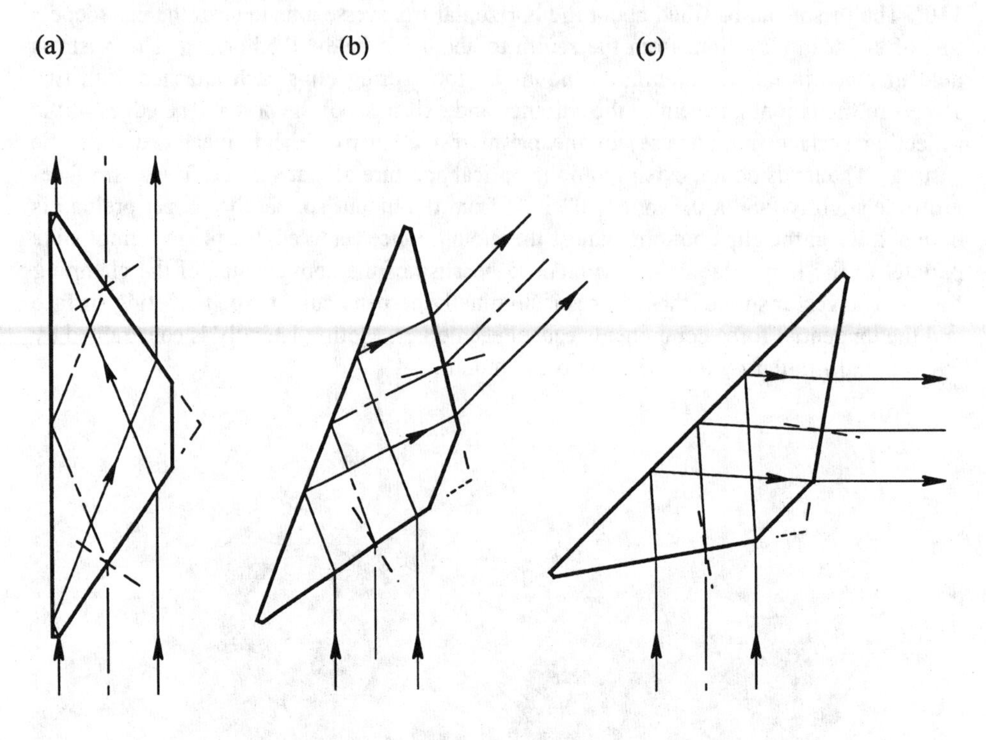

Fig. 4.8 Typical elevation scanning function of a Dove prism

The most popular types of derotation prisms are the Dove, double Dove, Pechan, and delta. In order to function successfully all these prisms must be mounted securely yet be capable of adjustment at the time of assembly so as to minimize image motion during operation. A design for an adjustable derotation prism mount will be discussed next.

In Fig. 4.9, we see a sectional view of a typical Pechan prism mount.[38] If used in a collimated beam it requres only angular adjustment of the optical axis relative to the rotation axis. Here, it was to be used in an uncollimated beam so both angular and lateral adjustments were needed. Bearing wobble would cause angular errors. To minimize this in the design considered here, Class 5 angular contact bearings, mounted back to back, were oriented with factory-identified high spots matched then preloaded. Runout over 180° motion was measured as about 0.0003 in. (7.6 μm). The bearing axis was adjusted laterally by fine thread screws (not shown) that permitted centration with respect to the optical system axis to better than 0.0005 in. (12.7 μm). The prism was adjusted laterally within the bearing housing in the plane of refraction by sliding it against a flat vertical reference surface with fine thread screws pressing against the reflecting surfaces through pressure pads. A spherical seat with center of rotation at the intersection of the hypotenuse face with the optic axis (to minimize axis cross-coupling) was provided for angular adjustment. This movement was controlled by the adjustment screws indicated.

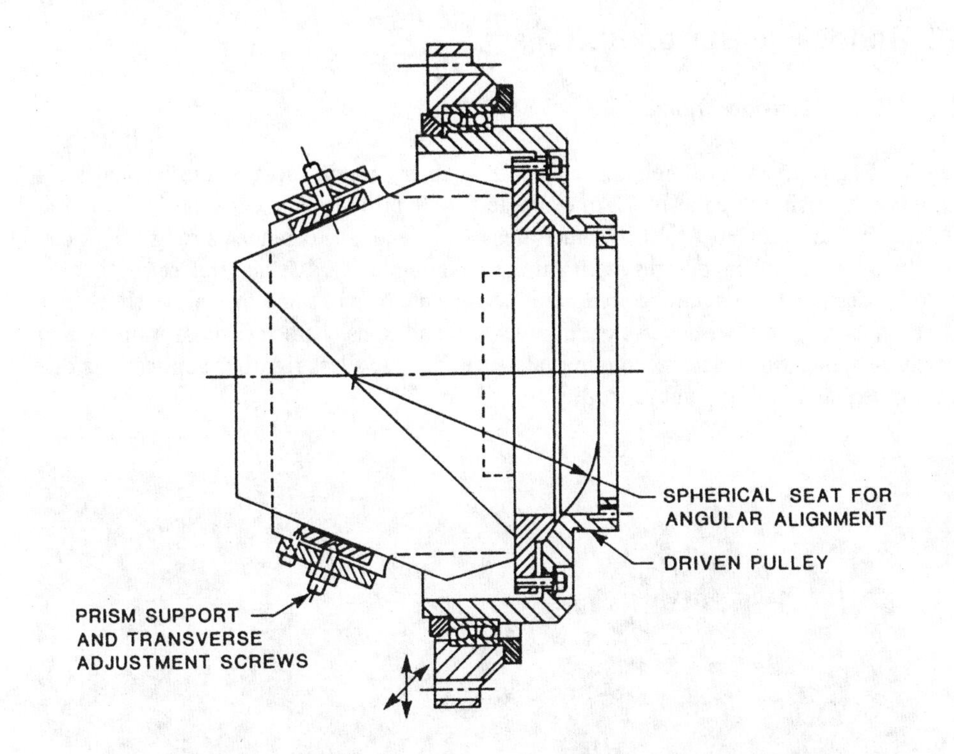

Fig. 4.9 A Pechan prism derotation assembly. (From Delgado[38])

In each of the last four design examples, the Porro, the Amici, the Dove-type, and the Pechan prisms are loaded against machined surfaces on the mounts. Since it is virtually impossible for the metal surfaces to be as flat as the interfacing glass surfaces, contact will occur first on the three highest points on the mount. Usually, these points are not directly opposite the springs that hold the prism in place. Moments are then applied to the glass, and surface deformations may occur. If the spring loading is large, the metal or the glass may bend enough for more point contacts to form. We then have a condition of uncontrolled overconstraint. Since the prisms are stiff and the instruments into which these designs have been incorporated have demonstrated long service life in relatively adverse environments, we conclude that these problems are usually tolerable. Adding small-area pads that can be lapped flat and coplanar and located opposite or nearly opposite the force points, such as are used in the designs shown in Fig. 4.1 through 4.3, reduces the likelihood of prism distortion due to applied constraints.

A potential problem occurs if, under shock or vibration, the prism loses contact with some of its interfacing positional references. When the driving force is dissipated, the prism may land in a new orientation. There it remains due to friction until disturbed again. This action introduces uncertainty into the location and/or orientation of the optical component that may affect performance. If the springs are strong enough that the optic always maintains contact with the references, this problem should not exist.

4.3 Bonded prism mountings

4.3.1 General considerations

Many prisms are mounted by bonding their ground faces to mechanical pads using epoxy or similar adhesives. An example from a military periscope used in armored vehicles is shown in Fig. 4.10. Contact areas large enough to render strong joints can usually be provided in designs with minimum complexity. Mechanical strength of a carefully designed and manufactured bond is adequate to withstand the severe shock and vibration as well as other adverse environmental conditions characteristic of military and aerospace applications. The technique is also used in many less rigorous applications due to its inherent simplicity and reliability.

Fig. 4.10 Photographs of a typical bonded roof penta prism. (From Yoder[8] by courtesy of Marcel Dekker, Inc.)

The critical aspects of the design are characteristics of the adhesive, thickness of the adhesive layer, cleanliness of the surfaces to be bonded, dissimilarity of coefficients of expansion of the materials, area of the bond, environmental conditions, and care with which the parts are assembled. Several typical adhesives used for this purpose are listed in Table C14 of Appendix C. While the adhesive manufacturer's recomendations should be consulted, experimental verification of adequacy of the design, the materials to be used, the method of application, and cure temperature and duration are advisable in critical applications.

Guidelines for determining the appropriate bond area have appeared in the literature.[39] In general, the adhesive shear or tensile strength (usually approximately equal) is set equal to the product of prism weight and maximum expected acceleration, a_G, divided by the bond area, Q. If this ratio is greater than unity, some safety factor, f_S, exists. This factor should be at least 2 and possibly as large as 10 to allow for some unplanned, non-optimum conditions during processing. Most prism designs considered in Section 3.3 list equations for calculating the required circular- or racetrack- shaped bond area, Q, and the maximum bond area achievable ($Q_{MAX\,(C)}$ or $Q_{MAX\,(RT)}$, as appropriate) to simplify the interface design task. In those equations, $f_S = 2$ and the adhesive strength, J, is taken as 17.24 N/mm^2 (2500 lb/in.^{2}).

For maximum glass-to-metal bond strength, the adhesive layer should have a particular thickness. If 3M epoxy EC2216-B/A is used, experience indicates that a thickness of 0.075 to 0.125 mm (0.003 to 0.005 in.) is best. Some adhesive manufacturers recommend thicknesses as large as 0.4 mm (0.016 in.) while some users have found success with 0.05 mm (0.002 in.) thicknesses. A thin bond is stiffer than a thick one. A means for ensuring that a particular bond thickness is achieved is to place shims of the specified thickness at three locations in the interface between the glass and the metal. If possible, these should be arranged in a triangular pattern and lie outside the bonded area. The glass, mount, and shims must be held together to ensure contact. Adhesive should not be allowed to get onto the shims. Another technique for obtaining a uniform layer thickness is to mix small glass beads into the ahesive before applying it to the surfaces to be bonded. When the surfaces are clamped together, the larger beads contact both faces and hold them apart. Since such beads can be purchased with closely controlled diameters, achievement of correct bond thickness is relatively easy.

Since adhesive layers normally shrink by a few percent of each dimension during curing, it is advisable to keep these dimensions as small as possible while providing adequate strength. Many designs have the required bond area subdivided into three smaller areas preferably arranged in triangular fashion. See Fig. 4.11. This allows one to spread the bond over the maximum-sized geometrical pattern for maximum stability. The shrinkage is significantly reduced in such a design. Note that the bonds are on the larger prism only.

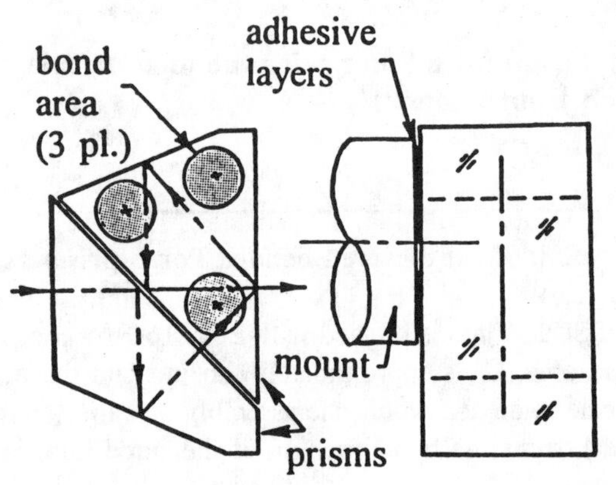

Fig. 4.11 An example of a triangular bond distribution on a Pechan prism subassembly. (Reprinted with permission from Yoder[64]. Copyright CRC Press.)

Obviously, bonding adhesive should be kept away from the airspace between separated prisms such as the Pechan shown in Fig. 4.11. In fact, even with components cemented together rather than airspaced, bridging the cemented joint with bonding adhesive can cause glass breakage at low temperature due to differential shrinkage of the adhesive trapped within the v-groove formed by bevels on the edges of the optics. Further, fillets of excess adhesive at the edges of glass-to-metal bonds should be permitted only if the adhesive is quite flexible. For this reason, urethane adhesives are better than the harder epoxies if fillets are inevitable due, for example, to the inaccessibility of the region for observation or removal of excess material.

4.3.2 Cantilevered techniques

Figures 4.10 and 4.11 illustrate prisms bonded on one ground surface in cantilevered fashion with respect to a mechanical mounting surface. Experience has shown this to provide adequate support for a variety of severe shock and vibration conditions including shocks of several hundred times gravity.[8] Figure 4.12 shows one such design which is known to have withstood a_G of >1200 without damage. Figure 4.13 shows the same prism mounted in the optical assembly. This prism and its mount are the subject of Numerical Example No. 8.

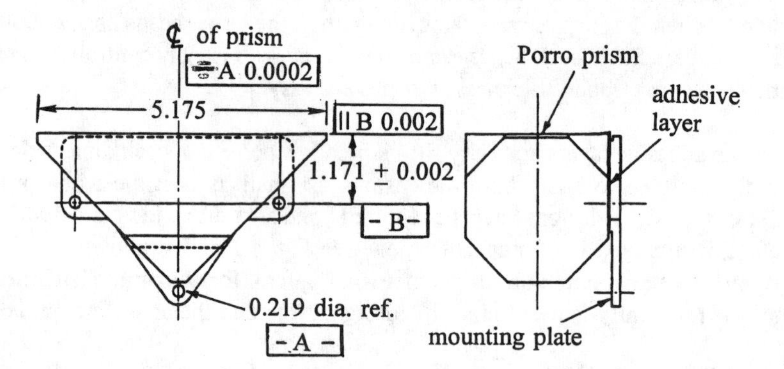

Fig. 4.12 Mounting design for a Porro prism bonded on one side. Dimensions are inches. (Adapted from Yoder[37])

Numerical Example No. 8: A cantilevered bonded Porro prism assembly.

A Porro prism made of SK16 glass is bonded with EC2216-B/A epoxy to a 416 stainless steel bracket. The bond area, Q, is approximately 5.6 in.^{2} and the prism weight, W, is about 2.2 lb. What acceleration, a_G, would the assembly be expected to withstand with a safety factor, f_S, of 2? Assume the bond strength of the cured joint is 2500 lb/in.^{2}.

The guidelines given above lead to the following equation for a_G:

$$a_G = JQ/Wf_S = (2500)(5.6) / (2.2)(2) = 3182 \text{ times gravity}$$

This indicates that the prism should withstand the 1200 G measured acceleration cited above with a safety factor of about 2.6.

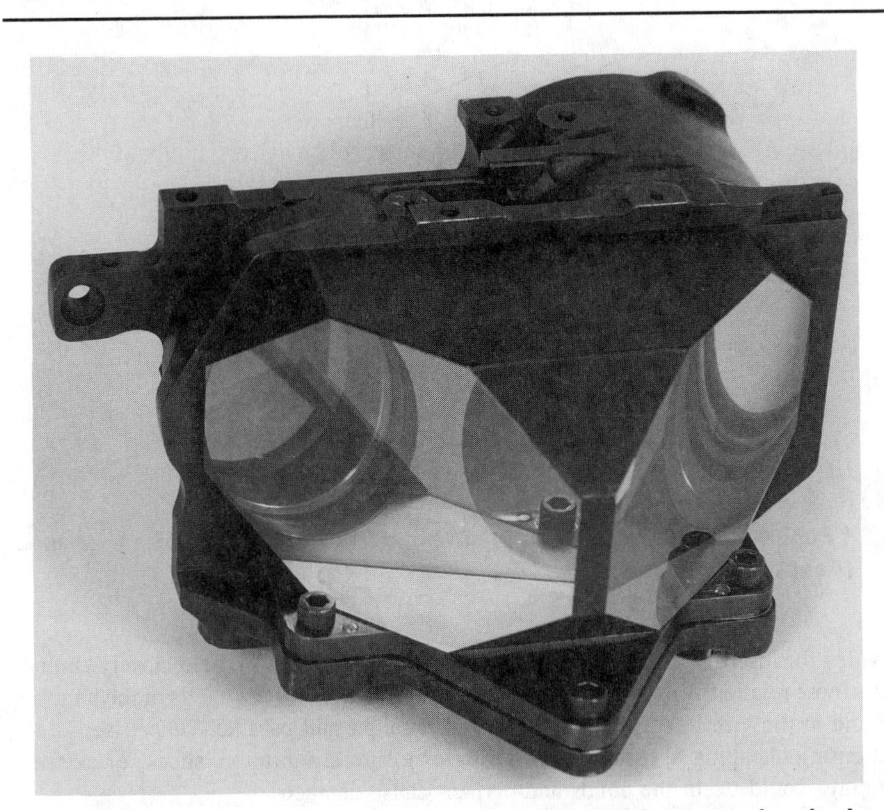

Fig. 4.13 Photograph of the Porro prism of Fig. 4.12 mounted to its bracket. (Courtesy of Raytheon Optical Systems Corp., Danbury, CT)

Since, during shipment or use, the mounting surface could be oriented in any direction with respect to gravity and environmental acceleration forces could be applied in any direction, the cantilevered mounting may not be adequate under extreme conditions. Additional support may be needed. This leads us to the following design variation.

4.3.3 Double-sided support techniques

Some designs for bonding prisms utilize multiple adhesive joints between the prism and structure. See, for example, Fig. 4.14. Here, increased bond area and support from both sides are provided to a right angle prism by bonding it to the ends of two metal stub shafts. The bonded subassembly rests in and is firmly clamped by two precision-machined pillow blocks of conventional split design. Ease of adjustment of the prism's alignment about the transverse axis is a key feature of this design. Once installed in the optical instrument and adjusted, the shafts are securely clamped.[35] It is necessary in such designs that the surfaces to be bonded be nearly parallel and that reasonable clearances

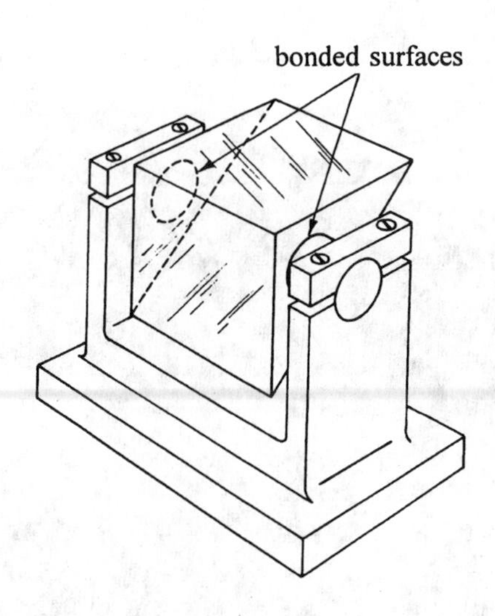

Fig. 4.14 Schematic of a right angle prism bonded to both sides of a U-shaped mount. (Adapted from Durie[35])

be provided for insertion of the adhesive layers. Tolerances must be held closely enough to ensure these relationships. Further, the the bearing surfaces for the shafts in the bonding fixture and in the instrument must be extremely straight and parallel. Otherwise, forces exerted during clamping at assembly or during exposure to vibration, shock, or extreme temperatures could strain the bonds and perhaps cause damage.

Potential problems with differential expansion between glass and metal at extreme temperatures in the design of Fig. 4.14 can be avoided by building flexibility into one support arm as indicated in Fig. 4.15. Units made to an earlier design without this flexure were damaged at low temperature when the aluminum mount contracted more than than the prism causing the arms to pivot about the bottom edge of the prism and pull away from the prism at the top of the bonds. Allowing the arm to bend slightly prevented such damage.[40] Hole "E" (2 pl.) allowed epoxy to be injected after the prism was aligned.

A potential problem in any design in which epoxy is injected through access holes to a bond cavity is that the "plug" of adhesive in the hole can shrink significantly at low temperature and perhaps pull the adjacent glass sufficiently to fracture it.[41] It is well known that well-made epoxy joints can be stronger than the tensile strength of glasses. Minimizing the length of the hole, and hence the length of the plug, or removing some epoxy from the hole before curing would help avoid this problem. To avoid this problem completely, some designs allow one to apply an appropriate amount of the adhesive to one bonding surface and then to bring the surfaces together with controlled spacing to give the proper bond thickness. This technique is especially applicable if the uncured adhesive is very viscous or if it carries glass beads to control bond thickness.

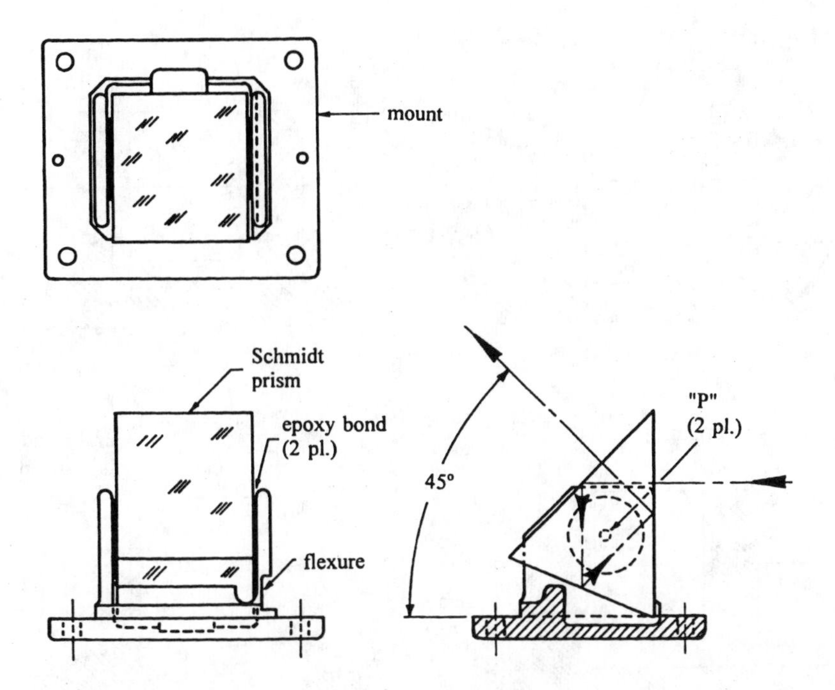

Fig. 4.15 A Schmidt prism bonded on both sides to a U-shaped mount. (From Willey[41])

Two versions of another design concept with support provided from two sides by arms forming a U-shaped mount are shown schematically in Figure 4.18.[41] In View (a), the crown glass prism is bonded to the ends of two cylindrical stainless steel plates passing through clearance holes in the arms. The entire mount is made of stainless steel so glass-to-metal thermal expansion mismatch is not a big problem. Alignment of the prism and the plates relative to the mount must be accomplished using mechanical references or optical fixturing prior to this first bonding step. After the first bonds have cured, the plates are epoxied to the arms as indicated in View (b). With this approach, tolerances on location and tilt of the surfaces to be bonded are relaxed since the plates align themselves to the prism in the clearance holes before they are bonded to the arms.

In Fig. 4.16(c), the prism is aligned to the mount and then is bonded to a raised pad on one support arm (left) and to the metal plug shown protruding through, but not attached to the second (right) arm. After these bonds have cured, the plug is bonded to the right arm to provide the required dual support. See View (d). Once again tolerances on the bonding surface locations and orientations are relatively loose.

Either of the last two design concepts can be extended to allow multiple plates or plugs of selected shapes (not necessarily round) to be passed through clearance holes in the support arms and bonded to opposite sides of the prism. After cure of the first bonds, these plates or plugs would be bonded to the support arms. This assembly process,

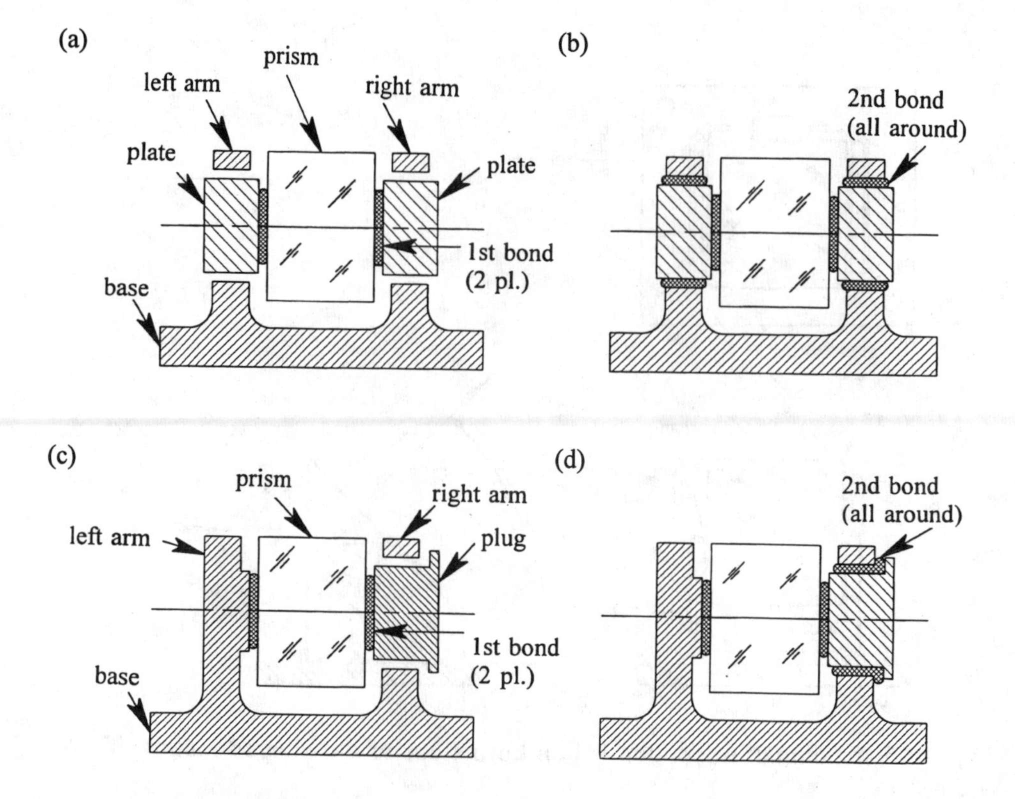

Fig. 4.16 Two concepts for double-sided bonding of a prism to a U-shaped mount. (Adapted from Beckmann[41])

"plastic pinning," is frequently used to lock aligned metal parts together in structures without the dangers associated with drilling and reaming for metal pinning. The beauty of the idea presented here in connection with constraining optical components is that precise fitting of parts is not required yet alignment established optically or by fixturing is retained after bonding. It also allows the benefits of separation of support points to increase mechanical rigidity.

A different optomechanical design involving bonding of prisms is illustrated in Fig. 4.17. Here, a Schmidt-Pechan roof prism (cemented subassembly) is inserted into a close-fitting seat molded into the filled-plastic housing of a commercial binocular. The prism subassembly is then provisionally secured in place with dabs of UV-curing adhesive applied through openings in the housing walls.[42,43] After confirming proper alignment, the prisms are secured by adding several beads of polyurethane adhesive through the same wall openings. The slight resiliency of the housing accomodates differential thermal expansion characteristics of the adjacent materials. With precision-molded structural members, adjustments are not required. Figure 4.18 shows some details of the internal configuration of such a prism mounting.

Fig. 4.17 A Schmidt Pechan erecting prism subassembly mounted in a plastic binocular housing. (From Seil[42])

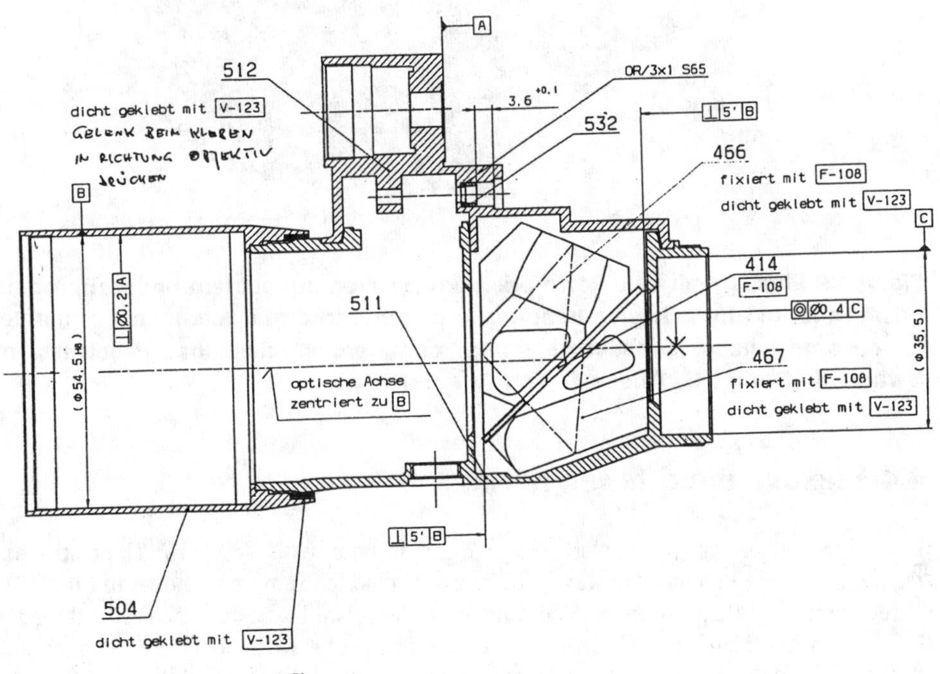

Fig. 4.18 Drawing of a roof prism mounted in the manner of Fig. 4.17 (Courtesy of Swarovski Optik KG, Absam/Tyrol, Austria)

Figure 4.19 is a photograph of an assembly comprising a Porro prism erecting system and a rhomboid prism mounted by the same general technique as just described for constraining roof prisms.[43] In this design, one Porro prism is attached with adhesive to its plastic bracket. This bracket then slides on two parallel metal rods to provide axial movement of the prism for focus adjustment of an optical instrument. The adhesive beads are more clearly shown in Fig. 4.20. Minimization of components and ease of assembly are prime features of this design. Customer experience and acceptance of products made by this technique have demonstrated the durability and adequacy of optomechanical performance achievable with this type of assembly.

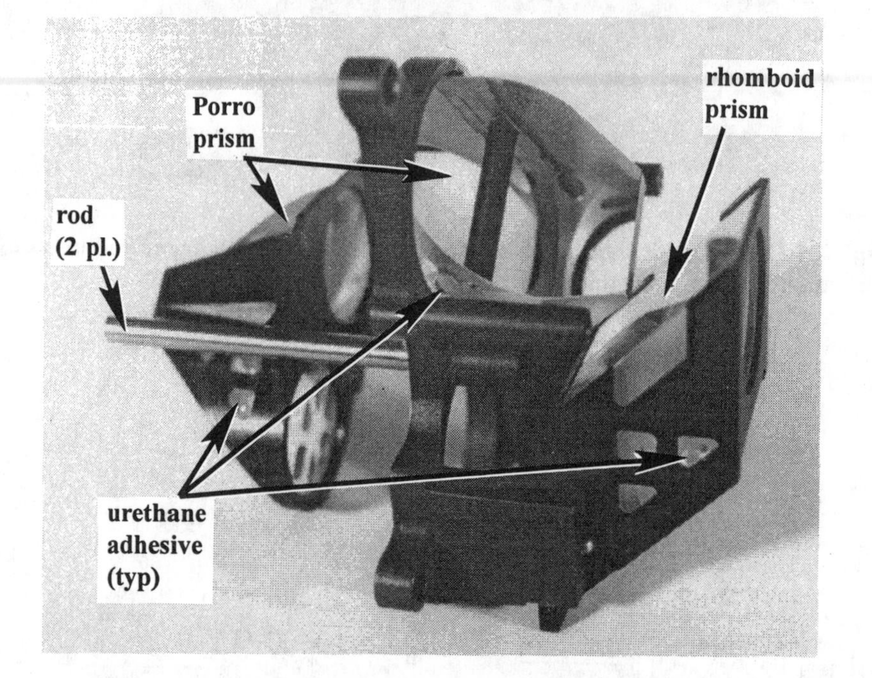

Fig. 4.19 Photograph of a Porro prism image erecting system and a rhomboid prism mounted by adhesive-bonding to plastic structural members for use as an optomechanical subassembly in a commercial telescope. (Courtesy of Swarovski Optik KG, Absam/Tyrol, Austria)

4.4 Flexure mounts for prisms

Some prisms (particularly large ones or ones with very critical positioning requirements) are mounted by way of flexures. A generic example is shown in Fig. 4.21. Three compound flexures are bonded directly to the prism base and attached by threaded joints to a baseplate (not shown). To reduce strain due to temperature changes and differential expansion between the prism material and the baseplate, all three flexures are designed to bend in several directions. They are very stiff axially. Flexure No. 1 locates the prism horizontally at a fixed point. It has a "universal joint" at its top to allow for angular misalignment at the bonded joint. The second flexure constrains rotation about the

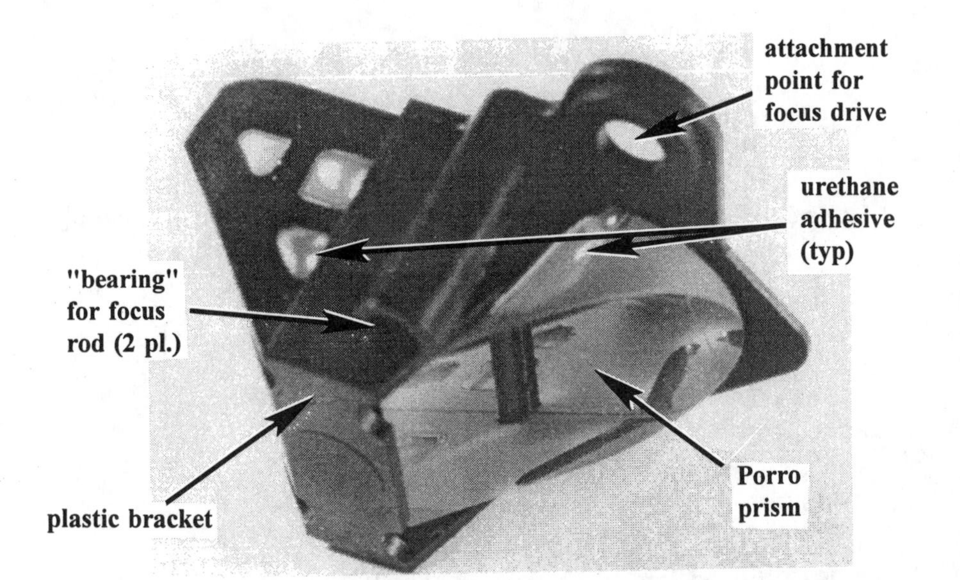

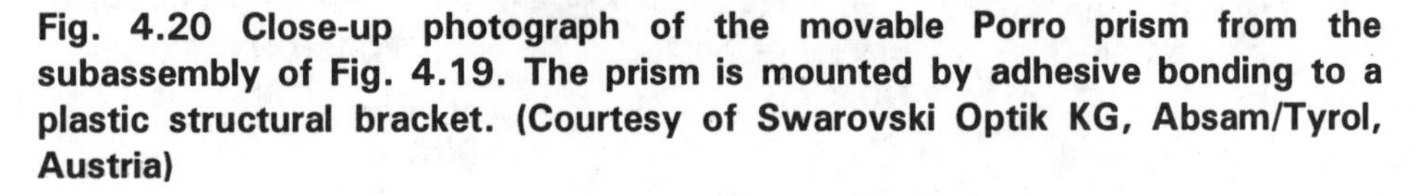
Fig. 4.20 Close-up photograph of the movable Porro prism from the subassembly of Fig. 4.19. The prism is mounted by adhesive bonding to a plastic structural bracket. (Courtesy of Swarovski Optik KG, Absam/Tyrol, Austria)

fixed point (first flexure), but allows relative expansion along a line connecting the first and second flexures because of its universal joint at top and single axis joint at the bottom. The third flexure has a universal joint at each end; it thus supports its share of the prism's weight and prevents rotation about the line connecting the other two flexures. The third flexure does not constrain the prism transversely. All three flexures have torsional compliance. Small differences in lengths of the flexures are accomodated through compliance of the three top universal joints. Because of the flexures, the prism remains fixed in space without distorting or disturbing the structure to which it is attached even in the presence of significant temperature changes and differential expansion of the prism and the mounting structure.[8] Specific examples of flexure mountings for large prisms are discussed in Sections 9.3 and 9.4.

A different type of flexure mounting is shown in Fig. 4.22. The prism is a small [25.4 mm (1.00 in.)] OD internally reflecting axicon with a $\pm 60°$ conical entrance/exit face. As explained in Section 3.3.19, its function is to convert a circular laser beam into a larger annular beam.[44] The BK7 glass prism is bonded with a thin annular ring of elastomer into the bore of an aluminum cell. The cell is attached to its support structure in an instrument with three screws passing through holes in three integral tangentially-oriented flexure blades. The interface to structure is plastic pinned to prevent lateral motion relative to the structure. The subassembly is relatively insensitive to temperature changes because the flexures will bend symmetrically and keep the axicon apex centered with respect to the laser beam. Fine adjustments of the subassembly in angle and lateral translation are accomplished by external means.

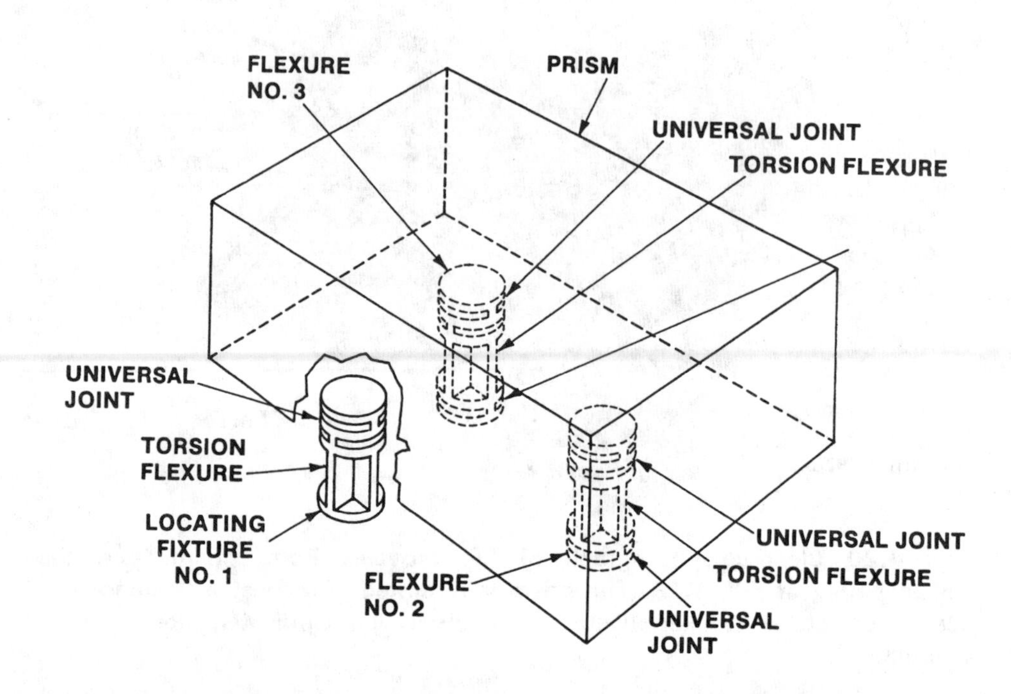

Fig. 4.21 Conceptual sketch of a flexure mounting for a large prism. (From Yoder[8] by courtesy of Marcel Dekker, Inc.)

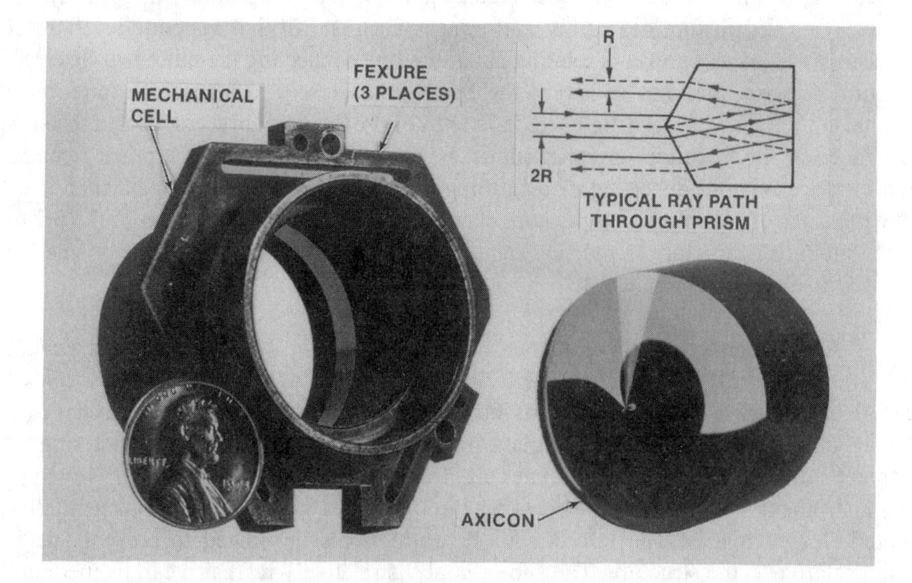

Fig. 4.22 Axicon and flexure-mounted cell. (Adapted from Yoder et al[44])

CHAPTER 5

ESTIMATION OF CONTACT STRESSES IN PRISMS

Force exerted over a small area on the surface of an optical component causes elastic deformation, i.e., strain, of the local region and, hence, proportional stress within that region. If the stress is so large as to exceed the damage threshold of the optical material, damage might well occur. Deterministic calculations of damage thresholds for glass-type materials by fracture mechanics methods are complex. Statistical methods can determine the probability of failure under specified conditions.[36] As a short cut, we here apply rule-of-thumb values for the stress in glass that might cause damage. In compression, this is 50,000 $lb/in.^2$ (3.4×10^8 Pa) while, in tension, it is 1000 $lb/in.^2$ (6.9×10^6 Pa).

Similarly, stress builds up within the mechanical member that constrains the optical material. This must be compared to the yield strength of that material (generally taken as that resulting in a deformation of 2×10^{-3}) to see if damage could occur there.

In this chapter, we outline ways to use the previously discussed estimates of forces needed to hold a prism under acceleration, i.e., shock and vibration, to calculate the deflections needed in constraining members such as leaf springs. We then see how to estimate the contact stress within the regions of the prisms compressed by these forces using various shapes of contacting mechanical members. The bending stresses in the spring blades also are considered. Next, we consider the use of resilient pads as springs in prism constraints. Finally, tensile stress in a single-sided bonded joint is discussed. We show why glass can fracture at low temperatures in such a design.

5.1 Compressive stress in clamped prisms

5.1.1 Constraint with multiple cantilevered leaf springs

As was illustrated by Numerical Example No. 4, it is easy to calculate the force, P_i, needed at each spring to ensure that the prism stays in contact with the reference surfaces under maximum expected acceleration. Equation (4.1). We now need a relationship for calculating from the spring dimensions and material characteristics how much deflection from the "at rest" condition is needed to provide the desired force. This can be done by using standard design formulas for the chosen type of spring. Since flat leaf springs are the most common types used to hold prisms, we limit our considerations here to that type. Figures 5.1(a) and 5.1(b) illustrate common types.

The cantilevered leaf spring is a simple beam, usually with rectangular cross-section, fixed to structure at one end and pressing against the prism at the other end. Equation (5.1) can be used to calculate the beam deflection, Δy, of each spring from its relaxed (undeflected) condition required to hold the prism with a force, P_i, as determined from Eq. (4.1) for given (or calculated) prism weight, w; maximum acceleration, a_G; and number of springs, N_i[45]. We assume that the force is divided equally among the springs.

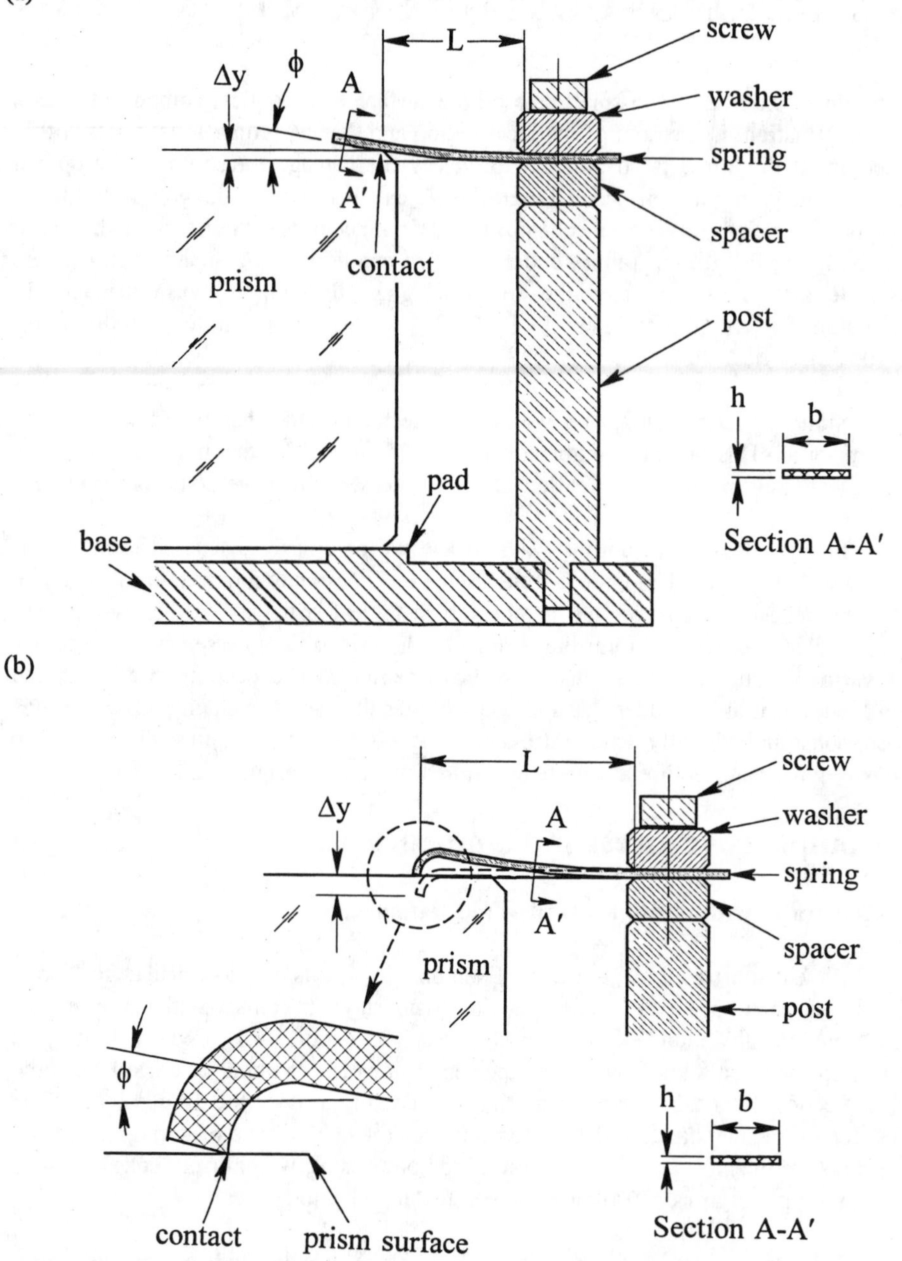

Fig. 5.1 Two concepts for cantilevered flat spring constraints for a prism: (a) contact at the prism edge, (b) contact at a sharp corner of a bent spring. In each case, line contact occurs across the full width of the spring and the prism is pressed against pads on the bottom of the mount directly opposite the spring contacts

$$\Delta y = (1-\upsilon_M^2)(4L^3P_i)/E_M bh^3, \tag{5.1}$$

where L is the free (cantilevered) length of the spring (in mm or in.)
P_i is the required force per spring (in N or lb)
E_M is Young's modulus for the spring material (in N/mm² or lb/in.²)
υ_M is Poisson's ratio for the spring material
b is the width of the spring (in mm or in.),
h is the thickness of the spring (in mm or in.), and b >> h.

The tilt of the free end of the spring, shown as ϕ in Figs. 5.1(a) and 5.1(b), can be calculated from Equation (5.2). This angle is in radians.

$$\phi = (1-\upsilon_M^2)(6L^2P_i)/E_M bh^3. \tag{5.2}$$

Numerical Example No. 9: Cantilevered spring penta prism mount.
Assume a prism weighing 1.25 lb (0.567 kg) is to be constrained at a_G = 10 times gravity by three titanium springs with dimensions as follows: L = 0.625 in. (15.875 mm), b = 0.637 in. (16.180 mm), and h = 0.035 in. (0.889 mm). Use the model of Fig. 5.1 and Eqs. (4.1) and (5.1) to find out how much each spring should be deflected from its relaxed condition in order to provide the proper constraining force. Applying Eq. (5.2), what is the angle of the spring end?

From Table C12, $E_M = 16.5\times10^6$ lb/in.² (1.14×10^5 Pa), $\upsilon_M = 0.34$

By Eq. (4.1): $P_i = (1.25)(10) / 3 = 4.167$ lb (18.534 N)
By Eq. (5.1): $\Delta y = (1 - 0.34^2)(4)(0.625)^3(4.167) / (1.65\times10^7)(0.637)(0.035)^3$
$= 0.0080$ in. (0.203 mm)
By Eq. (5.2): $\phi = (1 - 0.34^2)(6)(0.625)^2(4.167) / (1.65\times10^7)(0.637)(0.035)^3$
$= 0.0192$ radian $=1.099°$

Numerical Example No. 10: Effect of error in spring deflection.
In the design of Numerical Example No. 9, what are the variations in force and spring tip angle if the spring deflection is in error by 0.001 in. (0.025 mm)?

Let $\Delta x = 0.0090 + 0.0010 = 0.0100$ in.
$= (4)(0.625)^3(P_i + \Delta P_i) / (1.65\times10^7)(0.637)(0.035)^3$
The new force $= P_i + \Delta P_i = 4.614$ lb, $\Delta P_i = 0.447$ lb (11% error)
Error in ϕ if P_i changes to 4.614 lb is:
$[(0.0217)(4.614) / (4.167)] - 0.0217 = 0.0023$ rad (0.106°)

This angle change seems small, but actually is the same percentage error as for Δy.

Another pertinent equation quantifies the bending stress, S_B, within the material of each cantilevered spring due to the imposed nominal deflection when holding the prism in place:

$$S_B = 6LP_i/bh^2. \tag{5.3}$$

This stress should not exceed the yield strength of the material used.

Numerical Example No. 11: Stress in a cantilevered spring.
Calculate the nominal stress in each of the springs used in Numerical Example No. 9.

From Eq. (5.3),

$$S_B = (6)(0.625)(4.167) / (0.637)(0.035)^2 = 20{,}024 \text{ lb/in.}^2 \text{ (138 MPa)}.$$

Note: From Table C12, the minimum yield strength of titanium is 1.20×10^5 lb/in.^{2} so a safety factor of 6 therefore exists.

The force from the spring-type constraint and the reaction force from the base pad opposite are applied over specific areas. These areas depend upon the designs of the glass-to-metal interfaces. Ideally, the interfaces are appropriately sized and the metal part conforms intimately with the glass part, e.g., a flat constraining surface contacts a flat prism surface. This is not the case in the designs shown in Fig. 5.1. In View (a), the fixed end of the spring is below the top of the prism so contact occurs at the beveled edge of the prism. It would be expected that the spring is sufficiently flexible in twist so as to establish contact along a line of length equal to the spring width, b. The free length of spring that can bend is L and extends from the contact line to the edge of the clamp near the top of the post. In View (b) of that figure, the spring is bent at about 90° near its free end so, when the free length bends to deflect that end by Δy, the corner of the spring touches a flat surface of the prism. Again, line contact occurs over the spring width b.

If the spring is clamped above the prism surface and tilted as shown in Fig. 5.2(a) by virtue of wedge angles built into the spacer and washer or by two sets of spherical washers (not shown) sandwiching the spring, a larger area contact can theoretically be achieved on the prism. Similarly, in View (b), a flat pad attached to (or machined integral with) the spring lies flat against the prism surface. Both these designs are subject to error if the contact area is not accurately tangent to the prism surface. If the spacers are too long or too short, the area contacts can degenerate into lines and free lengths, L, can vary from nominal.

A preferred interface is shown in Fig. 5.3(a). Starting with thick stock, a cylindrical pad is machined integrally into the bottom of the free end of the spring and the blade portion of the spring is machined from both sides to the desired thickness, h. Transition to the thickness of the clamped end is by way of machined cylindrical fillets which serve to reduce stress concentrations that would occur with sharp corners. The

(a)

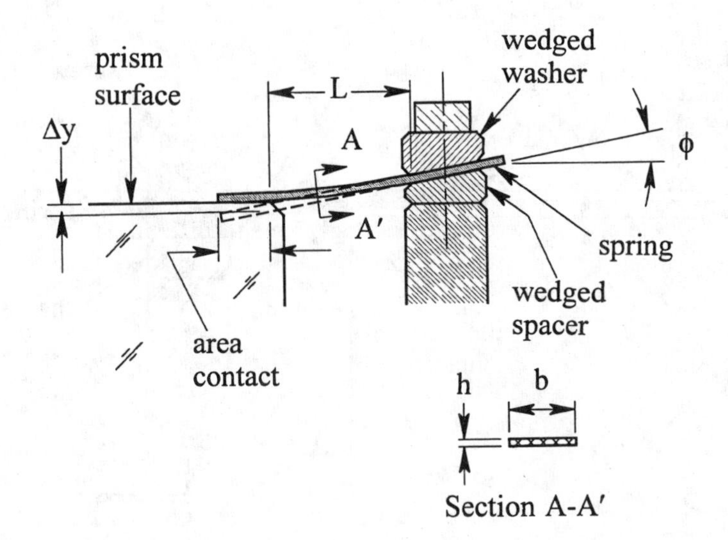

(b)

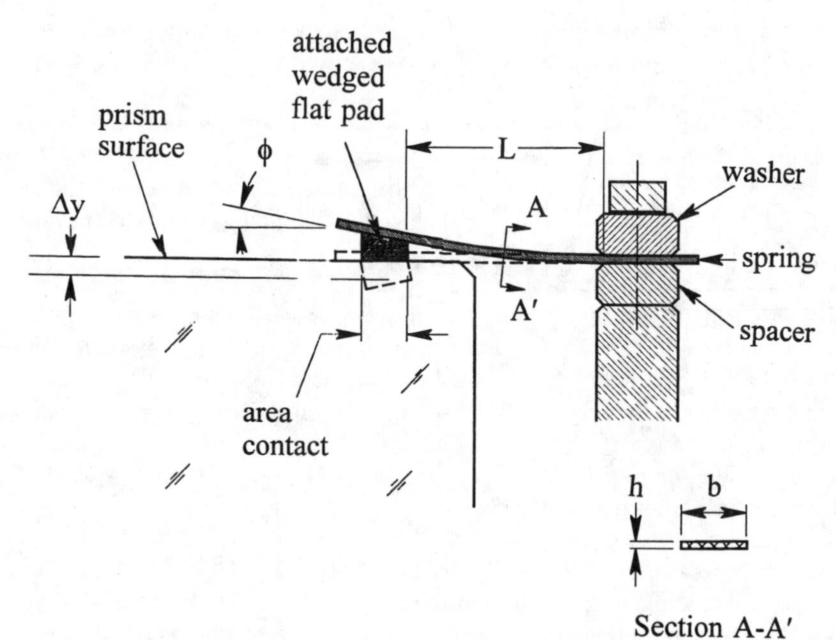

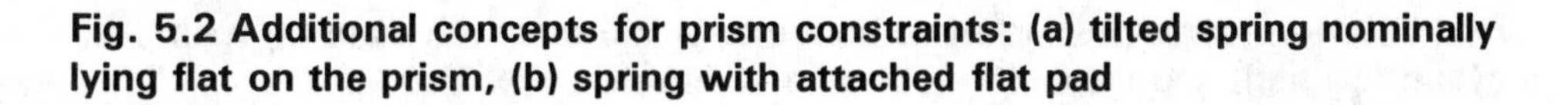

Fig. 5.2 Additional concepts for prism constraints: (a) tilted spring nominally lying flat on the prism, (b) spring with attached flat pad

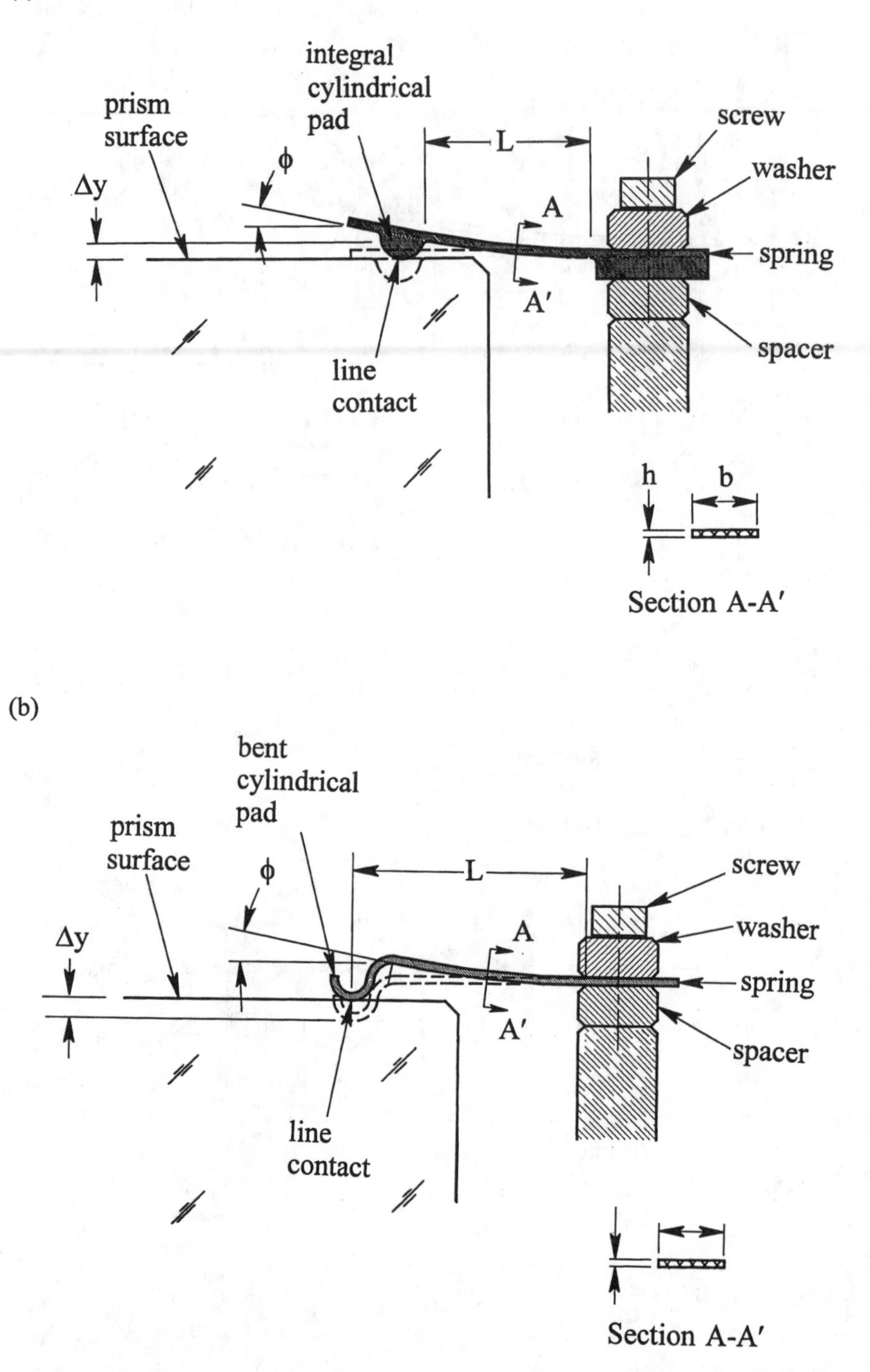

Fig. 5.3 More concepts for prism constraints: (a) monolithic spring with cylindrical pad, and (b) spring with bent cylindrical tip

height of the clamped end is adjusted (usually by grinding the thickness of the spacer) to produce a predetermined deflection of the spring. Line contact again occurs over the width of the pad which usually would equal the spring width. The radius of the cylindrical pad is a design variable. A less sophisticated version of this spring could be machined from uniform thickness stock with the pad attached with epoxy or screws. This type construction was indicated in Fig. 4.4.

Fig. 5.3(b) shows another possibility. Here, the free end of the spring is bent into a cylinder which contacts the prism at a line. The radius of the cylinder and the length L may be harder to control here than in the case shown in View (a) and uniform contact across the bent surface may not occur because of difficulties in bending metal to precise contours.

Another concept, not shown in Fig. 5.3, has a spherical interface with the prism surface achieved either by machining or by die-forming an integral "dimple" near the free end of the spring blade and allowing the convex side of the dimple to touch the prism. This concept probably would be subject to the same inherent fabrication difficulties as the design of Fig. 5.3(b).

5.1.2 Constraint with one leaf spring clamped at both ends

As illustrated in the discussion of semi-kinematic mounts for prisms in Section 4.1, and, especially, Figs. 4.3(d), 4.5, 4.6, and 4.7, it is sometimes appropriate for a single force to hold a prism in one direction against three opposite reference pads. It is quite likely that the spring would be a leaf spring clamped to the structure at both ends and straddling the prism. See Fig. 5.4.

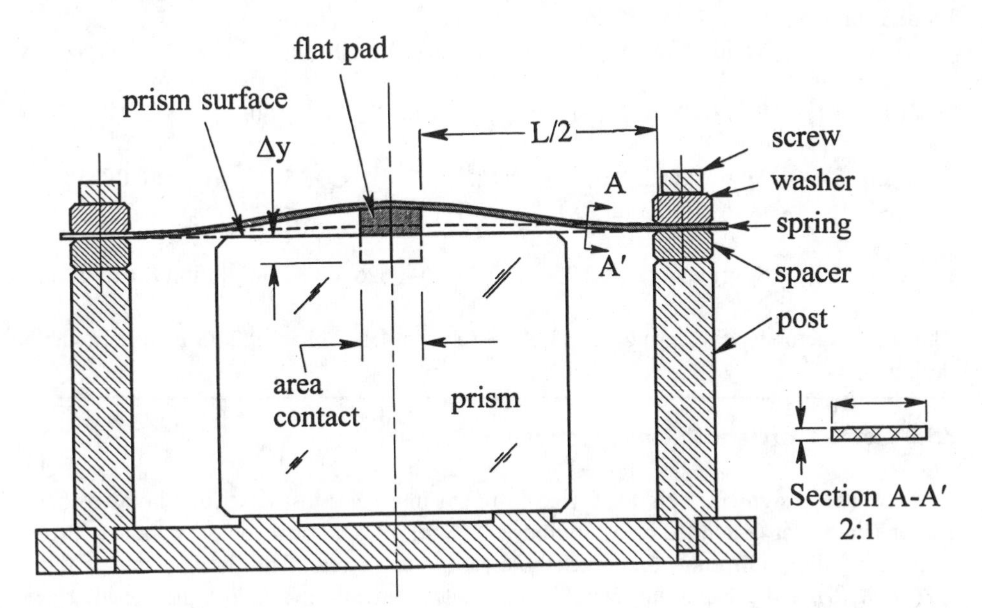

Fig. 5.4 Prism constrained with one leaf spring clamped at both ends

If b >> h, the following equations, adapted from Roark[45], can be used to calculate the spring displacement, Δy, at its center required to produce a preload of P_i and the bending stress, S_B, in the spring.

$$\Delta y = (0.0625)(1 - \upsilon_M^2)(P_i L^3)/E_M bh^3, \quad (5.4)$$

where L is the total free length of the spring from clamp to clamp (in mm or in.)
P_i is the required total force (in N or lb)
E_M is Young's modulus for the spring material (in N/mm^2 or lb/in/^{2})
b is the width of the spring (in mm or in.)
h is the thickness of the spring (in mm or in.).

$$S_B = (3/2)P_i L/bh^2, \quad (5.5)$$

where all terms are as defined above.

Numerical Example No. 12: Prism constrained with one straddling leaf spring. Assume that the 1.25 lb prism of Numerical Example No. 9 is to be constrained against locating pads under 10 times gravity by one spring force, P_i, from a straddling leaf spring. The spring is beryllium copper and has dimensions L = 2.958 in. (75.133 mm), b = 0.750 in. (19.050 mm) by h = 0.0625 in. (1.5875 mm). What force and spring deflection are needed and what is the bending stress in the blade?

From Table C8, $E_M = 18.5 \times 10^6$ lb/in.^{2}
Yield strength = 155,000 lb/in.^{2} (107×10^7 Pa)

From Eq. 4.1, $P_i = (w/N_i)(a_G) = (1.25 / 1)(10) = 12.5$ lb (55.60 N)

From Eq. 5.4, $\Delta y = (0.0625)(1 - 0.35^2)(12.5)(2.958^3) / (18.5 \times 10^6)(0.75)(0.0625^3)$
$= 0.0052$ in. (0.1330 mm)

From Eq. 5.5, $S_B = (1.5)(12.5)(2.958) / (0.375)(0.0625^2) = 37{,}863$ lb/in.^{2} (261,061 Pa)

The safety factor for the spring is 155,000/37,863 = 4.1. This seems to be a reasonable design.

5.1.3 Contact stress with curved interfaces

The design concepts for spring-to-prism interfaces just described involve sharp corner (line), flat (extended area), cylindrical (line), or spherical (point) contacts. From a stress-buildup viewpoint, the larger the area, the less stress. The flat contacts of Figs. 5.2(a), 5.2(b), and 5.4 therefore would seem best. As mentioned earlier, the flat interface is subject to degeneration into line contact if tolerances are not held closely so its potential stress advantage may not be achieved. By choosing the contact radius carefully, the stress with a given preload and a cylindrical or spherical interface can be significantly reduced from that which occurs in a sharp-corner interface at the same preload. The curved interface offers the advantage that tilt of the spring end as the spring deflects will not change the nature of the contact nor appreciably affect the free length, L.

In section view, either a convex spherical or cylindrical surface on a flat prism surface would appear essentially as indicated schematically in Fig. 5.3(a). Neglecting friction, if the spring tilts as it deflects, the preload is still delivered normal to the prism surface. We next consider means for estimating the stress developed in the glass region near the contact due to this preload.

If the mechanical contact interface is cylindrical, the geometry of Fig. 5.5 applies.

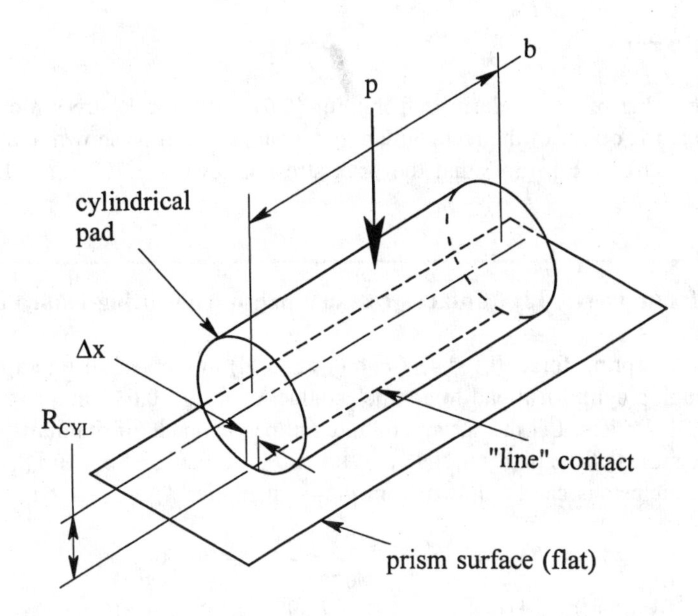

Fig. 5.5 Geometric model of a cylindrical interface on a flat prism surface

Adapting an equation from Rourk[45] as applied by Yoder,[8] the stress, $S_{P\,cyl}$, is:

$$S_{Pcyl} = 0.564(p/K_2R_{cyl})^{1/2}, \tag{5.6}$$

where: p is the linear preload = P_i/b (in N/mm or lb/in.)
b is the width of contact (in mm or in.)
R_{cyl} is the cylindrical radius of the surface contacting the prism (in mm or in.)
K_2 is given by Eq. (5.7).

$$K_2 = [(1 - \upsilon_G^2)/E_G] + [(1 - \upsilon_M^2)/E_M], \tag{5.7}$$

where: υ_G and E_G are Poisson's ratio and Young's modulus for the glass (prism)

υ_M and E_M are Poisson's ratio and Young's modulus for the metal (pad).

Since the prism and the pad are assumed to be elastic in this analysis, the force that presses them together causes each material to deform slightly along the line of contact. The deformed region is a rectangle of size b by Δx as shown in Fig. 5.5. Equation (5.8) is used to calculate Δx.

$$\Delta x = 2.263 (K_2 p R_{cyl})^{1/2}. \tag{5.8}$$

If we divide the total contact preload by the contact area we obtain the average stress in the contact region as indicated in Eq. (5.9).

$$S_{Pavg} = P_i / b \Delta x. \tag{5.9}$$

The value of $S_{P\ cyl}$ derived from Eq. (5.6) is the peak stress along the line parallel to b at the center of the rectangular contact area. It can be shown analytically by dividing Eq. (5.6) by Eq. (5.9) that the peak stress is always 1.277 times the average stress.

Numerical Example No. 13: Contact stress in a prism with spring-loaded cylindrical interface.

Assume that the spring force, P_i, = 4.167 lb (18.535 N) in Numerical Example No. 9 is applied through a cylindrical pad to a "line" contact of length 0.637 in. (15.875 mm) at the top of a BK7 prism. Let the spring and integral pad be made of titanium. Assume, in turn, that the radius of the pad is 0.1, 1, 10, and 100 in. Calculate $S_{P\ cyl}$ and $S_{P\ avg}$ for each case. What conclusions can be drawn from these calculations?

R_{CYL}	given	(in.)	0.1	1.0	10	100
υ_G	Table C1		0.208	0.208	0.208	0.208
E_G	Table C1	(lb/in.^{2})	1.17×10^7	1.17×10^7	1.17×10^7	1.17×10^7
υ_M	Table C12		0.340	0.340	0.340	0.340
E_M	Table C12	(lb/in.^{2})	1.65×10^7	1.65×10^7	1.65×10^7	1.65×10^7
K_2	Eq. (5.7)	(in.^{2}/lb)	1.354×10^{-7}	1.354×10^{-7}	1.354×10^{-7}	1.354×10^{-7}
P_i	given	(lb)	4.167	4.167	4.167	4.167
p	P_i/b	(lb/in.)	6.5411	6.5411	6.5411	6.5411
$S_{P\ cyl}$	Eq. (5.6)	(lb/in.^{2})	12,396	3920	1240	392
Δx	Eq. (5.8)	(in.)	6.734×10^{-4}	2.129×10^{-3}	6.734×10^{-3}	2.129×10^{-2}
$S_{P\ avg}$	Eq. (5.9)	(lb/in.^{2})	9713	3071	971	307
$S_{P\ cyl} / S_{P\ avg}$			1.276	1.276	1.277	1.277

Conclusions: (1) longer R_{cyl} give lower stresses, (2) stresses shown here do not pose any damage problems, and (3) the ratio of $S_{P\ cyl}$ to $S_{P\ avg}$ is as indicated in the text.

Note that the spring deflection required to produce the specified preload is not affected by change in pad radius.

Examination of Eq. (5.6) shows that an increase in R_{cyl} from one value to another would reduce $S_{P\,cyl}$ by $[(R_{cyl})_2/(R_{cyl})_1]^{1/2}$. The changes in Numerical Example No. 13 were by factors of 10 so we would expect the corresponding changes in $S_{P\,cyl}$ to be by factors of $10^{1/2} = 3.162$. This is exactly the case. We might then extrapolate those changes in the direction of reducing the cylindrical radius further. For $R_{cyl} = 0.001$ in., S_{cyl} would increase to 123,980 lb/in.^{2} which would greatly exceed the rule-of-thumb compression strength of 50,000 lb/in.^{2}. Hence, "line" contact such as is depicted in Figs. 5.1(a) and 5.1(b) would be dangerous because the radius of a sharp edge is extremely small. This situation might be remedied for the case of Fig. 5.1(a) by a simple change in the design of the prism where the protective bevel touched by the spring is fine ground to a cylindrical radius such as 1 mm. The design of Fig. 5.1(b) could also be improved if the spring corner that touches the prism is rounded to a cylindrical radius of the order of 1 mm.

If the curved mechanical interface with the flat prism surface is changed to a spherical contour and all other design features remain constant, the following equation would give the contact stress in the glass region surrounding the "point" contact.[45]

$$S_{Psph} = 0.578[P_i/(R_{sph}^2 K_2^2)]^{1/3}. \quad (5.10)$$

The radius, r_{sph}, of the elastically deformed circular contact region is given by Eq. (5.11).

$$r_{sph} = 0.908(P_i R_{sph} K_2)^{1/3}. \quad (5.11)$$

The average stress in this region is as given by Eq. (5.12). In this case, the ratio of peak stress to average stress can be shown to be 1.497.

$$S_{Pavg} = P_i/\pi r_{sph}^2, \quad (5.12)$$

where all terms are as defined above.

Numerical Example No. 14: Contact stress in a prism with spring-loaded spherical interface.

Repeat the last numerical example assuming spherical interfaces with $R_{sph} = 0.1$, 1.0, and 10.0 in. What conclusions can be drawn?

R_{sph}	given	(in.)	0.1	1.0	10.0
υ_G	Table C1		0.208	0.208	0.208
E_G	Table C1	(lb/in.^{2})	1.17×10^7	1.17×10^7	1.17×10^7
υ_M	Table C12		0.340	0.340	0.340
E_M	Table C12	(lb/in.^{2})	1.65×10^7	1.65×10^7	1.65×10^7
K_2	Eq. (5.7)	(in.^{2}/lb)	1.354×10^{-7}	1.354×10^{-7}	1.354×10^{-7}
P_i	given	(lb)	4.1667	4.1667	4.1667
$S_{P\,sph}$	Eq. (5.10)	(lb/in.^{2})	163,520	35,234	7592
r_{sph}	Eq. (5.11)	(in.)	0.0035	0.0075	0.0162
$S_{P\,avg}$	Eq. (5.12)	(lb/in.^{2})	108,270	23,579	5054
$S_{P\,cyl}/S_{P\,avg}$			1.510	1.494	1.502

We see that (1) longer R_{sph} give lower stresses, (2) the contact stresses do not pose severe glass damage problems for R_{sph} greater than about 1 in., and (3) the ratio of $S_{P\,sph}$ to $S_{P\,avg}$ for each case is as indicated in the text.

Note from Eq. (5.10) that changing R_{sph} by some factor, f_R, changes $S_{P\,sph}$ by $f_R^{-3/2}$ so, if the factor is 10 as in Numerical Example No. 14, we would expect the $S_{P\,sph}$ changes from column to column to be $10^{2/3}$ or 4.641. This is borne out by the calculations in the example.

5.1.4 Contact stress at prism locating pads

In many prism mounting designs, the width, d_P, of each locating pad touching the base of the prism (see Fig. 5.1) is set equal the width, b, of the spring and each pad is circular. In such designs, the localized contact stress, S_{pad}, is estimated from Eq. (5.13). This stress should be compared to the rule-of-thumb glass compressive strength of 50,000 lb/in.^{2} to determine what safety factor, if any, exists in the design.

$$S_{Pavg} = 4P_i / \pi b^2. \tag{5.13}$$

This equation can easily be adjusted to accommodate other pad shapes.

The same stress is created in the pad, but the metal is more tolerable of compression so this is of little concern.

Numerical Example No. 15: Contact stress in a prism at a locating pad.
Assume that the spring forces in Numerical Example No. 9 are applied normally to three coplanar circular flat areas of diameters = 0.637 in. (16.180 mm) at the bottom of the prism. Assume uniform force distribution and, using data from that example, calculate the average stress, S_{Pad}, in the prism adjacent to one pad.

From Eq. (5.13), $S_{Pad} = (4)(4.167) / (\pi)(0.637)^2 = 13.075$ lb/in.^{2} (90,155 Pa).

As might be expected, this stress is negligible.

5.1.5 Spring-loading prisms through non-rigid pads

If a prism is clamped between rigid plates with only a resilient pad to provide spring force to resist shock and vibration as indicated conceptually by Fig. 4.3(c), we can design the interface only if the elastic characteristic of the pad material is known. The pertinent material property is the "spring constant," C_P, defined as the load required to be applied normal to the pad surface to produce a unit deflection. See Eq. (5.14).

$$P_i = C_P \Delta y. \tag{5.14}$$

Most resilient materials have a limited elastic range, tend to creep over time, and typically take a permanent set under sustained high compressive load, i.e., one greater than that for which the material acts elastically. For these reasons, these materials might be considered unreliable for use in the manner suggested here. Since they are sometimes used, we discuss one type of material as it might be used in a prism mount.

Typically, a material such as a visco-elastic, thermoset, polyether base polyurethane behaves as indicated in Fig. 5.6 for different durometers and three deflections as percentages of the pad thickness. The material depicted is commonly used

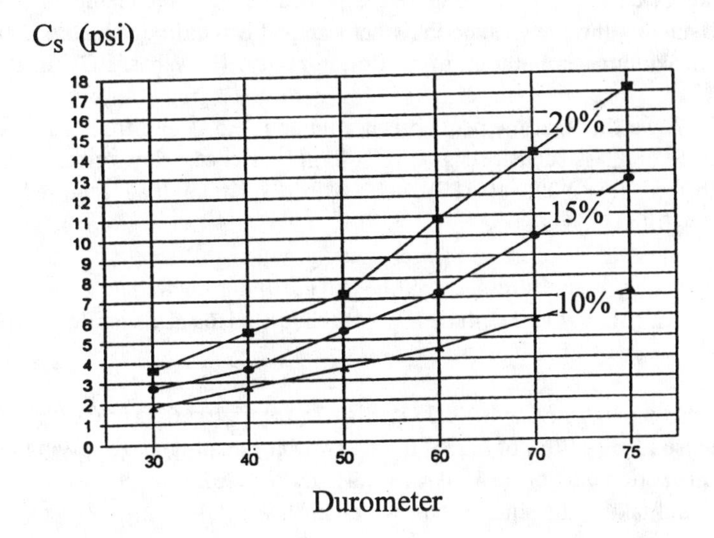

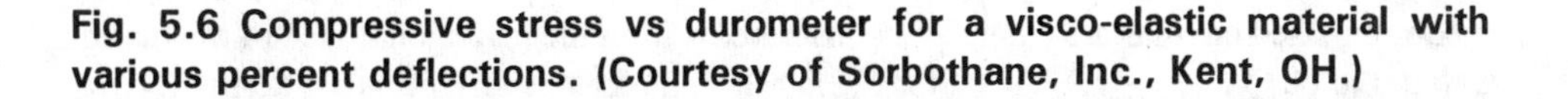
Fig. 5.6 Compressive stress vs durometer for a visco-elastic material with various percent deflections. (Courtesy of Sorbothane, Inc., Kent, OH.)

in vibration isolators and behaves elastically if the change in thickness is between 10% and 25% of the total thickness.[46] Obviously, a softer material deflects more per unit load. Since we should design the interface for the maximum applied force (which would occur at maximum acceleration), we might well choose the 20% deflection curve of Fig. 5.6 so deflections under conditions of lesser severity would lie within the linear range of the material. The manufacturer's literature suggests that deflection, Δy, is related to load, P_i, as follows:

$$\Delta y = (0.15)(P_i)(t_P)/[(C_S)(1+2\gamma^2)(A_P)]. \tag{5.15}$$

where: P_i is the required force per pad (in N or lb)

t_P is the pad thickness (in mm or in.)
γ is a "shape factor" = D_P / $4t_P$ for a square or circular shape
D_P is the pad width or diameter (in mm or in.)
A_P is the pad area (in mm or in.) = D_P^2 for a square or $\pi D_P^2/4$ for a circular pad.

The following example illustrates the design of a resilient interface using such a material. Similar calculations can be applied to other resilient materials if their elastic properties are available.

Numerical Example No. 16: Prism clamped through a resilient pad.
A SF6 penta prism with aperture A = 2.000 in. is to be clamped against a baseplate by a rigid plate in a manner similar to that of Fig. 4.3(c). A circular pad with thickness 3/8 in. made of material characterized by Fig. 5.6 is placed between the clamping plate and the prism. (a) Using 30 durometer material, what size pad is required if a 20% deflection is to occur at a maximum acceleration of 10 times gravity? (b) What is C_P for the pad?

(a) From Fig. 3.17, we find that the penta prism volume is $1.5A^3$ and from Table C1, we determine that SF6 glass density is 0.187 lb/in.^{3}. Hence, the prism weight = (volume)(density) = $(1.5)(2.000^3)(0.187)$ = 2.44 lb. The total load required = P_i = (2.44)(10) = 22.440 lb.

From Fig. 5.6, the compressive stress, C_S, in the pad @ 30 durometer and 20% deflection is 3.7 lb/in.^{2}. We calculate γ as $D_P/4t_P = D_P/(4)(0.375) = 0.667D_P$ and $A_P = \pi D_P^2/4 = 0.785D_P^2$.

From Eq. (5.15), $\Delta y = (0.15)(22.440)(0.375) / [(3.7)(1 + (2)(0.667D_P)^2)(0.785D_P^2)]$.
This deflection also is to be 20% of 0.375 in. or 0.075 in. Equating and applying algebra, we obtain the quadratic equation: $0.194D_P^4 + 0.218D_P^2 - 1.262 = 0$.
Solving for D_P^2 and taking the square root, we obtain D_P = 1.431 in.

We need to see if this pad will fit onto the side of the prism. From Fig. 3.17, the maximum sized circular area that can be inscribed within the pentagonal face of the prism is found to be $Q_{MAX(C)} = 1.13A^2 = (1.13)(2.000)^2 = 4.520$ in.^{2}. From this we derive $D_{P\,MAX} = [(4)(4.520)/\pi]^{1/2} = 2.399$ in. This shows that the pad will easily fit onto the prism.

(b) The spring constant of the pad is C_P = 22.440 / 0.075 = 299 lb/in.

With a little thought we conclude that a stiffer pad (such as 70 durometer) would be smaller than the one in this example. A smaller pad also would result if the thickness t_P were reduced.

5.2 Tensile stress in the single-sided bonded interface

A prism bonded on one side must depend upon the tensile and shear strength of the adhesive joint for integrity of the bond. If it is bonded on two opposite sides, significantly greater reliability exists because an opposing force is felt at the second interface whenever the prism tries to pull away from the first interface. We here consider means for estimating the stress in the bonding material and in the optical material under acceleration or shock loading that tends to separate the optical component from the mechanical one in a design typified by Fig. 4.11.

From Newton's Second Law, $F_N = ma_G = \sigma Q/f_S$, where F_N is the force normal to the interface, $m = w/g$ is the mass of the prism, g is the acceleration due to gravity, a_G is the acceleration load factor measured as "times gravity," σ is the tensile yield strength of the weaker of the prism material or the bond, Q is the area of the bond, and f_S is a safety factor.

In a simple prism such as a cube, right angle, Porro, penta, etc., having no roof or other non-symmetrical feature and in which the bond covers the entire base area, Q, of the prism polyhedron, w = (base area)(height)(density) = $Qh\rho$. Usually, the prism optical aperture is round and equals A. Hence, $F_N = QA\rho a_G = \sigma Q/f_S$, and

$$a_G = \sigma / A \rho f_S. \tag{5.16}$$

In a more complex prism design such as an Amici, a roof penta, or any other prism having a bond area not equal to the prism base area, $F_N = V\rho a_G = \sigma Q/f_S$, where V is the prism volume and all other terms are as previously defined. So,

$$a_G = \sigma Q / V \rho f_S. \tag{5.17}$$

In either of the above cases, the adequacy of the bond itself under the calculated acceleration should be confirmed. To do so, we utilize Eqs. (5.16) or (5.17) and substitute σ_A and σ_G in turn. Here, σ_A has the appropriate value [such as 2500 lb/in.^{2} (17.2 MPa) of EC2216-B/A epoxy] for the adhesive used while σ_G = 1000 lb/in.^{2} (6.9 MPa) for glass-type materials.

Numerical Example No. 17: Stress in a prism and epoxy bond under acceleration
Assume the bonded roof penta prism shown in Fig. 4.12 is subjected to test vibration loads perpendicular to the bonded face equal to 1500 and 4200 times gravity. If the bond has a strength of 2500 lb/in.^{2}, the glass has a tensile yield strength of 1000 lb/in.^{2}, the prism weight is 0.25 lb, and the bond area is 0.785 in.^{2}, what are the safety factors for the glass and the adhesive under these loads?

(a) At 1500× gravity loading:

For the glass, from Eq. (5.17), a_G = 1500 = (1000)(0.785)/(0.25)(f_S) and f_S = 2.1

For the adhesive, a = 1500 = (2500)(0.785)/(0.25)(f_S) and f_S = 3.14

(b) At 4200× gravity loading:

For the glass, from Eq. (5.17), a_G = 4200 = (1000)(0.785)/(0.25)(f_S) and f_S = 0.75

For the adhesive, a = 4200 = (2500)(0.785)/(0.25)(f_S) and f_S = 1.87

Note that this explains why "well made" glass-to-metal bonded joints may fail under extreme shock or vibration by tearing glass from the optical component. The bond typically is stronger than glass!

CHAPTER 6
SMALL MIRROR DESIGN

As in the case of prisms, a clear understanding of certain important aspects of small mirror design is believed to be necessary before we consider different techniques for mounting those mirrors. This chapter therefore deals primarily with the mirror's geometric configuration. We define the relative advantages of first- and second-surface mirror types, provide guidance in specifying the appropriate aperture dimensions for the reflecting surface, and illustrate substrate configurations that might be employed to minimize component weight and self-weight deflection. Typical designs for metallic mirrors are then considered. The chapter closes with a few observations about the design and use of pellicles.

6.1 First- and second-surface mirrors

Most mirrors used in optical instruments have light-reflecting coatings made of metallic and/or nonmetallic thin films on their first optically polished surfaces. These are quite logically called "first-surface" mirrors. The metals most commonly used as coatings are aluminum, silver, and gold because of their high reflectivities in the UV, visible, and/or IR spectral regions. Protective coatings such as silicon monoxide or magnesium fluoride are frequently used on metallic coatings to increase their durability. Non-metallic films usually comprise multi-layer stacks of dielectric films of materials with high and low indices of refraction. Dielectric reflecting films usually function best over narrower spectral bands than the metals, but can have very high reflectivity at specific wavelengths. They are especially helpful in monochromatic systems using laser radiation. The state of polarization of the reflected beam may be significantly modified by all-dielectric stacks or dielectric overcoats when the beam angle of incidence differs from zero. Figures 6.1 and 6.2 show typical reflectance vs. wavelength curves for different first-surface reflecting coatings at normal and/or 45° incidence.

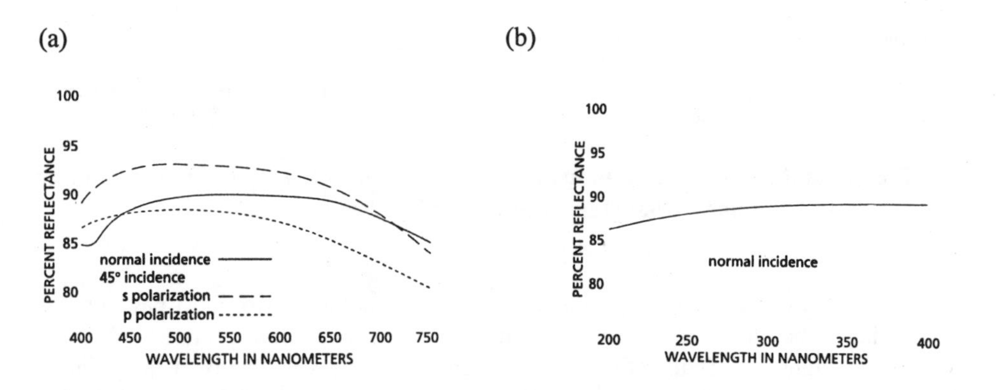

Fig. 6.1 Reflectance vs. wavelength for first-surface metallic coatings of (a) protected aluminum and (b) UV enhanced aluminum

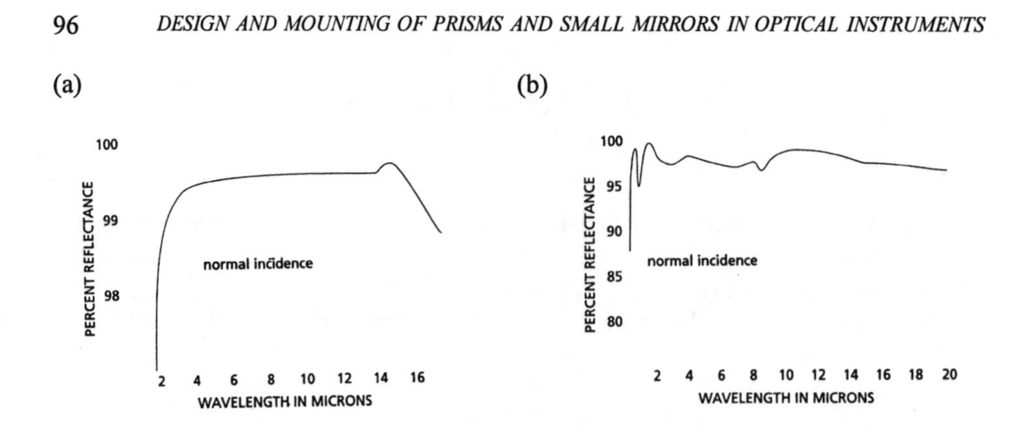

Fig. 6.2 Reflectance vs. wavelength for first-surface thin films of (a) protected gold and (b) protected silver

Figure 6.3(a) shows reflectance vs. wavelength for a typical multilayer dielectric film while Fig. 6.3(b) shows reflectance vs. wavelength for a second-surface coating of silver. The latter type reflecting coating is applied to the back, i.e., second surface of a mirror (or prism). This can be an advantage from the durability viewpoint because the film is protected from the outside environment and physical damage due to handling or use. A suitable protective coating such as electroplated copper plus enamel frequently is applied to the back of the thin film coating to protect it.

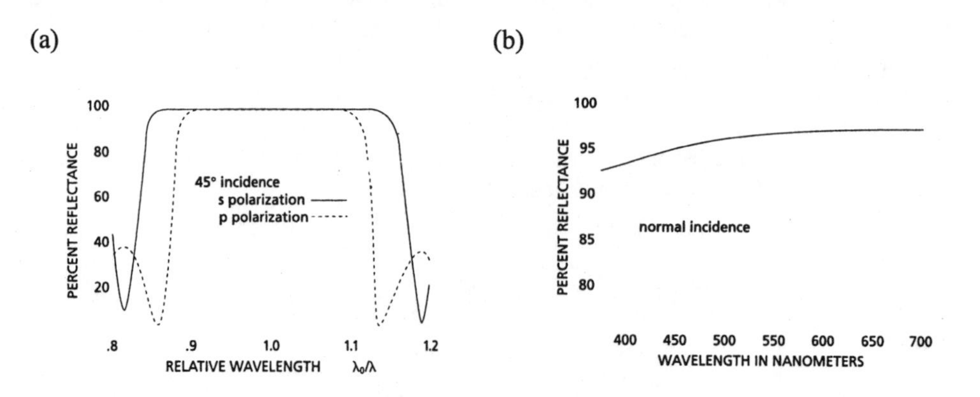

Fig. 6.3 Reflectance v.s wavelength for (a) a first-surface multi-layer dielectric thin film and (b) a second-surface thin film of silver

Since the radiation to be reflected by a second-surface mirror must pass through a refracting surface to get to the reflecting surface, a ghost reflection is formed at the first surface. The intensity of this reflection can be calculated from the index of refraction of the substrate using Fresnel's equations.[47,15] At normal incidence, the reflectance, R_λ, of an uncoated air-to-glass interface is:

$$R_\lambda = (n_\lambda - 1)^2/(n_\lambda + 1)^2. \tag{6.1}$$

For a typical optical glass (BaK2) with n_d = 1.540, $R_d = (0.540/2.540)^2$ = 4.5%.

Figure 6.4 illustrates a concave second-surface mirror and its function in forming a conventional image of a distant object and a ghost image of that object resulting from reflection at the first surface of the mirror. Light from the ghost is superimposed upon the conventional image as stray light and tends to reduce the contrast of the latter image. The ghost can be reduced in intensity by antireflection coating the first surface, but it is hard to eliminate it completely. The axial separation of the two images can be increased or decreased by careful choice of radii and mirror thickness. Typically, single-layer antireflection coatings reduce reflected intensity to about 1% to 2% of the incident intensity while multi-layer coatings typically reduce it to about 0.5%. High-efficiency, multilayer coatings for monochromatic applications can reduce the reflected intensity to significantly less than 0.5% over a narrow band.

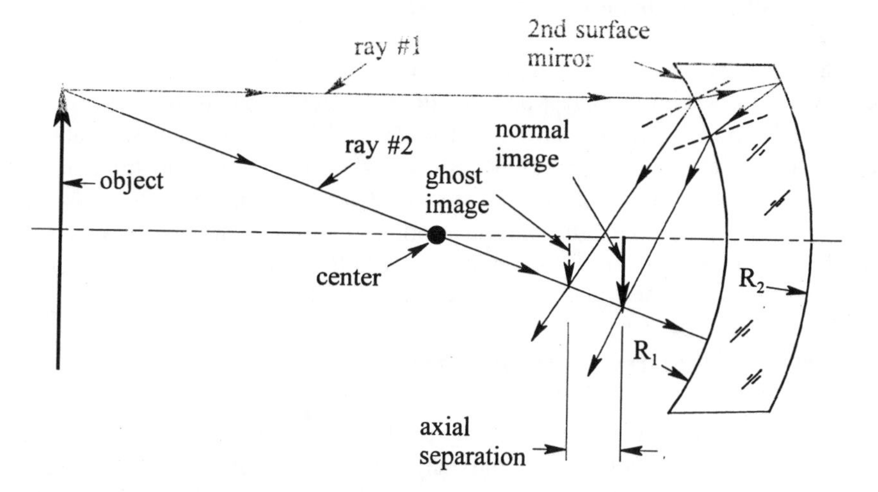

Fig. 6.4 Ghost image formation from the first surface of a second-surface mirror with concentric spherical surfaces. (Adapted from Kaspereit, ORDM 2-1)

Figure 6.5 shows ghost reflection formation from a flat second-surface mirror of thickness, t, oriented at 45° to the axis of an imaging system (lens). With a mirror at 45° in air, the ghost is displaced axially by a distance $d_A = (2t/n) + d_A$ and laterally by a distance $d_L = 2t/(2(2n^2 - 1))^{1/2}$. Once again, superimposition of the ghost onto the conventional image tends to reduce the contrast of the latter image. By antireflection coating the first surface, the ghost can be reduced in intensity, as quantified above.

An obvious difference between first- and second-surface mirrors is that a transparent substrate is needed for the latter, but not for the former. Tables C8 and C9 list mechanical properties and "figures of merit" for the common non-metallic and metallic mirror substrate materials. Of these, only fused silica has good refractive properties.

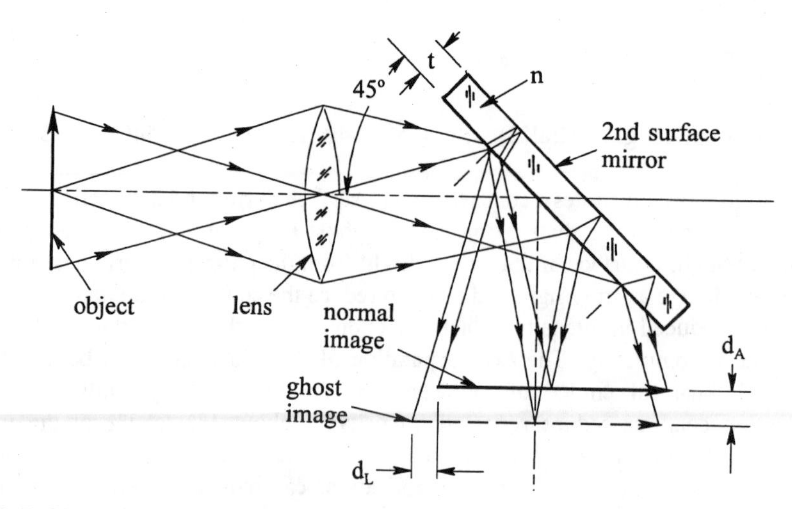

Fig. 6.5 Ghost image formation at a second-surface mirror inclined 45° to the axis. (Adapted from Kaspereit, ORDM 2-1)

Hence, second-surface mirrors must be made of that material, one of the optical glasses (see Tables C1 and C2), an optical plastic (see Table C3), or a crystal (see Tables C4 through C7). A related advantage of the second-surface mirror is that the radius and asphericity of the extra optical surface, an axial thickness, and an index are available for aberration control as well as for ghost image location control as mentioned above. Second-surface designs obviously do not work in mirrors with tapered or arched back surfaces or in built-up mirror substrates such as are discussed in Sections 6.2 and 6.3.

6.2 Mirror aperture determination

The size of a mirror is set primarily by the size and shape of the optical beam as it intercepts the reflecting surface plus any allowances that need to be made for mounting provisions, misalignment, and/or beam motion during operation. The, so called, "beamprint" can be approximated from a scaled layout of the optical system showing extreme rays of the light beam in at least two orthogonal meridians. This method is time-consuming to use and may be inaccurate due to compounded drafting errors. Modern computer-aided design methods have alleviated both these problems, especially with software that interfaces raytracing capability with creation of drawings or graphic renditions to any scale and in any perspective. In spite of these advances, some of us rely upon hand calculations, at least at early stages of system design. We therefore include here a set of equations (adapted from Schubert[48]) that allow minimum elliptical beamprint dimensions on a mirror which is tilted at an angle, θ, relative to the axis to be determined from the known beam diameter, D, at some axial reference plane positioned at a distance, L, from the axis intercept on the mirror, and the beam divergence angle, α, of the extreme off-axis ray to be reflected. Note that θ is 90° minus the angle of incidence of the axial ray at the mirror surface. All of these dimensions and the geometry are indicated in Fig. 6.6. A rotationally symmetric beam is assumed.

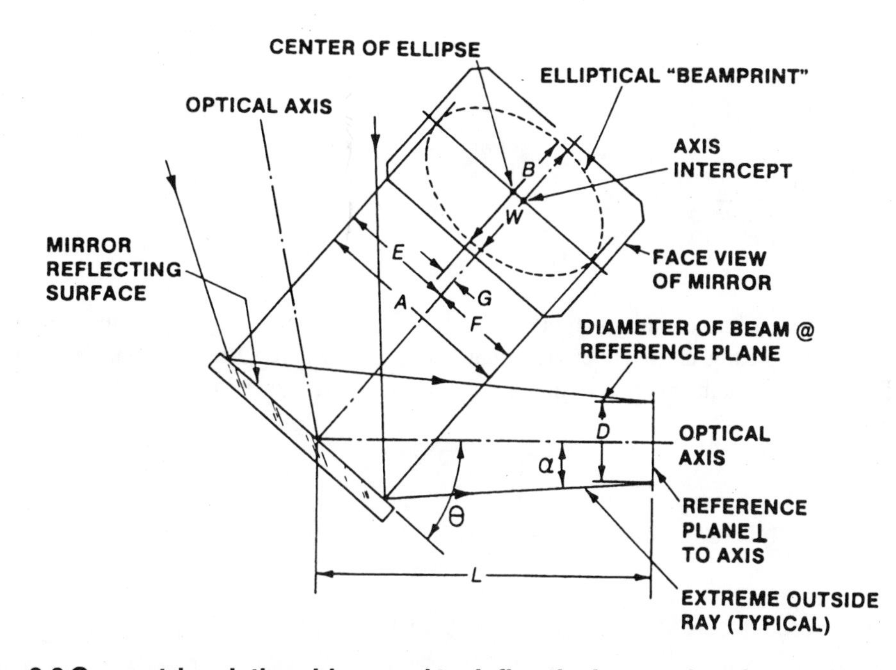

Fig. 6.6 Geometric relationships used to define the beamprint of a rotationally symmetric beam on a tilted mirror. (From Schubert[48])

Appropriate values are substituted into Eqs. (6.2) through (6.4) to find W, E, and F. Equations (6.5) through (6.7) are then used to find the lengths of the major and minor axes, A and B, of the elliptical beamprint and the offset, G, of the center of the ellipse from the axis intercept. All linear dimensions, except D and L, are measured in the mirror plane. The ellipse is assumed to be centered on the mirror in the minor-axis direction.

$$W = D + 2L\tan\alpha. \tag{6.2}$$

$$E = \frac{W\cos\alpha}{2\sin(\theta - \alpha)}. \tag{6.3}$$

$$F = \frac{W\cos\alpha}{2\sin(\theta + \alpha)}. \tag{6.4}$$

$$A = E + F. \tag{6.5}$$

$$G = (A/2) - F. \tag{6.6}$$

$$B = \frac{AW}{(A^2 - 4G^2)^{1/2}}. \tag{6.7}$$

These same equations apply regardless of the direction in which the beam is propagating as long as the reference plane is located where D is less than W. For a collimated beam, α and G are zero and the above equations reduce to the symmetrical case where:

$$B = W = D. \tag{6.8}$$

$$E = F = D/(2\sin\theta). \tag{6.9}$$

$$A = D/\sin\theta. \tag{6.10}$$

Numerical Example No. 18: Beam footprint on a tilted mirror.
Calculate the beamprint dimensions for a circular beam of D = 25.4 mm incident at L = 50 mm on a flat mirror tilted at an angle $\theta = 30^\circ$ to the axis. Assume the beam divergence to be (a) 0.5° and (b) zero.

(a) From Eq. (6.2), $W = 25.4 + (2)(50)(\tan 0.5^\circ) = 26.273$ mm
From Eq. (6.3), $E = (26.273)(\cos 0.5^\circ) / [2 \sin (30^\circ - 0.5^\circ)] = 26.677$ mm
From Eq. (6.4), $F = (26.273)(\cos 0.5^\circ) / [2 \sin (30^\circ + 0.5^\circ)] = 25.881$ mm
From Eq. (6.5), $A = 26.677 + 25.881 = 52.558$ mm
From Eq. (6.6), $G = (52.558 / 2) - 25.882 = 0.397$ mm
From Eq. (6.7), $B = (52.558)(26.273) / [52.558^2 - (4)(0.397^2)]^{1/2} = 26.276$ mm

(b) From Eq. (6.8), $B = W = D = 25.4$ mm
From Eq. (6.9), $E = F = 25.4 / (2)(\sin 30^\circ) = 25.4$ mm

As expected, the elliptical beamprint in Part (a) is slightly decentered with respect to the axis in the plane of reflection, but symmetrical to the axis in the orthogonal plane. In Part (b), the beamprint is symmetrical in both directions.

The dimensions of the reflecting surface should be increased somewhat from those calculated with the above equations to allow for the factors mentioned above (mounting clearance, beam motion, etc.) and for reasonable manufacturing tolerances.

6.3 Weight reduction techniques

The most common small mirror substrate shapes are the solid cylindrical disc and the solid rectangular plate. If possible, they should have thicknesses about 1/5th or 1/6th the largest in-plane dimension to ensure adequate mechanical stiffness. Thinner or thicker substrates are used as the application allows or demands.

Even in small and modest sized mirrors, weight minimization can prove advantageous or, in some cases, absolutely necessary. Given a chosen substrate material, reduction in mirror weight from that of the regular solid can be made only by changing the configuration. The usual means for doing this are to remove unnecessary material from a solid substrate or to combine separate pieces to create a built-up structure with a lot of empty spaces inside. No matter what technique is used to minimize mirror weight, the end product must be of high quality and capable of economic fabrication and test. Rodkevich and Robachevskaya correctly stated some time ago the fundamental requirements for precision mirrors and lightweight versions thereof.[49] These statements are paraphrased here as follows:

(1) The mirror material must be highly immune to outside mechanical and temperature influences, isotropic, and possess stable properties and dimensions,
(2) The mirror material must accept a high-quality polished surface and a coating having the required reflection coefficient,
(3) The mirror construction must be capable of being shaped to a specified optical surface contour and must retain this shape under operating conditions,
(4) Lightweight mirrors must have lower mass than those made to traditional designs while maintaining adequate stiffness and property homogeneity,
(5) Similar techniques should be used for fabricating conventional and lightweight mirrors, and
(6) Mounting and load-relief during testing and use should employ conventional techniques and should not increase mass and/or mechanism complexity.

These idealized principles would serve as useful guidelines for the design of any sized mirror. Material selection, fabrication methods, and configuration design are key to meeting these guidelines. Tables C8(a) and (b) list materials properties such as coefficient of thermal expansion, thermal conductivity, Young's modulus, etc., which relate to inherent behavior undr changing environmental conditions. Dimensional stability and property homogeneity differ from one material type to another. Further information on these topics may be found in other publications.[4,50] Typical fabrication methods, surface finishes, and coatings for mirrors made of various materials are listed in Table C11.[4] Mounting methods are considered in Chapt. 7 of this book.

6.3.1 Contoured-back configurations

The baseline configuration for mirrors with flat, concave, and convex first (reflecting) surfaces is the regular solid having a flat second surface. We concentrate on circular-aperture mirrors since they are the most common. In general, most of the following discussion applies also to rectangular or non-symmetric designs. Reduction of weight by thinning the baseline substrate also reduces stiffness and increases self-weight deflection and susceptibility to acceleration forces. Hence, that technique can be used only within limits. The simplest means for lightweighting front-surface mirrors is to contour the second (back) surface. Figure 6.7 illustrates this approach for a series of six concave mirrors with the same aperture, $2r_2$, and radius of curvature, R_1. Fabrication complexity increases from left to right. We will discuss each variation in turn and show how to calculate mirror volume which, when multiplied by the appropriate density, gives mirror weight. We also will work out typical examples in which mirror diameter, reflecting surface radius of curvature, material, and axial thicknesses are the same. This allows direct comparison of relative weights for the different designs.

Figure 6.7(a) shows the baseline plano-concave type mirror. Its axial and edge thicknesses are t_A and t_E respectively. The sagittal depth, S_1, is given by Eq. (6.11) while the mirror volume is given by Eq. (6.12).

$$S_1 = R_1 - (R_1^2 - r_2^2)^{1/2}. \tag{6.11}$$

$$V_{baseline} = \pi r_2^2 t_E - (\pi/3)(S_1^2)(3R_1 - S_1). \tag{6.12}$$

Numerical Example No. 19: Baseline flat-back concave mirror.
A concave ULE mirror per Fig. 6.7(a) of diameter 18 in. and axial thickness t_A = 18/6 = 3 in. has a radius of curvature R_1 of 72 in. Calculate the volume and weight of the mirror.

Per Fig. 6.7(a), r_2 = diameter / 2 = 9 in.
By Eq. (6.11), $S_1 = 72 - (72^2 - 9^2)^{1/2} = 0.565$ in.
$t_E = t_A + S_1 = 3.565$ in.
By Eq. (6.12), Volume = $(\pi)(9^2)(3.565) - (\pi / 3)(0.565^2)((3)(72) - 0.565)$
$= 907.18 - 72.018 = 835.16$ in.^{2}
From Table C8(a), ρ for ULE is 0.0788 lb/in.^{3}
Mirror weight = (0.0788)(835.16) = 65.811 lb.

We will use this weight as the baseline value for comparison in the following five numerical examples.

The simplest back-surface contour is tapered or conical as indicated in Fig. 6.7(b). Thickness varies linearly between some selected radius, r_1, and the rim. The flat region on the back surface within r_1 might be used to mount the mirror if the thin edge is thought to be too fragile. Equation (6.13) is used to calculate the mirror volume.

$$V_{tapered} = V_{baseline} - (\pi t_1/2)(r_2^2 - r_1^2). \tag{6.13}$$

Note that $V_{baseline}$ is the baseline mirror volume from Eq. (6.12).

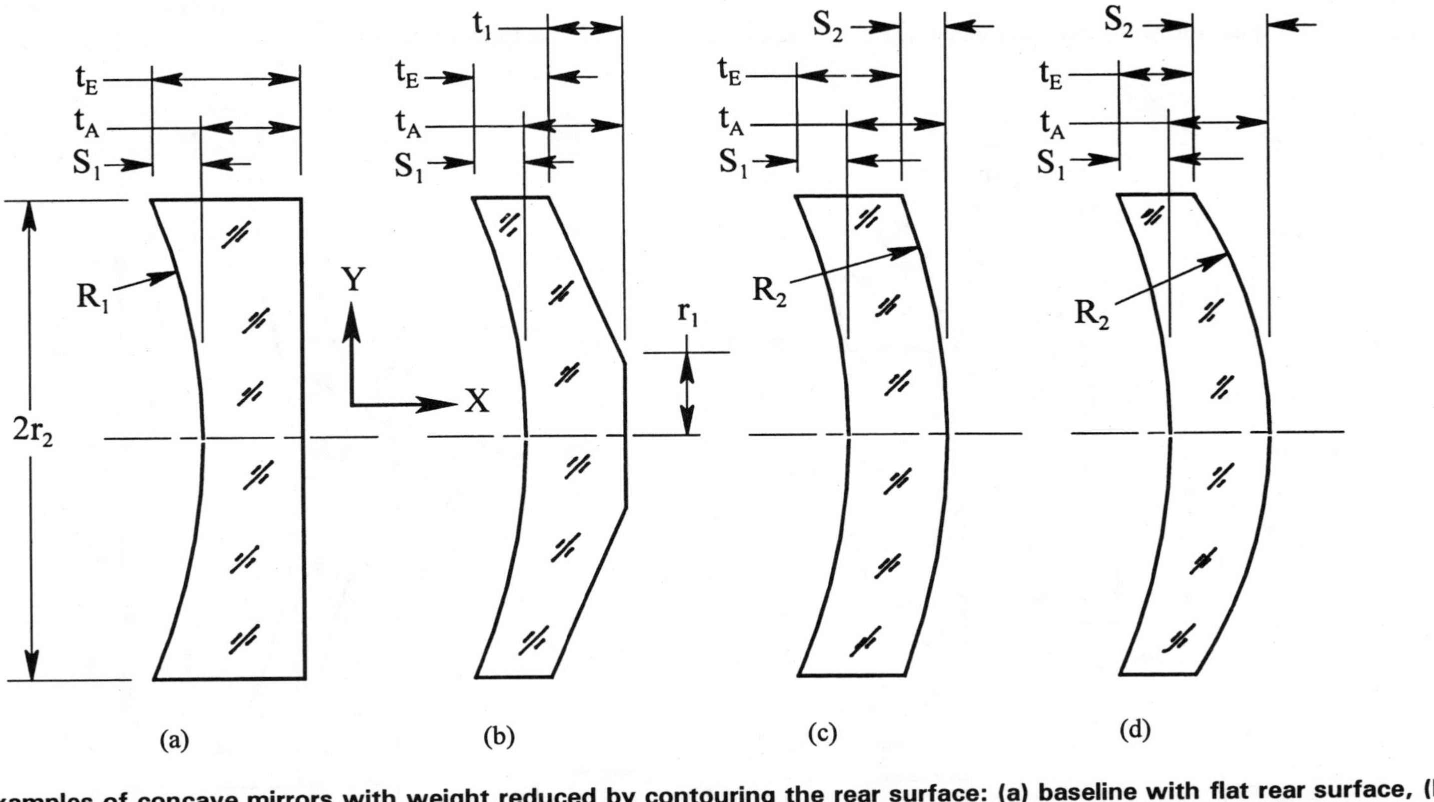

Fig. 6.7 Examples of concave mirrors with weight reduced by contouring the rear surface: (a) baseline with flat rear surface, (b) tapered (conical) rear surface, (c) concentric spherical front and rear surfaces with $R_2 = R_1 + t_1$, (d) spherical rear surface with $R_2 < R_1$. **(Continued on next page)**

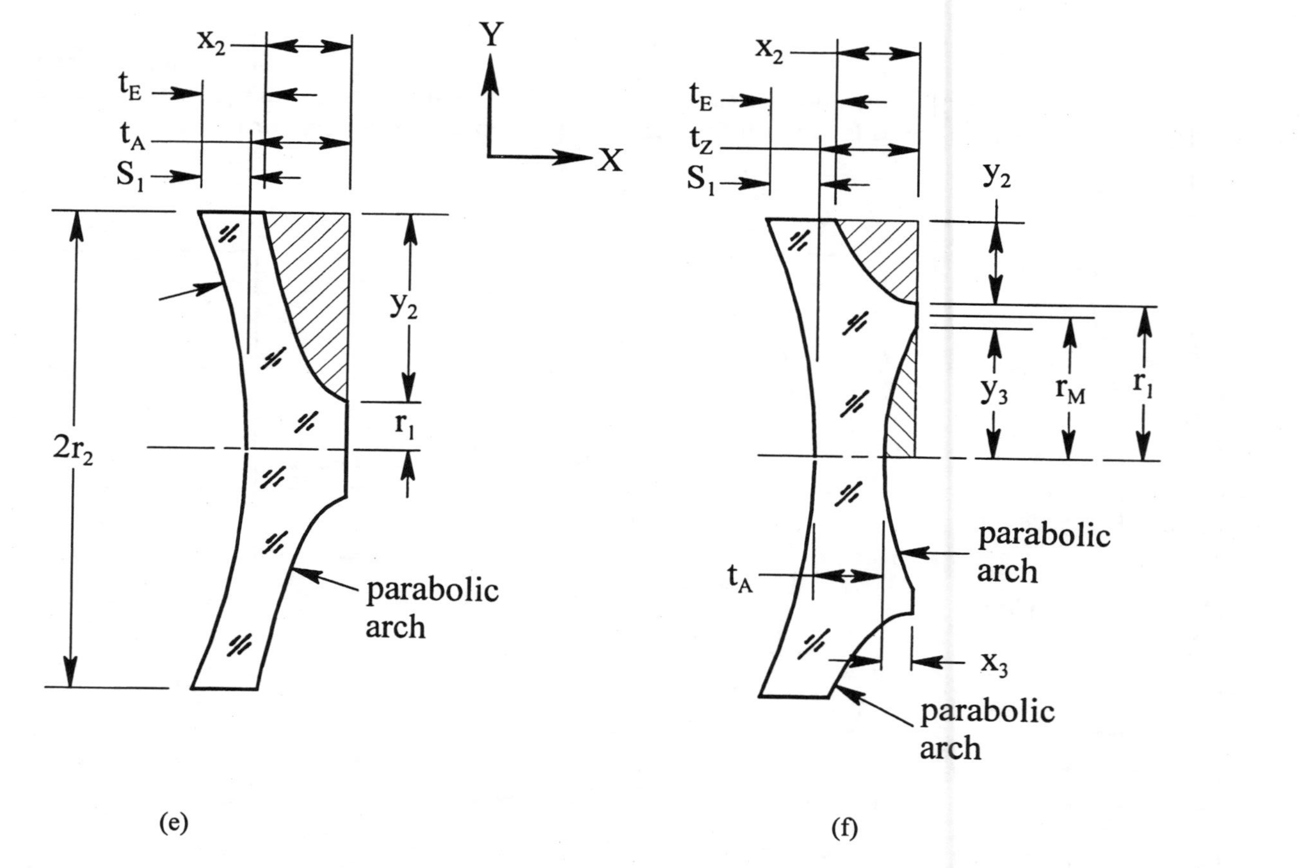

Fig. 6.7 (continued) Examples of concave mirrors with weight reduced by contouring the rear surface: (e) single arch configuration, and (f) double arch configuration.

Numerical Example No. 20: Tapered-back concave mirror.
Assume that the back surface of the mirror from Numerical Example No. 19 has a contoured back surface starting at a radius $r_1 = 1.5$ in. and tapering to an edge thickness, t_E, of 0.5 in. as shown in Fig. 6.7(b). What is the mirror weight and how does it compare to the baseline mirror weight?

$S_1 = 0.565$ in. (from Numerical Example No. 19)
$t_1 = t_A + S_1 - t_E = 3.000 + 0.565 - 0.500 = 3.065$ in.
From Eq. (6.13), $V_{tapered} = 835.16 - ((\pi)(3.065) / 2)(9^2 - 1.5^2)$
$= 835.16 - 379.14 = 456.02 \text{ in.}^3$

Mirror weight = (0.0788)(456.02) = 35.934 lb.

The mirror weight has been reduced to 35.934 / 65.811 = 55% of the baseline mirror weight.

In Views (c) and (d) of Fig. 6.7, we see meniscus-shaped mirrors with radii $R_2 = R_1 + t_1$ and $R_2 < R_1$, respectively. The first case has concentric spherical surfaces and uniform thickness over the aperture so only a slight reduction in weight from the plano-concave case [View (a)] is possible. The second case allows greater weight reduction because the rim is significantly thinner. The latter type mirror is usually mounted on a central hub passing through a perforation.

Equation (6.14) is used to estimate the meniscus-shaped mirror volume.

$$V_{concentric} = V_{baseline} - \pi r_2^2 s_2 + (\pi/3)(s_2^2)(3R_2 - s_2). \tag{6.14}$$

Numerical Example No. 21: Concentric meniscus solid mirror.
Assume that the back surface of the mirror from Numerical Example No. 19 has a spherical back surface with radius $R_2 = R_1 + t_A = 72 + 3 = 75$ in. Figure 6.7(c) applies. What are t_E and the mirror weight and how does the latter compare to the baseline mirror weight?

From Eq. (6.11), $S_2 = 75^2 - (75^2 - 9^2) = 0.542$ in.
$S_1 = 0.565$ in. (from Numerical Example No. 15)
$t_E = t_A + S_1 - S_2 = 3.000 + 0.565 - 0.542 = 3.023$ in.

From Eq. (6.14), $V_{concentric} = 835.16 - (\pi)(9^2)(0.542) + (\pi/3)(0.542^2)((3)(75) - 0.542)$
$= 835.16 - 137.92 + 69.05 = 766.29 \text{ in.}^3$
Mirror weight = (0.0788)(766.29) = 60.384 lb

The mirror weight has been reduced to 60.384 / 65.811 = 92% of the baseline mirror weight.

Numerical Example No. 22: Meniscus solid mirror with $R_2 < R_1$.
Assume that the back surface of the mirror from Numerical Example No. 19 has a spherical back surface with radius R_2 = 14.746 in. Figure 6.7(d) applies. What are t_E and the mirror weight and how does the latter compare to the baseline mirror weight?

From Eq. (6.11), $S_2 = 14.746 - (14.746^2 - 9^2)^{1/2} = 3.065$ in.
$S_1 = 0.565$ in. (from Numerical Example No. 19)
$t_E = t_A + S_1 - S_2 = 3.000 + 0.565 - 3.065 = 0.500$ in.

From Eq. (6.14),

$$V_{concentric} = 835.16 - (\pi)(9^2)(3.065) + (\pi/3)(3.065^2)((3)(14.746) - 3.065)$$
$$= 835.16 - 779.95 + 405.05 = 460.26 \text{ in.}^3$$

$$\text{Mirror weight} = (0.0788)(460.26) = 36.268 \text{ lb}$$

The mirror weight has been reduced to 36.268 / 65.811 = 55% of the baseline mirror weight. Note that this is the same as for the tapered-back mirror of Numerical Example No. 20.

The mirror design of Fig. 6.7(e) is called the single-arch configuration. The best way to mount such a mirror is on a central hub. The back surface could have parabolic or circular contour. In the former case, the axis of the parabola may be oriented parallel to the mirror axis (X) and decentered to locate the vertex at P_1 on the rim of the mirror (X-axis parabola) or oriented radially with the vertex at P_2 on the mirror back (Y-axis parabola). The circle might well be parallel to the reflecting surface at the mirror rim. It must pass through P_1 and P_2. These possibilities are drawn approximately to scale in Fig. 6.8 for the mirror considered in the numerical examples. The material volume removed by contouring the back surface with either parabolic or circular contour can be calculated by finding the sectional area to the right of the appropriate curve in Fig. 6.8 and bound at the right by the vertical line A-B then revolving that area around the mirror axis at the radius of the area centroid. It may be seen that the sectional area is smaller for the chosen circular contour than for either parabolic contour. This suggests that the mirror weight with a circular back contour will be greater than the same mirror with a parabolic contour.

The sectional area, A_P, for each half-parabola is given by Eq. (6.15) where x_2 and y_2 are as illustrated in Fig. 6.8. The greatest volume is removed with the X-axis parabola

$$A_P = (2/3)(x_2)(y_2). \tag{6.15}$$

because it has a slightly larger radius of revolution. Mirror volume for the X-axis and Y-axis cases are both given by Eq. (6.16).

$$V_{s\text{-}arch} = V_{baseline} - (A_p)(2\pi)(y_{centroid}). \tag{6.16}$$

where

$$y_{centroidY} = r_1 + (3/5)(y_2), \tag{6.17}$$

and

$$y_{centroidX} = r_1 + y_2 - (3/8)(y_2). \tag{6.18}$$

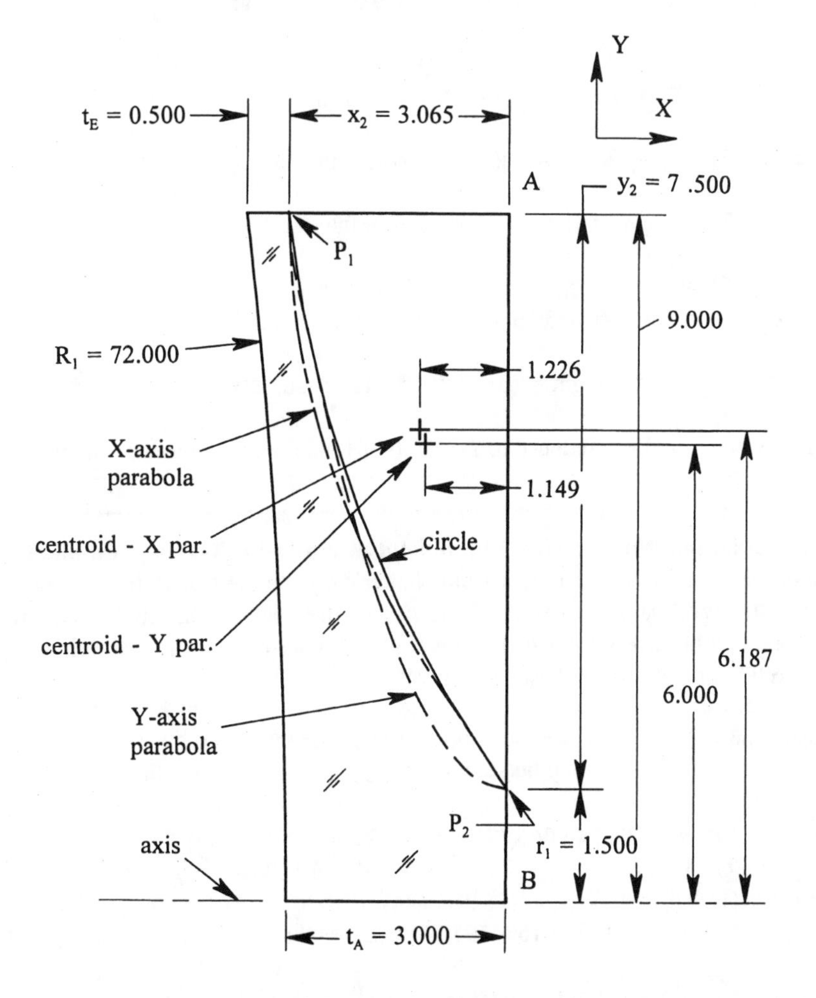

Fig. 6.8 Three possible contours (X-axis parabola, Y-axis parabola, and circle) for the single-arch mirror plotted to the same scale. Dimensions are inches and apply to Numerical Examples No. 23 and No. 24

The parabola with symmetry about the Y axis might be preferred since the thickness then decreases monotonically with increasing distance from the axis. This is not necessarily the case for the X-axis parabola; this is indicated in Fig. 6.8 where the minimum mirror thickness occurs well inside the rim.

Numerical Example No. 23: Single-arch solid mirror with Y-axis parabolic back. Change the back surface of the mirror from Numerical Example No. 19 to the single-arch, parabolic form with parabola vertex at the back surface of the mirror, i.e., a Y-axis parabola. Assume t_E is 0.500 in. Figures 6.7(e) and 6.8 apply. What is the mirror weight and how does the latter compare to the baseline mirror weight?

By observation, $x_2 = t_A + S_1 - t_E = 3.000 + 0.565 - 0.500 = 3.065$ in.
$y_2 = r_2 - r_1 = 9.000 - 1.5 = 7.500$ in.

From Eq. (6.15), $A_P = (2/3)(3.065)(7.5) = 15.325$ in.^{2}

From Eq. (6.17), $y_{\text{centroid Y}} = 1.5 + (3/5)(7.5) = 6.000$ in.

From Eq. (6.16), $V_{\text{s-arch}} = V_{\text{baseline}} - (15.325)(2\pi)(6.000)$
$= 835.16 - 577.739 = 257.42$ in.^{3}

Mirror weight = (0.0788)(257.42) = 20.28 lb

The mirror weight has been reduced to 20.28 / 65.81 = 31% of the baseline mirror weight.

Numerical Example No. 24: Single-arch solid mirror with X-axis parabolic back. Change the back surface of the mirror from Numerical Example No. 19 to the single-arch, parabolic form with parabola vertex at the rim of the mirror, i.e., an X-axis parabola. Assume t_E is 0.500 in. Figure 6.7(e) again applies. What is the mirror weight and how does the latter compare to the baseline mirror weight?

By observation, $x_2 = t_A + S_1 - t_E = 3.000 + 0.565 - 0.500 = 3.065$ in.
$y_2 = r_2 - r_1 = 9.000 - 1.5 = 7.500$ in.

From Eq. (6.15), $A_P = (2/3)(3.065)(7.5) = 15.325$ in.^{2}
From Eq. (6.17), $y_{\text{centroid X}} = 1.5 + 7.5 - (3/8)(7.5) = 6.187$ in.
From Eq. (6.16), $V_{\text{s-arch}} = V_{\text{baseline}} - (15.325)(2\pi)(6.187)$
$= 835.16 - 595.74 = 239.41$ in.^{3}

Mirror weight = (0.0788)(239.41) = 18.86 lb

The mirror weight has been reduced to 18.86 / 65.81 = 29% of the baseline mirror weight.

Comparison with Numerical Example No. 23 confirms the statement above that this design with a X-axis parabola gives slightly lower weight than the corresponding Y-axis parabola case.

The double-arch mirror of Fig. 6.7(f) is thickest at a zone chosen in the range 50 to 70 percent of the mirror diameter. It is supported at three or more points at this zone. The rear surface is shaped as two parabolic curves with minimum (and usually equal) axial thickness at the rim and center. The outer arch might well be a Y-axis parabola while the inner arch might be an X-axis parabola. The latter choice is made to avoid an inflection point in the inner arch surface at the axis. The sectional areas of the outer and inner arches can be calculated from Eqs. (6.15) and (6.19), respectively.

$$A_{P\text{-}inner} = (2/3)(x_3)(y_3), \tag{6.19}$$

where the x- and y-dimensions are as shown in Fig. 6.7.

The radius of revolution for the outer arch ($y_{centroidY}$) and the volume of that arch of the mirror are given by Eqs. (6.20) and (6.21) while those parameters for the inner arch are given by Eqs. (6.22) and (6.23).

$$y_{centroidY} = r_1 + (3/5)y_2. \tag{6.20}$$

$$V_{outer\,arch} = (A_{P\text{-}outer})(2\pi)(y_{centroidY}). \tag{6.21}$$

$$y_{centroidX} = (3/8)(y_3). \tag{6.22}$$

$$V_{inner\,arch} = (A_{P\text{-}inner})(2\pi)(y_{centroidX}). \tag{6.23}$$

The mirror volume is then:

$$V_{d\text{-}arch} = V_{baseline} - V_{outer\,arch} - V_{inner\,arch}. \tag{6.24}$$

Numerical Example No. 25: Double-arch solid mirror.

Recontour the back surface of the mirror from Numerical Example No. 19 to a double-arch configuration with Y-axis and X-axis parabolas for the outer and inner arches, respectively. Figure 6.7(f) applies. Assume $t_E = t_A = 0.5$ in. Let $t_Z = 3$ in., $S_1 = 0.565$ in., $r_m = 0.6r_2 = 5.4$ in., and the annular zone width = 0.6 in. What is the mirror weight and how does it compare with the baseline mirror weight?

$r_1 = 5.4 + 0.3 = 5.7$ in.
$y_3 = 5.4 - 0.3 = 5.1$ in.
$y_2 = r_2 - r_1 = 9 - 5.7 = 3.3$ in.

By observation, $t_A = 0.5$ in. $= t_Z - x_3 = 3 - x_3$, so $x_3 = 2.5$ in.
$t_E = 0.5$ in. $= t_Z + S_1 - x_2 = 3 + 0.565 - x_2$,

so $x_2 = 3.565 - 0.5 = 3.065$

By Eq. (6.15), $A_{P\text{-outer}} = (2/3)(3.065)(3.3) = 6.743$ in.^{2}
By Eq. (6.19), $A_{P\text{-inner}} = (2/3)(2.5)(5.1) = 8.5$ in.^{2}
By Eq. (6.20), $y_{\text{centroid Y}} = 5.7 + (3/5)(3.3) = 7.680$ in.
By Eq. (6.21), $y_{\text{centroid X}} = (3/8)(5.1) = 1.912$ in.
By Eq. (6.22), $V_{\text{outer-arch}} = (6.743)(2\pi)(7.680) = 325.38$ in.^{3}
By Eq. (6.23), $V_{\text{inner arch}} = (8.5)(2\pi)(1.912) = 102.11$ in.^{3}
By Eq. (6.24), $V_{\text{d-arch}} = 835.16 - 325.38 - 102.11 = 407.66$ in.^{3}

Mirror weight = (0.0788)(407.66) = 32.12 lb.

The mirror weight has been reduced to 32.12 / 65.811 = 49% of the baseline mirror weight.

It is noted that others have configured double-arch mirrors with weight relative to the equivalent flat-back version as low as about 30%.[36]

The symmetrical concave mirror configuration is not capable of reducing substrate weight, but is included here for completeness. It is generally used only when the axis is nominally horizontal or nearly so because gravity deflections then are symmetrical about the mid-plane so are smaller than with non-symmetrical configurations. This mirror suffers excessively from surface deformation when the axis is vertical.[36] This also applies to the symmetrical convex-surfaced mirror.

The volume of this type mirror is:

$$V_{symm} = V_{baseline} + (\pi/3)(S_2^2)(3R_2 - S_2). \qquad (6.25)$$

Note the resemblance of the second term to that in Eq. (6.12).

Numerical Example No. 26: The symmetrical concave solid mirror.

Change the back surface of the mirror from Numerical Example No. 19 to the symmetrical concave form. What is the weight and how does it compare to that of the baseline version?

Referring back to Numerical Example No. 19, we observe that the volume of the plano-concave mirror version was calculated as 907.18 - 72.018 in.^{3} It is obvious then that the volume in question here is 907.18 + 72.018 = 979.198 in.^{3}

The mirror weight is then (0.0788)(979.198) = 77.16 lb.

The mirror weight has been increased by a factor of 77.16 / 65.81 = 117%.

6.3.2 Machined and built-up structural configurations

Although conventionally used in mirror substrates larger than the 20 in. (61 cm) limit imposed arbitrarily for for mirrors considered in this book, the following lightweight design/fabrication techniques are sometimes applied to mirrors within this limit. We therefore include brief considerations for the information of the interested reader.

Figure 6.9 shows various construction configurations for machined and built-up lightweight mirrors.[51] These include symmetrical and non-symmetrical sandwiches, partly and fully open ("waffle") back designs, and fused-fiber- or foam-filled sandwich constructions. Each of these would have a characteristic areal density depending on material type, material distribution, thickness of members (face-plate, back-plate, and core webs), etc. In some designs, the core is integral with the front and back facesheets of the structure while, in others, the parts are separate and completely or partially attached together. Attachment means include: thermal fusing, adhesive and frit bonding, and, for metal mirrors, brazing or welding. The pattern of cells in the core has a strong influence on the mirror's weight and stiffness. Triangular, square, circular, and hexagonal shapes are most commonly used in cells.

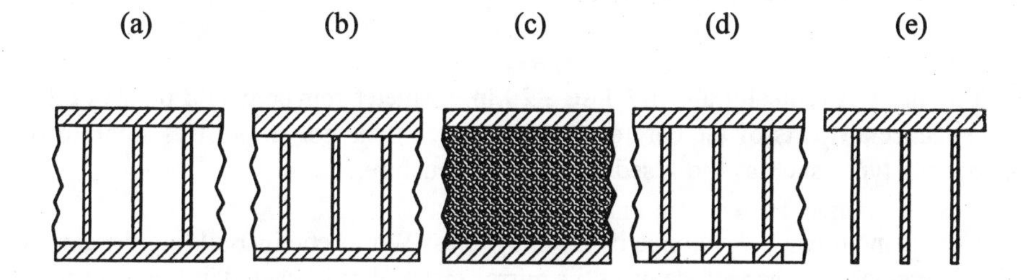

Fig. 6.9 Cross-sections for machined and built-up mirror substrates: (a) symmetrical sandwich, (b) non-symmetrical sandwich, (c) foam or fused fiber core sandwich, (d) partially open back, and (e) open back. (Adapted from Seibert[51])

A mirror lightweighted by removing non-essential material from within the substrate envelope is structurally more efficient than the equivalent-sized solid mirror. Since the material near the neutral plane contributes little to bending stiffness, it can safely be eliminated. This reduces weight so the desirable high stiffness-to-weight ratio can be provided. This accomplishment results in some reduction of shear resistance. The manner in which the mirror is supported contributes strongly to the effects of gravity and externally applied accelerations.

Figure 6.10 shows one classical type of built-up construction; it is called an "egg-crate" configuration. The core is created as cellular "webs" made of thin slotted strips that interlock but are not attached to each other. Front and back faceplates are fused to the top and bottom ends of the core to form the mirror substrate. The diameter-to-thickness ratio

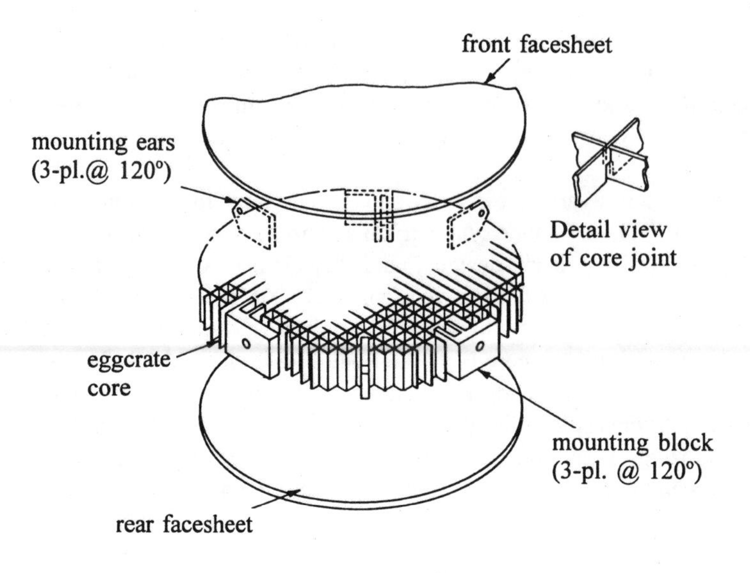

Fig. 6.10 The "eggcrate" mirror construction.

of such a mirror is typically about 7:1 so a 20 in. diameter mirror would be about 2.85 in. thick. Since all parts of the core are not connected, it is not as stiff as some of the more modern types such as the fused monolithic structure.

A fusing process developed by Corning Glass Works entails building the core by attaching together "ell-shaped" strips of the mirror material as indicated in Fig. 6.11 using multiple torches to locally soften the material so adjacent regions fuse.[53] In some designs, cylindrical rings are fused to the outer rim of the core to enclose it and to increase its stiffness. If the mirror is perforated, a ring may be fused into the central hole for the same reasons. See, for example, Fig. 6.12. When the entire core has been created, its top and bottom ends are usually ground flat and parallel. In some cases they may be generated or ground spherical to form a meniscus shape. In either case, facesheets are located on the core and the entirety is heated in a furnace until fused together. If fused on a curved mandrel, the structure can be sagged to the meniscus shape desired for minimum glass removal. The substrate so created is monolithic and has the characteristics of the bulk material throughout. During the fusing operation, the softened material usually distorts resulting in shape defects as indicated schematically in Fig. 6.13(a). The desired optical surface must be created within the front facesheet after carefully inspecting the blank to find the region with minimum internal defects (bubbles and included impurities) by grinding away the unwanted material as indicated in Fig. 6.13(b) to locate the optical surface within this, so called, critical zone.

A better mirror blank can be formed by making a core and attaching it to the facesheets using an assembly process similar to brazing in which all parts are attached to each other by "frit." Figure 6.14 shows a design for a plano-concave closed sandwich.

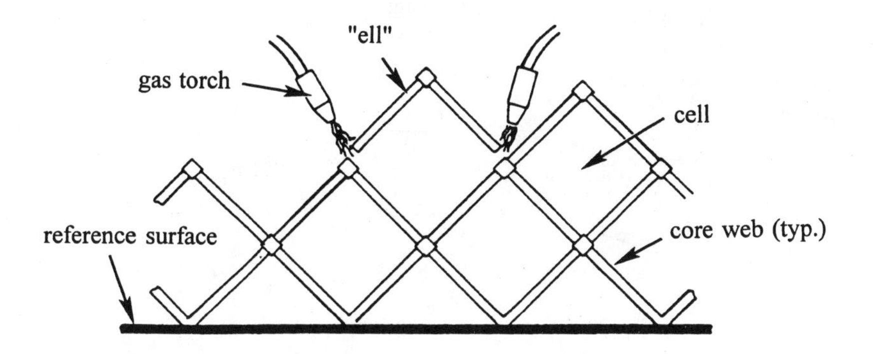

Fig. 6.11 Attaching 90° "ells" by torch welding to form a monolithic mirror core. (Adapted from Lewis[53])

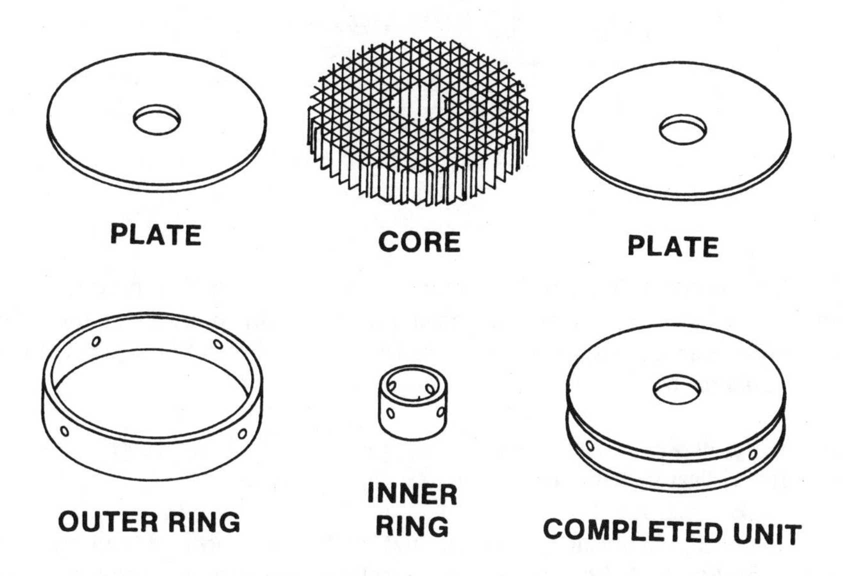

Fig. 6.12 Basic parts of a typical perforated fused monolithic mirror substrate

Frit is a "adhesive" made of an organic vehicle and powdered glass that melts at a temperature lower than the softening point of the mirror material. It works especially well with Corning ULE. The CTE of the frit is controlled so as not to introduce excessive stress into the blank during or subsequent to application. The resulting blank is free of the defects shown in Fig. 6.13(a) because the mirror structural elements never reach the softening temperature. Mirror blanks made by this process can have thinner webs and a higher diameter-to-thickness ratio (typically 167%) than monolithic fused blanks. The frit-

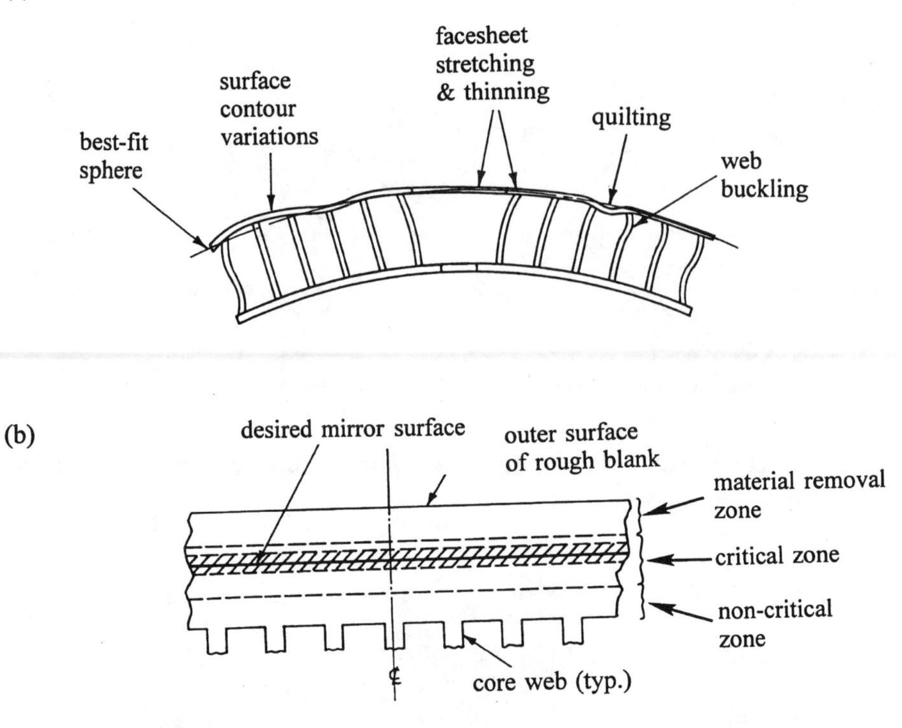

Fig. 6.13 Details regarding the monolithic fused mirror blank: (a) typical defects caused by heating ULE above the softening point, (b) location of the mirror surface within the best region of the front facesheet. (From Yoder[8] by courtesy of Marcel Dekker, Inc.)

bonded mirror will weigh significantly less (typically 27%), yet have greater rigidity (typically 164%) than the conventional fused monolithic type.[52]

A fused silica lightweight mirror substrate with a core machined from a solid disc is illustrated by Fig. 6.15. This has a symmetrical concave form and comprises a core machined with through holes of various shapes and ground concave on both sides fused to preformed meniscus-shaped facesheets. The 20 in. (50.8 cm) diameter mirror of Fig. 6.15 weighed about 16 lb (7.3 kg). Figure 6.16 shows the pattern of holes produced by drilling and grinding with annular diamond-bonded core and end-mill tools. Cusps remaining after core drilling were removed with diamond cutters. The wall thickness after machining was 1 to 3 mm.[54] This mirror shape is best used with its axis horizontal since its self-weight deflection with axis vertical is relatively large.

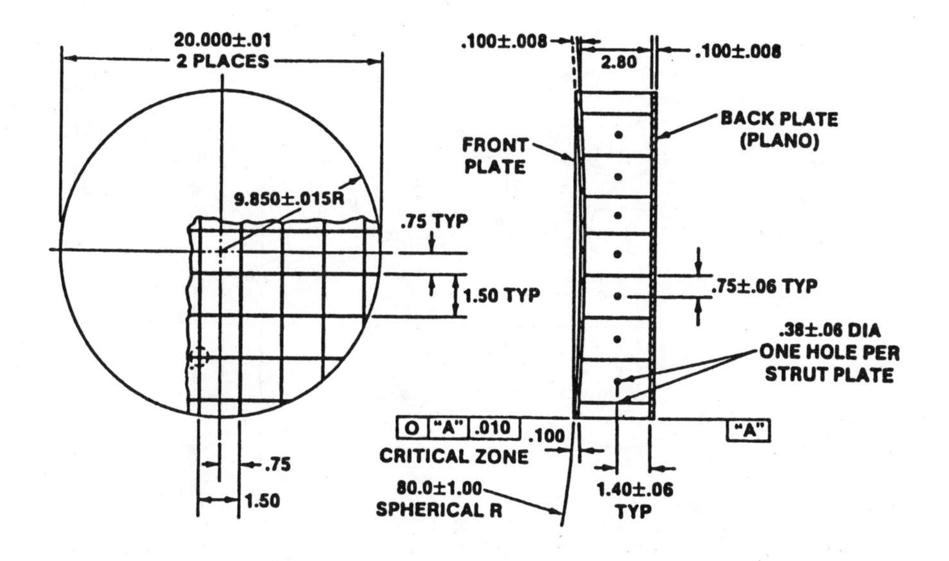

Fig. 6.14 A mirror design suitable for frit-bonding. (From Fitzsimmons and Crowe[52])

Yet another technique for forming small to moderate sized lightweight mirrors is illustrated in Fig. 6.17. Here, short tubes of glass are placed on end between glass facesheets to form a sandwich configuration. The rear facesheet is perforated with one hole per tube. The assembly is heated to the softening point in a furnace while air or other gas is pumped into the holes to pneumatically force the softened tubes into contact and to fuse together so as to form a square [View (a)] or hexagonal [View (b)] cell pattern. The result is a monolithic structure. Examples are shown in Fig. 6.18. Materials typically used in these mirrors are Corning Pyrex 7740, Schott Tempax, Vycor, and fused silica. Weight is considerably reduced from that of the built-up fused monolith because the pressure support allows thinner walls and facesheets to be used.[55]

6.4 Metallic mirrors

Metals commonly used to make mirrors are listed with their key mechanical properties in Table C8b. The most common are wrought aluminum and beryllium; the latter being most popular in cryogenic space applications. Mirrors for use in high-energy laser applications or high-power light sources need to be cooled. This frequently is done by circulating coolant through tubular passages machined into the mirror substrate. These frequently are made of oxygen free high conductivity (OFHC) copper or TZM, an alloy of titanium, zirconium, and molybdenum.

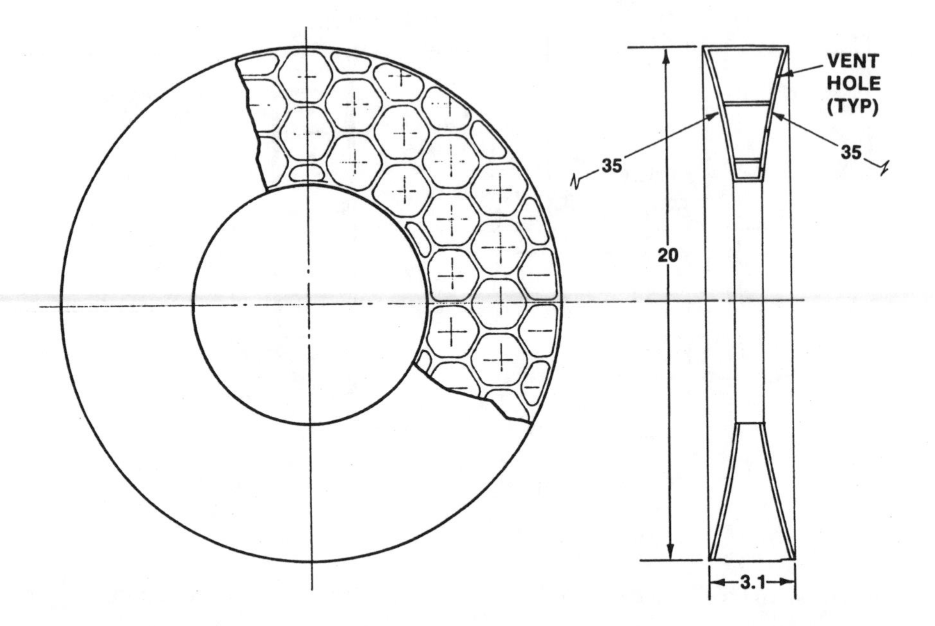

Fig. 6.15 Diagram of a symmetrical concave mirror with core machined with through holes then fused to facesheets. (From Pepi and Wollensak[54])

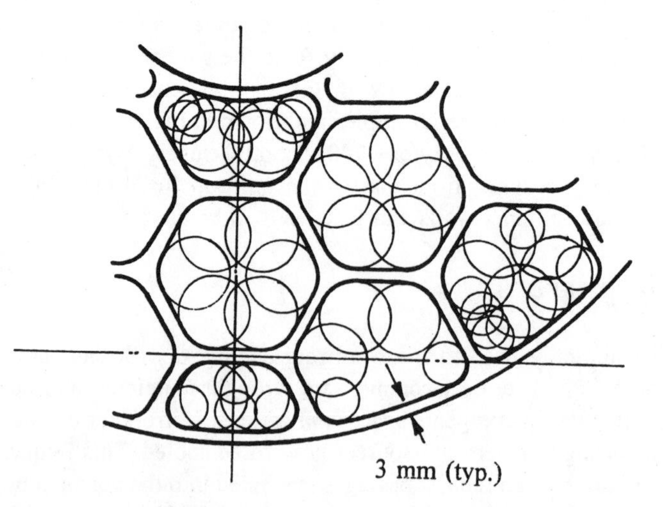

Fig. 6.16 Typical core machining pattern for a mirror such as is shown in Fig. 6.14. (From Pepi and Wollensak[54])

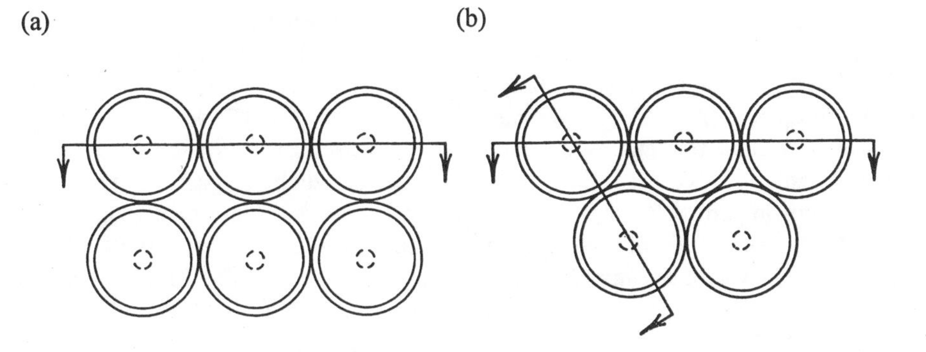

Fig. 6.17 Tube placement prior to heating and pneumatic expansion in the Hextek manufacturing process: (a) square cells, (b) hexagonal cells. (Courtesy of Hextek Corporation, Tucson, AZ)

Fig. 6.18 Photographs of two fused monolithic mirror blanks made by the Hextek process. (Courtesy of Hextek Corporation, Tucson, AZ)

Fabrication of metallic mirrors typically involves the following steps: formation of the blank, geometric shaping, stress relieving, plating (usually with electroless nickel), optical finishing, and optical coating. Many materials can be cast; others are welded or brazed from components. Single point diamond turning (SPDT) has found great application in creating fine quality optical surfaces in metals such as aluminum, brass, copper, gold, silver, electroless nickel (coating), and beryllium copper. Purity of the material is very important.[56] Surface finish of metal surfaces is inferior to glass-type materials, but is adequate for infrared systems and many visible-light applications. Figure 6.19 shows a variety of metal mirrors made by one manufacturer using SPDT methods.

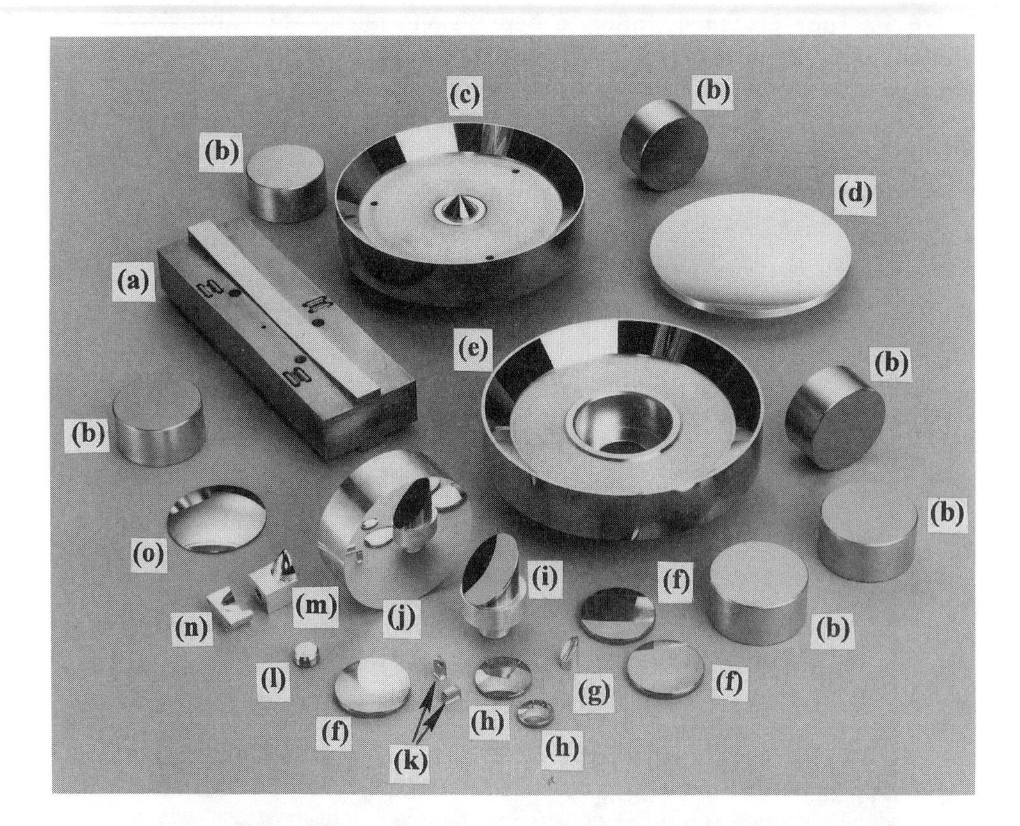

Fig. 6.19 Photograph of a variety of optical components made by the SPDT process. Shown are (a) CO_2 slab laser mirror, (b) water cooled CO_2 laser mirror, (c) Cu waxicon, (d) on-axis parabolic mirror, (e) copper axicon, (f) ZnSe aspheric windows, (g) parabolic reflector, (h) aspheric ZnSe lens, (i) Ni-plated Be mirror, (j) spherical Al mirror, (k) ZnSe aspheric retroreflector, (l) Cu "button" mirror, (m) parabloic replica mold, (n) electron microscope collimating mirror, and (o) aspheric Ge lens. (Courtesy of II-VI, Incorporated, Saxonburg, PA)

Figure 6.20 shows the back side of a typical metal mirror. It is the 7.3 in. (18.5 cm) diameter by 0.7 in. (1.78 cm) thick secondary mirror used in the infrared telescope of NASA's Kuiper Airborne Observatory.[57] Lightness of weight and low inertia are essential to the success of this equipment since the mirror moves mechanically in oscillatory tipping fashion to rapidly switch the field of view of the telescope from target of interest to background for calibration purposes.

Fig. 6.20 Photograph of the lightweighted aluminum scanning secondary mirror used in the Kuiper Airborne Observatory. (From Downey et al[57])

The 7:1 diameter-to-thickness Type 5083-O aluminum substrate was lightened by machining pockets into a solid blank. Total weight is 1.1 lb (0.5 kg) representing a 70% reduction from a solid. The convex hyperboloidal optical surface (which was not electroless nickel plated) was created by SPDT machining to final figure. Quality of the figure was about $\lambda/1.5$ P-V at 633 nm wavelength over 90% of the aperture. The final surface was coated with aluminum and silicon monoxide films. The mounting surfaces seen in section at the center of the mirror were diamond turned to facilitate accurate turning of the optical surface later. Surface figure at operating temperature (-40°C) is $\lambda/2$.

The mirror mounted on its drive mechanism is shown in Fig. 6.21. The square-wave response of the mirror and its drive mechanism for beam tilt angles up to $\pm$ 23 arc minute is about 40 Hz. It is driven in orthogonal tilts by four electromagnetic actuators

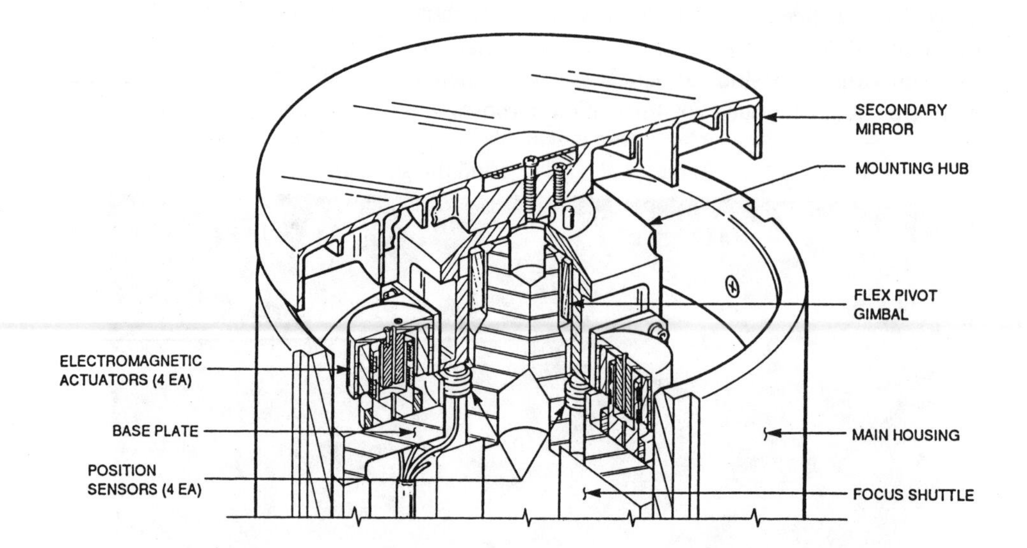

Fig. 6.21 Mirror from Fig. 6.20 mounted on its drive mechanism. (From Downey et al.[57])

located symmetrically at the back of the mirror. The moving assembly (weight about 2 lb) tilts about its center of gravity on two-axis flex pivot gimbals. The actuator coils are mounted to a stationary baseplate that provides a conductive path for temperature control. The entire moving assembly can be moved axially by a motor driven ballscrew through a range of $\pm$ 1.3 cm for focus adjustment during flight.

Many other lightweight beryllium mirrors have been fabricated by techniques similar to those just discussed. Usually these were used in space applications, although some have been used in high speed scanning applications where high stiffness and low weight are required to prevent surface distortion due to centrifugal force. For wavelengths beyond about 3 μm in the infrared, polished bare beryllium has high reflectance so it is not necessary to apply electroless nickel plating. This avoids thermal problems due to bimetallic effects from CTE mismatch.[58]

One very successful means for making beryllium mirrors is an improved powder metallurgy technique patented by Gould[59] and described by Paquin et al.[60] and Paquin[61]. In this process, high-purity beryllium powder is subjected to hot isostatic pressing (HIP) under high temperature and pressure. This yields blanks of near net shape with low porosity and few inclusions. The process improvement included forming internal lightweighting pockets by HIPing the material around void-formers made of leachable material (copper) that could be removed after compacting. Figure 6.22 shows such a mirror. It is a 9.5 in. (24.1 cm) diameter by 1.2 in. (2.8 cm) monolithic closed sandwich weighing 2.16 lb (0.98 kg). Hexagonal cells measuring 1 in. (2.5 cm) and webs 0.05 in.

(1.3 mm) thick were formed in this mirror. In the foreground of the figure the back face of another mirror made by the same process shows access holes for supporting the void-formers and removing them later. After polishing, the front facesheet had a figure of $\lambda/25$ P-V at 633 nm wavelength.

These experimental mirrors were extremely stiff with first resonance at about 8700 Hz. The manufacturing process is scalable to larger mirror sizes. At 17 in. (43.2 cm) diameter, the weight would be 7.1 lb (3.3 kg) and the structure would be stiff enough for mounting on a three-point support.

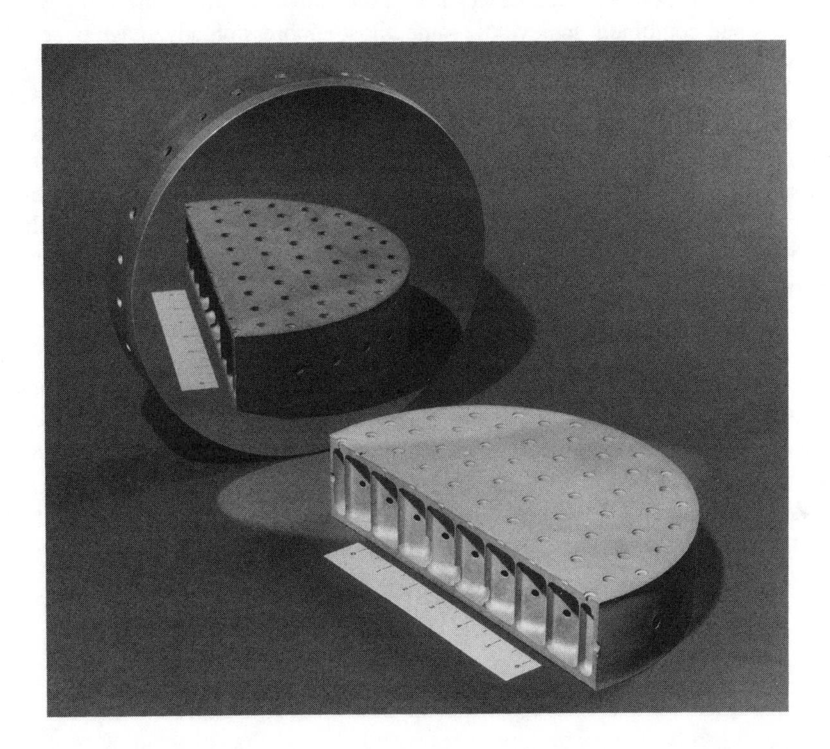

Fig. 6.22 Photograph of a 9.5 in. (24.1 cm) diameter monolithic, closed sandwich beryllium mirror made by the HIP process. (From Paquin[61])

6.5 Pellicles

Very thin mirrors, beamsplitters, and beam-combiners can be made from films of material such as nitrocellulose, polyester, or polyethylene. Thickness typically is 5 μm (0.0002 in.) ± 10%, although 2 μm (0.00008 in.) ± 10% thick films and special ones up to 20 μm (0.0008 in.) thick also are available. Surface quality of standard varieties typically is better than 40/20 scratch and dig while figure typically is 0.5 to 2λ per inch. The base material transmits well (>90%) from 0.35 to 2.4 μm, but has numerous deep absorbing regions beyond 2.4 μm.[64] See Fig. 6.23 for a simplified representation of the transmission characteristics of a typical standard type pelicle. Pellicles can be coated to reflect, split, or combine light beams in the visible to near IR region with conventional or custom- designed coatings. Standard antireflection (A/R) coatings can be applied to the back side of the films. Figure 6.24 shows mounted pellicle products as supplied by one manufacturer.

A prime feature of the pellicle is the absence of ghost imaging since the 1st and 2nd surface reflections at 45° incidence (see Fig. 6.5) are so close together they appear superimposed. Interference effects are frequently seen. A pellicle with uncoated front and A/R coated back surfaces at 45° incidence serves as a 4% beam sampler. If both surfaces are uncoated, it has about 8% total reflectance in the above-described spectral pass-band. Beamsplitters and combiners achieve the same ghost-suppression advantage over thicker conventional glass components.

Pellicles are supported by circular, square, or rectangular frames with beveled and lapped front surfaces to which the stretched film is attached. Frames typically are black anodized aluminum and have threaded holes for mounting. Special units can be made of stainless steel or ceramic. Figure 6.24 shows mounted pellicle products as supplied by one manufacturer. Figure 6.25 shows the configurations and dimensions of standard pellicle frames as supplied by one manufacturer.

Since pellicles are thin, they are more fragile than conventional plane-parallel optics. They are susceptible to acoustic vibration of adjacent air columns, but work well in a vacuum. Some thicker varieties (notably ones made of polyester films) can be used under water. Temperature range of usefulness is about -40° to +90° C. They can tolerate relative humidity to 95%. Pellicles must be mounted so as not to distort the frames because that would distort the optical surfaces. Custom frames can be designed in stiffer configurations so as to withstand significant mounting forces.

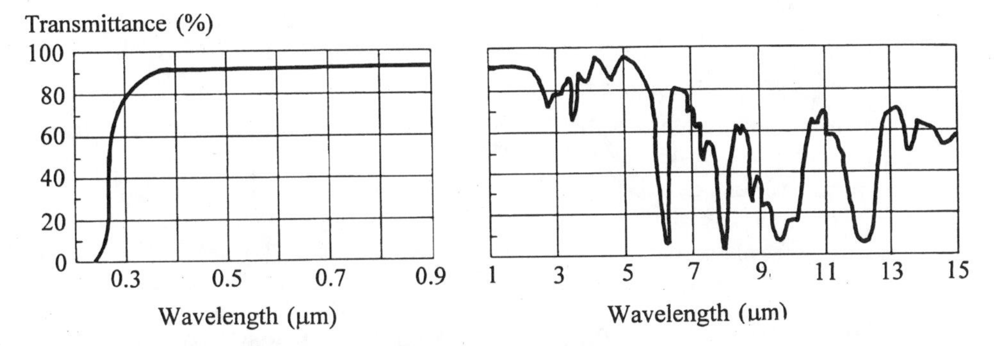

Fig. 6.23 Simplified transmission characteristics of a standard nitrocellulose pellicle in the visible and infrared regions. (Courtesy of National Photocolor Corporation, Mamaroneck, NY)

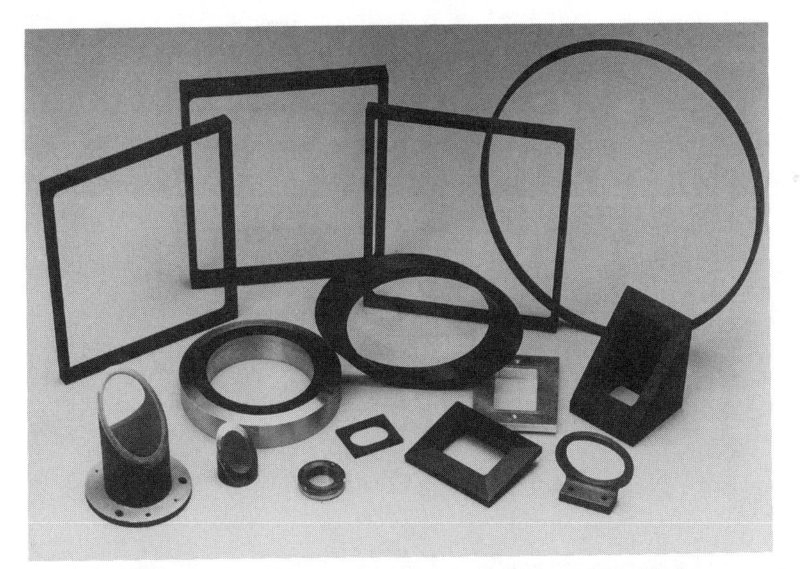

Fig. 6.24 Photograph of a variety of standard mounted pellicles as supplied by one manufacturer. (Courtesy of National Photocolor Corporation, Mamaroneck, NY)

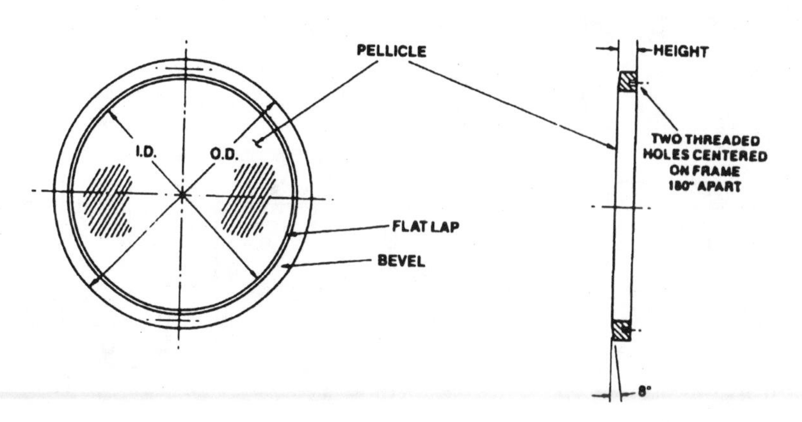

Size	I.D.	O.D.	Height	Mounting Holes
1"	1" (25.4mm)	$1^{3}/_{8}$" (34.9mm)	$^{3}/_{16}$" (4.8mm)	#2-56 thd. x $^{1}/_{8}$" dp.
2"	2" (50.8mm)	$2^{3}/_{8}$" (60.3mm)	$^{3}/_{16}$" (4.8mm)	#2-56 thd. x $^{1}/_{8}$" dp.
3"	3" (76.2mm)	$3^{1}/_{2}$" (88.9mm)	$^{1}/_{4}$" (6.4mm)	#6-32 thd. x $^{1}/_{8}$" dp.
4"	4" (101.6mm)	$4^{1}/_{2}$" (114.3mm)	$^{1}/_{4}$" (6.4mm)	#6-32 thd. x $^{1}/_{8}$" dp.
5"	5" (127.0mm)	$5^{1}/_{2}$" (139.7mm)	$^{5}/_{16}$" (7.9mm)	#6-32 thd. x $^{1}/_{8}$" dp.
6"	6" (152.4mm)	$6^{1}/_{2}$" (165.1mm)	$^{3}/_{8}$" (9.5mm)	#6-32 thd. x $^{3}/_{16}$" dp.

(b)

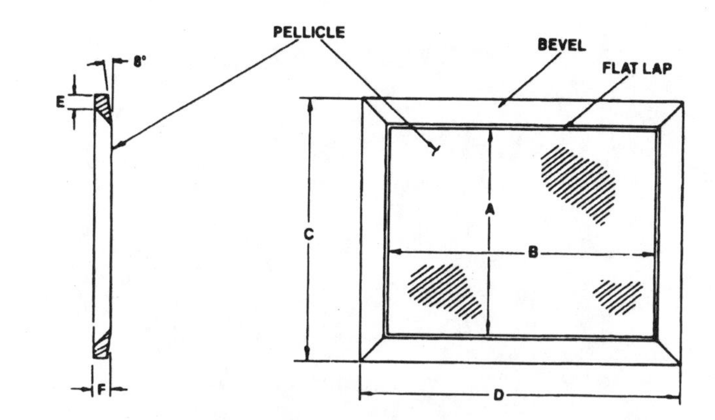

Nominal Dimensions

	Inches	Millimeters
A	5	127.0
B	7	177.8
C	$6^{5}/_{8}$	168.3
D	$8^{3}/_{8}$	212.7
E	$^{5}/_{16}$	7.9
F	$^{7}/_{16}$	11.1

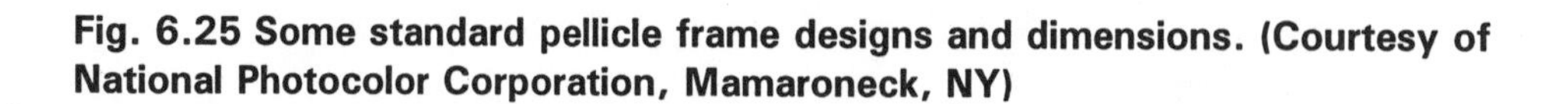

Fig. 6.25 Some standard pellicle frame designs and dimensions. (Courtesy of National Photocolor Corporation, Mamaroneck, NY)

CHAPTER 7
SMALL MIRROR MOUNTING TECHNIQUES

The appropriateness of a mechanical mounting for a mirror depends on a variety of factors, including:

- the inherent rigidity of the optic;
- the tolerable movement and distortion of the reflecting surface or surfaces;
- the magnitudes, locations, and orientations of the steady-state forces holding the optic against its mounting surfaces during operation;
- the transient forces driving the optic against, away from, or transversely to the mounting surfaces during exposure to extreme shock and vibration;
- the effects of steady-state and changing temperatures;
- the shape of the mounting interface on the optic;
- the size, shape, and orientation of the corresponding mounting surfaces (typically pads) on the mount;
- the rigidity and long-term stability of the mount;
- assembly, adjustment, maintenance, package size, weight, and configuration constraints; and

affordability in the context of cost of the entire instrument.

In this chapter we address a variety of techniques commonly used to constrain mirrors in the diameter range from a few cm to about 24 in. (61 cm). At the small end of this range, the mounts tend to be very simple; techniques typically used for mounting lenses may suffice. As would be expected, complexity increases with mirror size. We intentionally omit mountings for larger mirrors such as would be used in state-of-the-art astronomical applications because proper treatment of that aspect of mirror mounting would far exceed page limitations. General techniques considered comprise mechanically clamping, elastomeric bonding, optically contacting, and mounting on flexures. Mountings appropriate for non-metallic and metallic mirror substrates are included. In general, we progress from smaller to larger sized optics. It is pointed out that many mounting problems sometimes thought to exist only with the largest mirrors (i.e., those beyond the scope of this book) actually exist with small mirrors; the difference is one of scale. In some contemporary designs involving "small" size, but high performance, these same problems are of sufficient magnitude to warrant special consideration.

7.1 Mechanically clamped mirror mountings

Circular, rectangular, or non-symmetrically-shaped mirrors can often be mounted in the same manner as a lens. Circular ones up to, perhaps, 4 in. (10.2 cm) in diameter can be held with threaded retaining rings while they or non-circular ones can be held with clamping flanges. The OD limit for the threaded mount is set primarily by the increasing difficulty of machining thin circular rings with larger diameters.

Figure 7.1 shows the retaining ring concept. The mirror (here shown as convex although one with flat and concave surfaces can be mounted similarly) is held against a shoulder in the mount by axial preload exerted by tightening the retainer. The ring typically has a loose fit (Class-1 or -2 per ANSI Publication B1.1-1982) into the mount ID. Contact occurs on the polished surface of the mirror to encourage precise centering of curved-surfaced optics to the mechanical axis of the mount due to the radial component of the axial force. Precise edging or close tolerances on the OD of the mirror are not required. To minimize bending of the mirror, contact should occur approximately at the same height from the axis as the center of the contact area opposite. Sharp-corner contact on the polished surface is shown in Fig. 7.1. Since this maximizes contact stress in the mirror, another shape for the interfacing surface might well be utilized to advantage. A tangent (conical) or toroidal (donut-shaped) interface would be preferred. Chapter 3 of Ref. 1 gives details of the various types of interfaces for lenses which are equally applicable to some mirrors.

Pairs of holes are sometimes drilled into the exposed face of the retainer to accept pins on the end of a cylindrical wrench used to tighten the ring. Alternatively, a diametrical slot may be cut across the face of the retainer for this purpose. A flat plate that spans across the retainer is used as the wrench in this case. The edge of the wrench closest to the mirror face should be shaped to provide necessary clearance to prevent damaging the reflecting surface.

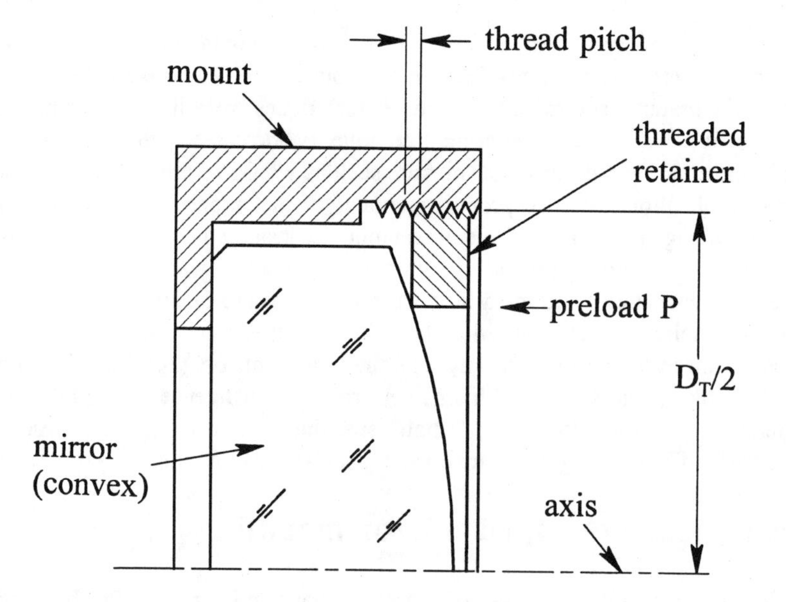

Fig. 7.1 Typical configuration of a convex mirror secured in its mount with a threaded retaining ring

The magnitude of the total preload (P) developed in a threaded retainer lens mount design with a specific torque (Q) applied to the ring at a fixed temperature can be estimated by the following equation:

$$P = 5Q/D_T, \tag{7.1}$$

where D_T is the pitch diameter of the thread as shown in Fig. 7.1.

Note that the accuracy of this equation depends largely upon the coefficient of friction in the threaded joint; this is quite variable in real life. This means that the torque applied to a threaded retainer cannot be relied upon to produce a specific preload.[63]

Numerical Example No. 27: Axial preload obtained from a torqued retainer. Assume that a 3.00 in. (7.62 cm) OD mirror is to be clamped with a total preload of 50 lb (222.4 N) delivered by a retainer screwed into a mount on a thread of pitch diameter 3.25 in. (8.25 cm). What torque should be applied?

Rearranging Eq. (7.1), $Q = PD_T/5 = (50)(3.25) / 5 = 31.5$ lb-in. (3.7 N-m).

A typical design for a circular mirror mounting involving a continuous-type clamping flange retainer is shown in Fig. 7.2. This type of retainer is most frequently used with mirrors larger than typically would be held with threaded rings or if more precise preload is needed in the application. Several close-fitting locating pads around the rim of the mirror help to center it to the mechanical axis of the mount. An annular location pad is used between the flat-ground mirror back surface and the mount shoulder to localize the restoring force directly opposite the clamping (preload) force. The shoulder and the latter pad typically should be stiff (i.e., metal) and lapped flat to minimize distortion of

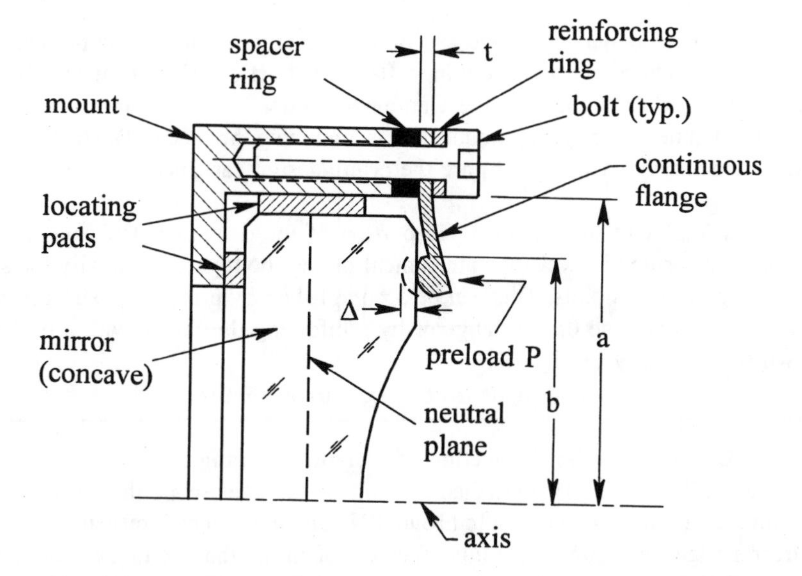

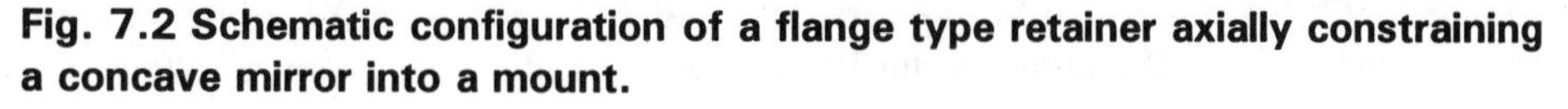

Fig. 7.2 Schematic configuration of a flange type retainer axially constraining a concave mirror into a mount.

the mirror surface. The interface to the flat bevel on the mirror is shown as toroidal to reduce contact stress. A flat pad would work well if aligned exactly to the bevel, but a sharp corner contact and thus increased stress could result from machining errors or temperature changes.

The function of the clamping flange is the same as the threaded retainer described above. The magnitude of the preload exerted thereby can be determined fairly closely using Eq. (7.2) which, acording to Roark[45], applies to a perforated circular plate with outer edge fixed and axially directed load applied uniformly along the inner edge to deflect that edge:

$$\Delta = K_{P1}(a^2 - b^2 + K_{P2}), \tag{7.2}$$

where

$$K_{P1} = 3P(m^2 - 1)/4\pi m^2 E_M t^3, \tag{7.3}$$

$$K_{P2} = \frac{2mb^2(a^2 - b^2) - 8ma^2b^2\ln(a/b) + 4a^2b^2(m+1)(\ln(a/b))^2}{a^2(m+1) + b^2(m+1)}, \tag{7.4}$$

and a = outer radius of cantilevered section
b = inner radius of cantilevered section
t = plate thickness
m = reciprocal of υ (Poisson's ratio) of plate
E_M = Young's modulus of plate
P = total preload.

The spacer under the flange can be ground at assembly to the particular axial thickness that produces the predetermined flange deflection when firm metal-to-metal contact is achieved by tightening the clamping screws. Variations in as-manufactured mirror edge thicknesses are accommodated by customizing the spacer. The flange material, thickness, and annular width (a - b) are the prime design variables.

It is important that the clamped portion of the flange be stiff enough that the deflections Δ measured between the attachment points (bolts) are essentially the same as those existing at those points. This can be accomplished by providing extra thickness at the clamped annular zone of the flange or by reinforcing the flange with a backup ring as shown schematically in Fig. 7.2.

Numerical Example No. 28: Deflection of a clamping flange.
Consider a 15.75 in. (400.05 mm) diameter mirror for a telescope that is to be held in place with a total preload P of 380 lb (1690.3 N) distributed uniformly around and near the mirror's edge by a 6061 aluminum flange that limits the aperture of the mirror to 15.000 in. (381.000 mm). A radial clearance of 0.010 in. (0.254 mm) is to be provided between the mirror OD and the mount ID. The pertinent dimensions are as follows:

$a = (15.750/2) + 0.010 = 7.885$ in. (200.279 mm)
$b = 15.000/2 = 7.750$ in. (196.850 mm)
$t = 0.010$ in. (0.254 mm)
$m = 1/0.332 = 3.012$
$E_M = 9.9\times10^6$ lb/in.^{2} (6.82×10^6 N/m^2)

To calculate the required flange edge displacement, Δ:

$$K_{P1} = (3)(380)(3.012^2 - 1) / (4\pi)(3.012^2)(9.9\times10^6)(0.010^3) = 8.1536$$

$$K_{P2} = \frac{[\,(2)(3.012)(7.75^2)(7.885^2 - 7.750^2) - (8)(93.012)(7.885^2)(7.750^2)(\ln(7.885/7.750)) + (4)(7.885^2)(7.750^2)(4.012)(\ln(7.885/7.750))^2\,]}{(7.885^2)(2.012) + (7.750^2)(4.012)}$$

$$= -2.1098$$

$$\Delta = (8.1536)(7.885^2 - 7.750^2 - 2.1098) = 0.0077 \text{ in. } (0.195 \text{ mm}).$$

This displacement is large enough to be measured with reasonable accuracy so the preload can be expected to be close to the desired value.

Figure 7.3 shows a simple means for attaching a nonmetallic- or metallic-type mirror configured as a plane-parallel plate to a metal surface. The reflecting surface is pressed against three flat machined (lapped) pads by three spring clips. The spring contacts are directly opposite the pads so as to minimize bending moments. This design constrains one translation and two tilts. The spacers that position the clips are machined to the proper length for the clips to exert clamping forces (preload) of controlled magnitude normal to the mirror surface. The spring clips should be strong enough to restrain the mirror against the shock and vibration to which it may be subjected.

The ends of the spring clips shown as bearing on the mirror in Fig. 7.3 are shaped as cylindrical pads. They are similar to those shown in Fig. 5.3 for a similar prism constraint. Line contact occurs. Stress generation along these lines is discussed in the next chapter. Other spring-end shapes are possible, of course. The discussion of effects of contact shape and geometric errors in orientation and/or location of contacts in Sect. 5.1 apply here too.

It is highly desirable for the rim of the mirror's reflecting surface to contact fixed reference pads rather than for the back side of the mirror to do so. This is because any wedge in the mirror substrate will then not affect pointing of the reflected beam. Obviously, the reference pads and the portion of the mirror contacting them must be clean; dirt between those surfaces could tilt the mirror. The pads must be coplanar and located directly opposite the applied forces so the clamping preload will not exert bending moments on the optic. Figure 7.4 an irregularity on a pad surface. A particle of sand or other particulate matter in the interface would act similarly. Orientation and figure of the reflecting surface could change if that defect were to change size or if the particle shifts

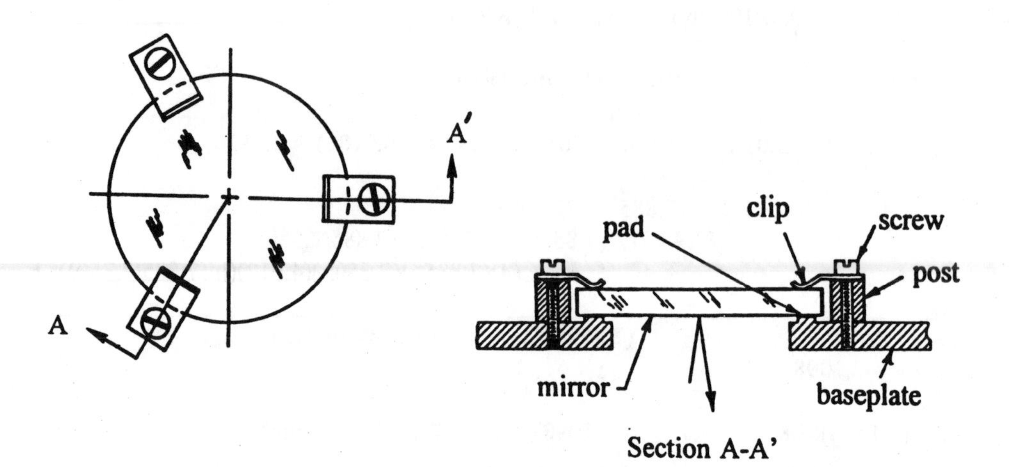

Fig. 7.3 A simple clamped mirror mount. (Adapted from Durie[35])

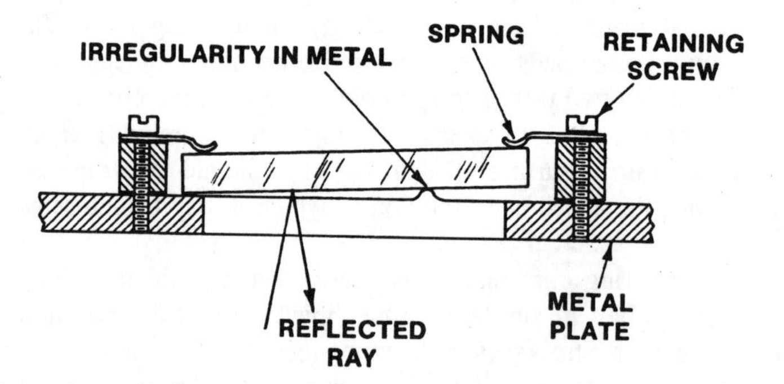

Fig. 7.4 Representation of pad irregularity or dirt particle in a mirror mount. (Adapted from Durie[35])

during vibration or shock.

In Fig. 7.5 we show a variation on the above design in which the spring force is supplied by compressed resilient pads. The reflecting surface bears against rigid clamps that are in contact with the main portion of the mount. The chief objection to this design is that the soft material may compress and lose its resiliency with time so preload is not necessarily constant in the long term.

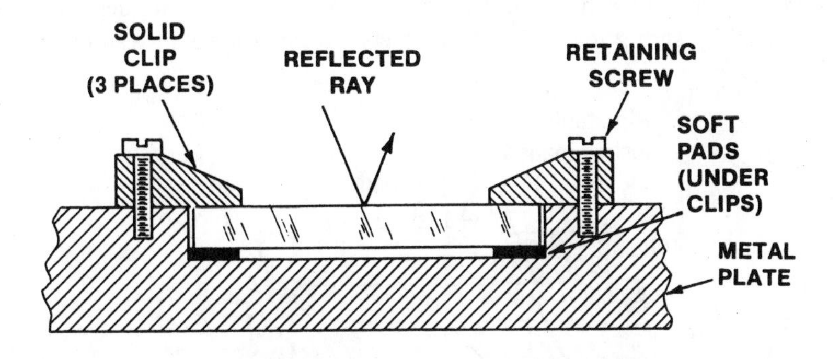

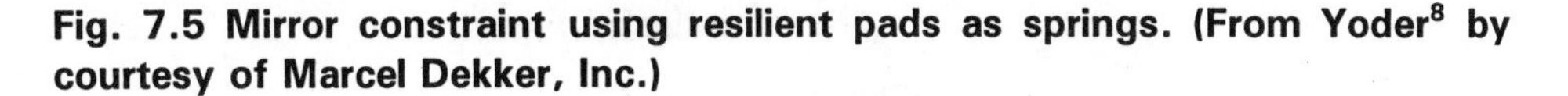

Fig. 7.5 Mirror constraint using resilient pads as springs. (From Yoder[8] by courtesy of Marcel Dekker, Inc.)

A mount for a beamsplitter plate is illustrated in Fig. 7.6. As in the discussion of a cube-shaped beamsplitter in connection with Figs. 4.1 and 4.2, this plate registers against fixed "points" and is spring-loaded directly opposite these points. Here, and in any design with hard contacts against the reflecting side of the mirror, the location and orientation of that surface do not change with temperature of the optic.

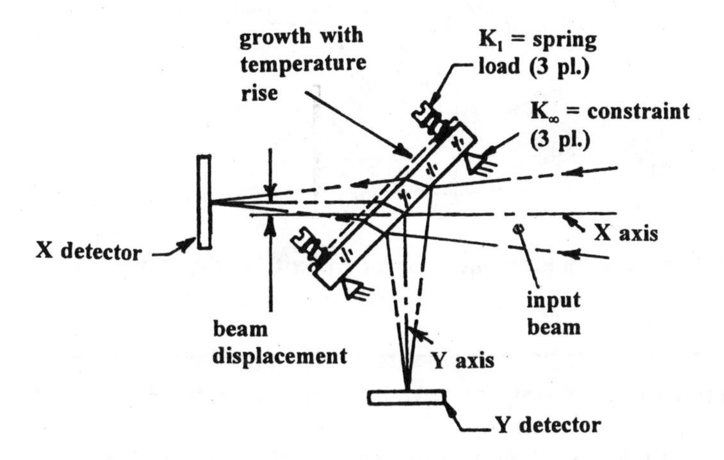

Fig. 7.6 Mounting for a beamsplitter plate. (Adapted from Lipshutz[34])

The optic is not constrained laterally other than by friction in the designs represented in Figs. 7.3 through 7.6. This may be acceptable because performance of a flat mirror is insensitive to these motions. Excessive lateral movement of the optic can be prevented spring loading it against stops. CTE differences must be considered if the mirror touches hard stops without springs.

A design concept including spring-loaded constraints in the plane of the reflecting surface is illustrated in Fig. 7.7. While compression coil springs are shown, cantilevered clips could be employed. This mount is semi-kinematic since all six degrees of freedom are constrained by spring loads and the contacts are small areas instead of points.

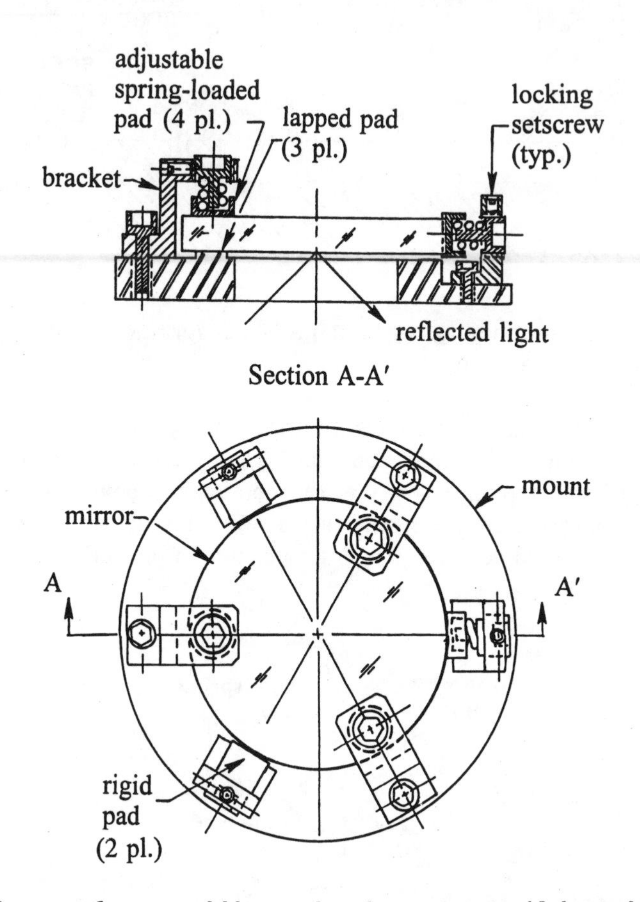

Fig. 7.7 Concept for a semi-kinematic mirror mount. (Adapted from Yoder[64])

Numerical Example No. 29: Clamping force to constrain a mirror

What normal force is required of each of three spring clips to constrain a flat round mirror weighing 0.09 lb (0.04 kg) with a safety factor of 2 under accelerations of 15 times gravity?

$P_i = w a_G f_S / N = (0.09)(15)(2) / 3 = 0.9$ lb (4 N)

A non-kinematic mount for a 16 in. (40.6 cm) spherical primary mirror that was used in a Schmidt telescope[65] is shown in Fig. 7.8. The mirror rim was ground with a spherical contour, centered at the mirror's face, to avoid chipping when installed or removed for recoating. A narrow [0.25 in. (6.35 mm) annular width] rim ground on the mirror's face was pressed against three steel pads, or lugs, inside an Invar mirror cell. This cell was attached to a perforated Invar tube. The pads were previously filed until they defined a plane perpendicular to the axis of the telescope tube. Three shims with aggregate thickness of about 0.09 in. (2.3 mm) were inserted between the spherical ground rim of the mirror and the ID of the cell. These shims caused the cell to spring very slightly out of round. By successive adjustments of the thicknesses and locations of the shims, the mirror's central normal was brought parallel to the telescope axis. The decentration thus produced was equivalent to tipping the mirror.

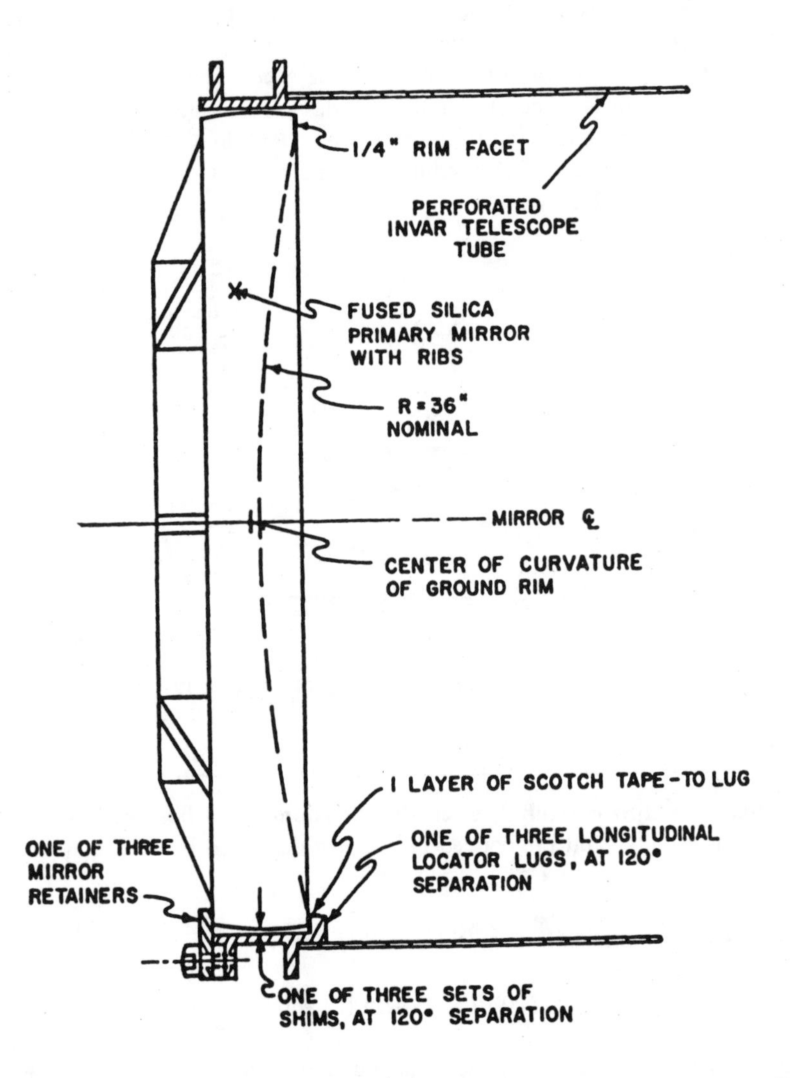

Fig. 7.8 Non-kinematic rim mounting for a telescope primary mirror. (From Strong[65])

Strong's mirror was restrained axially and prevented from rotating about the telescope axis by friction provided by three retaining spring clips that clamped the mirror against the three axial constraining lugs. One layer of thin plastic tape (Scotch tape) was used to isolate the mirror from the lugs. This increased the mount's resistance to mechanical shock and provided a slight thermal isolation.

7.2 Bonded mirror mountings

First-surface mirrors of apertures typically 6 in. (15.2 cm) or less can be bonded directly to a mechanical support in much the same manner as described earlier for prisms. The ratio of largest face dimension to thickness should be less than 10:1 and preferably no more than 6:1 in order for dimensional changes in the adhesive during cure or under temperature changes not distort the mirror surface excessively. Figure 7.9 illustrates such a design. The mirror is made of Schott BK7 glass, is 2 in. (5.1 cm) in diameter, and is 0.33 in. (0.84 cm) thick (6:1 ratio). It weighs about 0.09 lb (0.04 kg). The mounting base is stainless steel type 416. The bonding land is circular and has an area of 0.50 in.^{2}. It is bonded with 3M 2216 epoxy.

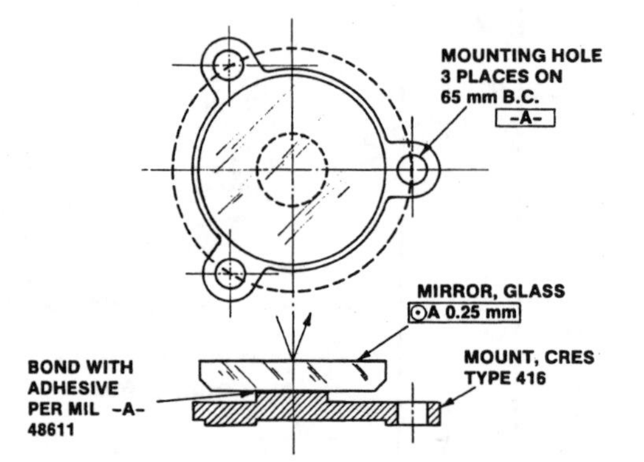

Fig. 7.9 A typical bonded first-surface mirror assemply. (From Yoder[37])

Numerical Example No. 30: Acceleration resistance of a bonded mirror.
What acceleration should the bonded mirror just described be able to withstand with a safety factor of 2?

As indicated in Numerical Example No. 8:

$$a_G = JQ / Wf_S = (2500)(0.5) / (0.09)(2) = 6944 \text{ times gravity}$$

We conclude that this design should be adequate for most applications. In fact, the glass should fail before the bond [at a_G = (6944)(1000) / 2500 = 2778 times gravity].

As pointed out for prism bonding (Sect. 4.3), the adhesive should, if possible, be distributed in small separated areas with total area equal to the calculated minimum value for the anticipated g-loading. This minimizes shrinkage and thermal expansion problems and helps secure the mirror in a more kinematic fashion with wider lever arm to resist in-plane torques.

Numerical Example No. 31: Distributed bond size required to secure a mirror.
A circular Zerodur mirror with diameter 4.25 in and thickness 0.708 in. is to withstand 200 times gravity accelerations with a safety factor of 4 when bonded with epoxy of strength J = 2500 lb/in.^{2} arranged in a triangular pattern of three equal circular areas on the mirror back. What should be the individual bond diameters?

From Table C8a, the density of Zerodur is 0.091 lb/in.^{2}
The mirror weight is then: $W = \pi r^2 t\rho = (\pi)(4.25 / 2)^2(0.708)(0.091) = 0.914$ lb

The minimum total bond area required is $Q_M = Wa_G f_S / J = (0.914)(200)(4) / 2500$
$= 0.292$ in.^{2}
The diameter if a single circular bond is used $= 2\sqrt{(Q_{MIN} / \pi)}$
$= (2)(0.292 / \pi)^{1/2} = 0.610$ in.

The minimum diameter of each of the three bonds is:
$2\sqrt{(Q_{MIN} / 3\pi)} = (2)(0.292 / 3\pi)^{1/2} = 0.352$ in.

The single and multiple areas would then appear as shown here

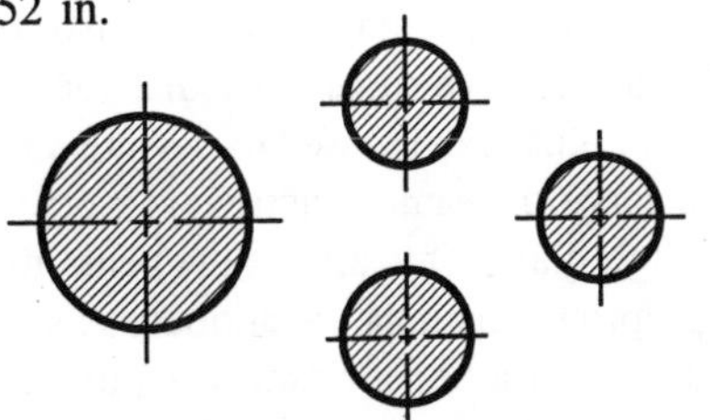

Another quite different technique for bonding a mirror to a mechanical support is sketched in Fig. 7.10. Here, a round mirror is bonded to three leaf springs that are, in turn, attached mechanically by screws, rivets, or adhesive to a circular plate of essentially the same diameter as the mirror. The springs are flat so they can flex radially to accommodate differences in thermal expansion. They are of the same length and material so that thermally induced tilts are minimized. The local areas on both the mirror and mount where the springs are attached are flattened in order to obtain adequate contact area for bonding and to prevent cupping of the springs. The springs should be as light and flexible as is consistent with vibration and shock requirements. Høg[66] discussed a design of this general type. Other flexure mountings for mirrors are discussed in Sect. 7.4.

Although discussed here in the context of mountings for non-image forming optics, the mounting arrangement of Fig. 7.10 may also be used to support image-forming mirrors since the design serves well to keep the optic centered in spite of temperature changes.

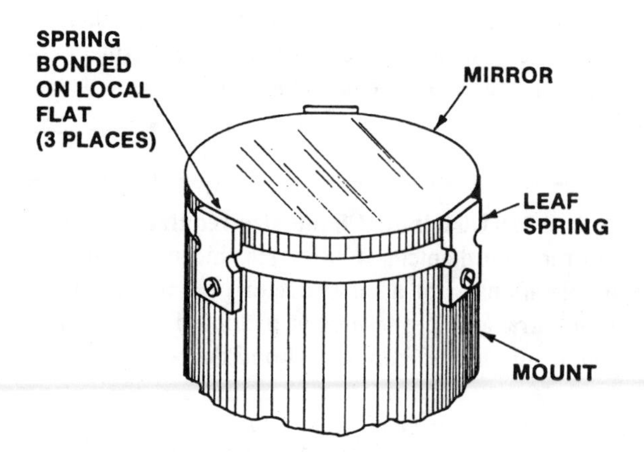

Fig. 7.10 Sketch of a mirror mount with radially compliant springs. (Adapted from Høg[66])

7.3 Multiple Mirror Mountings

Although mirrors are generally thought of as single optical elements, it is occasionally advantageous to use two or more such optics in opto-mechanical assemblies in order to serve some particular function. For example, two flat mirrors oriented at 45° to each other can be used to deviate a light beam by 90°. If rigidly attached together, they will serve the same function as a penta prism, but not require transmission of the beam through glass. This allows the penta mirror device to be used in the UV or IR, assuming appropriate coatings are applied. Also, the weight of the penta mirror is generally lower than that of a penta prism of equivalent aperture.

The problem in the design and fabrication of the penta mirror is how to hold the mirrors in the proper relative orientation so as to maintain long-term alignment stability and not distort the optical surfaces. One approach that has been used is to clamp the mirrors individually to a metal block or built-up structure providing the 45° dihedral angle. See Fig. 7.11. Here two rounded-end rectangular gold-coated mirrors are held by three screws to carefully lapped pads on either side of an aluminum casting. The screws each compress two Belleville washers that provide preload. This hardware was used as part of an automatic theodolite system for pre-launch azimuth alignment of the Saturn Space Vehicles, so was used in a generally stable environment inside a concrete bunker at Cape Canaveral.[67]

A type of penta mirror used successfully in military optical range finders had the glass mirrors bonded on edge to a glass base plate, which was in turn attached to the optical bar of the range finder.[68] Figure 7.12 shows an assembly of this general type. The base plate in this example is metal. The useful aperture exceeds 50 mm (1.97 in.).

Figure 7.13 shows another approach for creating a penta mirror assembly that

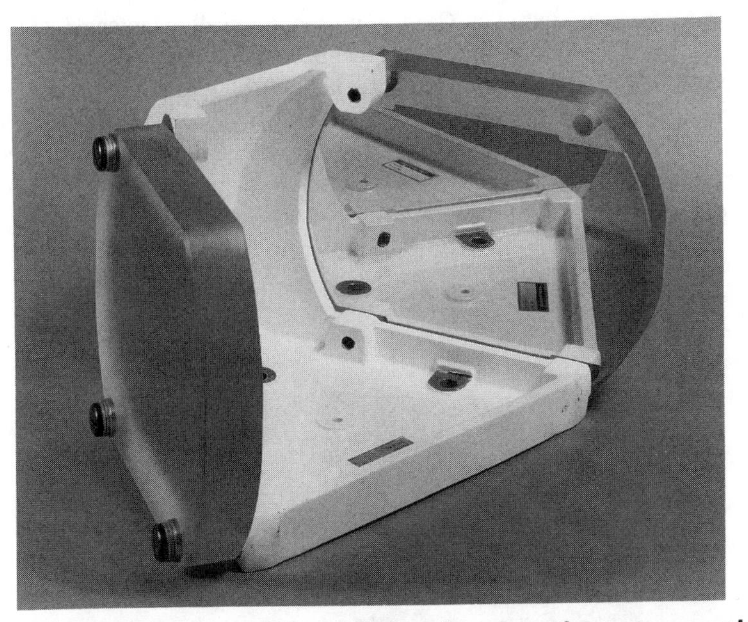

Fig. 7.11 Penta mirror assembly made by clamping mirrors to a precision metal casting. (Courtesy of NASA Marshall Space Flight Center, Huntsville, AL)

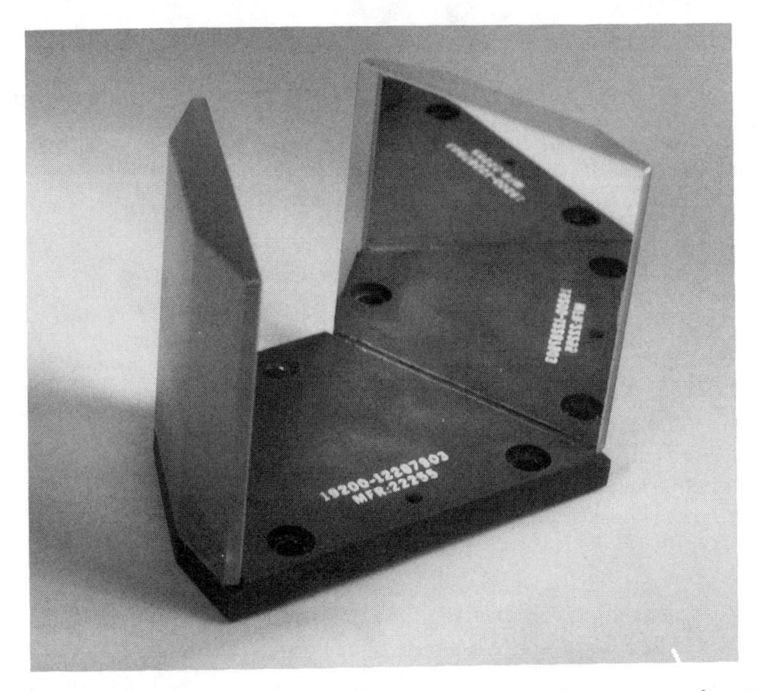

Fig. 7.12 Penta mirror made by bonding glass mirrors on edge to a metal bracket. (Courtesy of PLX Corporation, Deer Park, NY)

had the polished faces of two flat Cer-Vit mirrors optically contacted along their innermost edges to a Cer-Vit angle block that had been ground and polished to within 1 arc-sec of the nominal 45°. The angle block was hollowed out to reduce weight without reducing strength. Triangular Cer-Vit cover plates were then attached with optical cement to both the top and bottom of the assembly and a rectangular cover plate was cemented across its back. These three plates served not only as mechanical braces, but also to seal the exposed edges of the contacted joints. An Invar plate was bonded to one of these cover

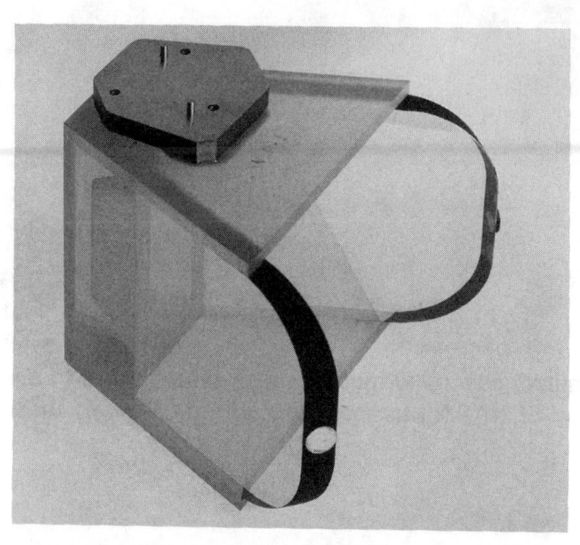

Fig. 7.13 A 10 cm (3.94 in.) aperture penta mirror assembly made by optically contacting Cer-Vit components. (From Yoder[69])

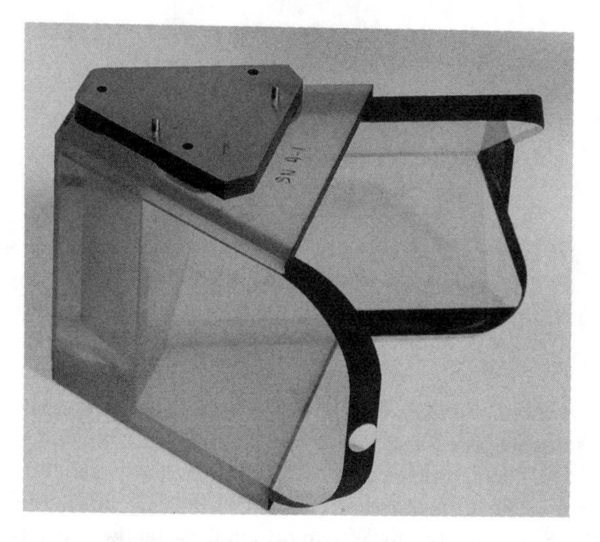

Fig. 7.14 A 10 cm (3.94 in.) aperture roof penta mirror assembly made by optically contacting Cer-Vit components. (From Yoder[69])

plates to serve as a mounting interface. With mirror plates measuring approximately 11 by 16 by 1.3 cm (4.33 by 6.30 by 0.51 in.), the assembly had a clear aperture of 10 cm (3.94 in.). A roof penta mirror assembly, of similar construction and size (see Fig. 7.14) also has been described in the literature.[69]

Fig. 7.15 Photograph of the penta mirror assembly of Fig. 7.13 mounted in its Invar housing. (Courtesy of Raytheon Optical Systems Corp., Danbury, CT)

To verify these optically-contacted designs, a prototype of the penta mirror and mount assembly (see Fig. 7.15) was subjected to adverse thermal, vibration, and shock environments. First, it was temperature-cycled several times from -2 to +68°C (28 to 154°F) while monitoring the reflected wavefront interferometrically. The test setup was capable of detecting changes of $\lambda/30$ and had an inherent error of less than $\lambda/15$ for λ = 633 nm. The maximum thermally induced P-V wavefront distortion measured $\lambda/4$. The assembly was then vibrated without failure at loadings up to 5 times gravity and frequencies of 5 to 500 Hz along each of three orthogonal axes. Shock testing at up to 28 times gravity peak loading in 8 msec in two directions also caused no damage to the test item.

A roof mirror functionally equivalent to a Porro prism is shown in Fig. 7.16. This assembly has an aperture of slightly over 1.75 by 4.0 in. (4.4 by 10.2 cm). Its mirrors are 0.5 in. (12.7 mm) thick Pyrex. These mirrors are epoxied on one long edge to a Pyrex keel that is, in turn, bonded to a 0.125-in. (3.2 mm) thick stainless steel mounting plate. The end plates are made with nominal 90° angles. Each plate is cemented to the top of one mirror and the end of the other mirror. Cementing is done in a jig that aligns the mirrors to 90° within tolerances as small as 0.5 arc-sec. The tolerance on the mirror figure is typically λ/10 at 633 nm.

Figure 7.17 shows front and back views of a hollow retroreflector which is the mirror equivalent of the cube corner prism (see Figs. 3.29 and 3.30). It comprises three

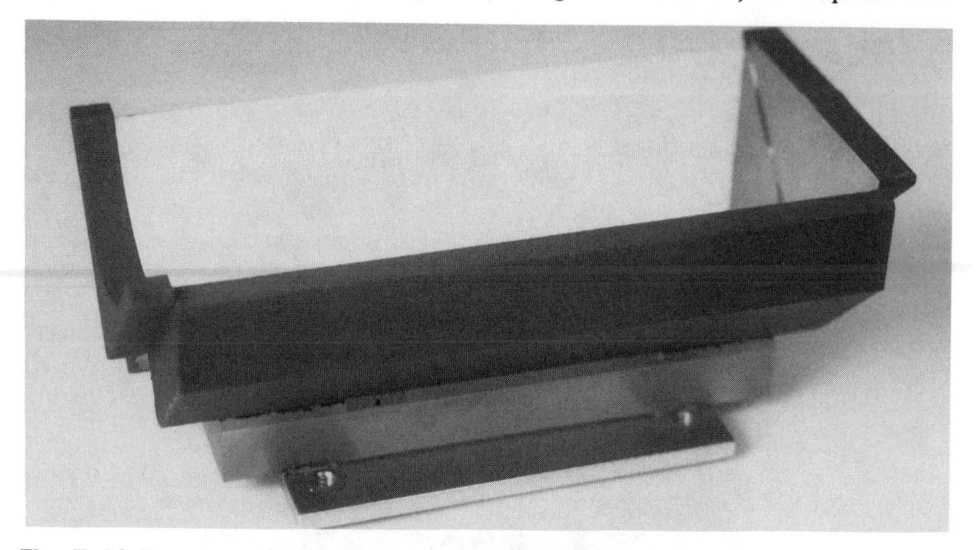

Fig. 7.16 Photograph of a Porro-type roof mirror made by bonding two flat mirrors at 90°. (Courtesy of PLX Corporation, Deer Park, NY)

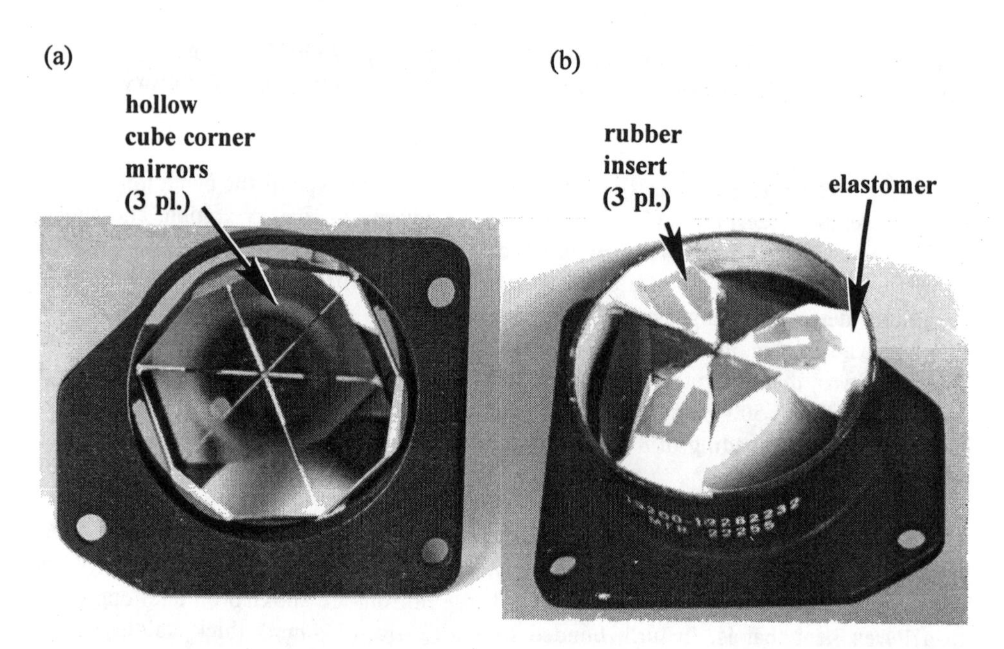

Fig. 7.17 Front (a) and back (b) photographs of a hollow corner retroreflector made by bonding three flat square mirrors in mutually-perpendicular fashion. (Courtesy of PLX Corporation, Deer Park, NY)

nominally square-faced Pyrex mirrors. The aperture of this unit is approximately 45 mm (1.77 in.). The mirrors are "potted" into an aluminum housing configured for ease of mounting in a military application using an elastomeric material (white) surrounding three rubbery inserts (gray). Accuracy of the nominally 180° light deviation typically between 0.5 arc-sec and 5 arc-min is achieved by jigging during elastomer cure. Apertures as large as about 5 in. (127 mm) are commercially available.

A variation of the hollow retroreflector is the, so called, lateral-transfer hollow retroreflector. An example of one commercially available unit is shown in Fig. 7.18. It has one flat mirror mounted at 45° at the end of an elongated box housing and a roof mirror at the other end of that box. The three mirrors are nominally mutually perpendicular so the assembly functions as a transverse section of a hollow cube corner retroreflector. Devices of this type with apertures up to 2 in. (5.1 cm) and lateral offsets of the axis of more than 30 in. (76 cm) are commercially available. A similar 6 in. (15.2 cm) offset device made entirely of beryllium also has been described.[70] That unit had a beamsplitter at one end instead of a conventional mirror. The Be unit was lighter than the equivalent glass/metal version would be.

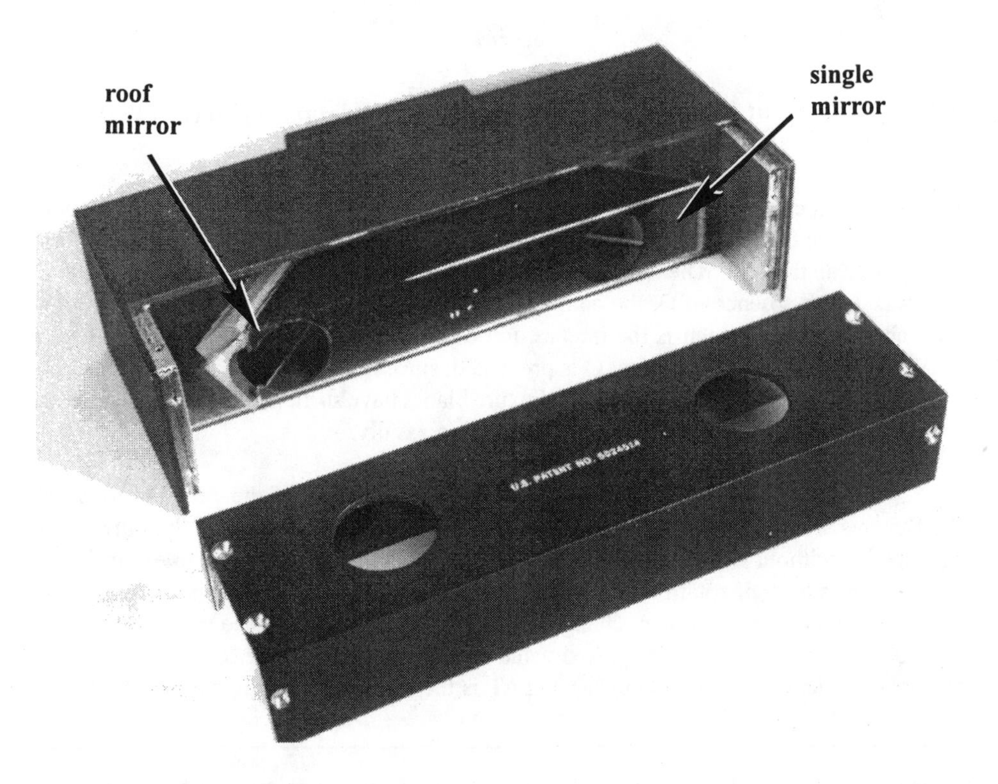

Fig. 7.18 Photograph of a partially disassembled lateral-transfer hollow retroreflector. (Courtesy of PLX Corporation, Deer Park, NY)

7.4 Flexure mountings for mirrors

The principle of one type of flexure mounting for mirrors is illustrated by Fig. 7.19. The mirror is circular and mounted in a cell. That cell is suspended from three thin flexure blades. The single direction of allowable motion for each flexure acting alone is

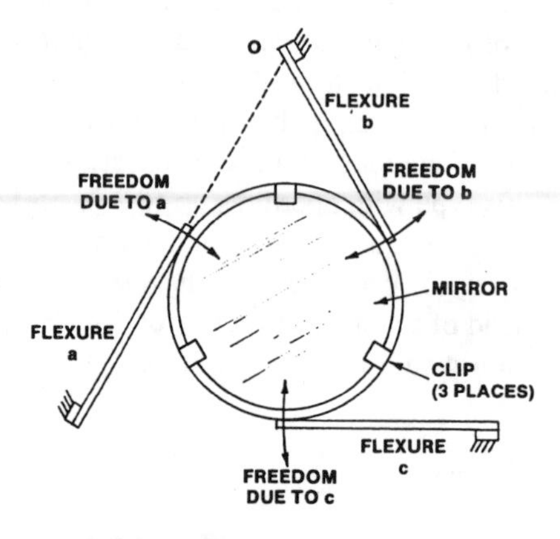

Fig. 7.19 Concept for a flexure mounting for a circular mirror. (From Yoder[8] by courtesy of Marcel Dekker, Inc.)

indicated by a curved arrow. Ideally, these lines of freedom should intersect at a point, the flexure lengths should be equal, and the fixed ends of the three flexures should form an equilateral triangle. The function of this system of flexures may be explained as follows. In the absence of C, the combination of flexures A and B will permit rotation only about point O, which is the intersection of flexure B with a line extending flexure A. With C in place, rotation about O is prevented, since flexure C is stiff in that direction. Although not apparent in the figure, the flexure blades have sufficient depth perpendicular to the page to prevent the mirror from translating axially.

Even if temperature changes cause thermal expansion of the structure to which the flexures are attached or of the mirror/cell assembly, radial motion of the mirror will be impeded without stressing the mirror. The only permitted motion due to expansion or contraction is a small rotation about the normal through the intersection of the lines of freedom. This occurs because of changes in lengths of the flexures. The magnitude of this rotation, θ, in radians, may be approximated as $\theta = (3\alpha)(\Delta T)$, where α is the CTE of the flexure material (typically in ppm/°F) and ΔT is the temperature change (typically in °F).

Numerical Example No. 32: Rotation of a flexure-mounted mirror with temperature change.

If the flexures of Fig. 7.19 are beryllium copper with a CTE of 8.0 ppm/°F and ΔT is 20°F, what is the mirror rotation, θ?

From the above text, $\theta = (3)(8.0 \times 10^{-6})(20) = 0.027$ deg = 1.6 arc-min

This rotation probably is inconsequential in most any application.

Figure 7.20 shows a variation of the above flexure mounting concept in which a mirror of rectangular shape is supported in a cell attached to three deep flexure blades. The dashed lines indicate the directions of freedom (approximated as straight lines). The intersection of these lines, which is stationary, does not coincide with the geometric center of the mirror or the center of gravity of this particular mirror/cell combination. By changing the angles of the corner bevels and relocating the flexures, the intersection point could be centralized and the design improved from a dynamic viewpoint. In either case, differential thermal expansion at the mount-to-structure interface can occur without stressing the mirror. Further, axial movement of the mirror is prevented by the large stiffness of the blades in that direction.

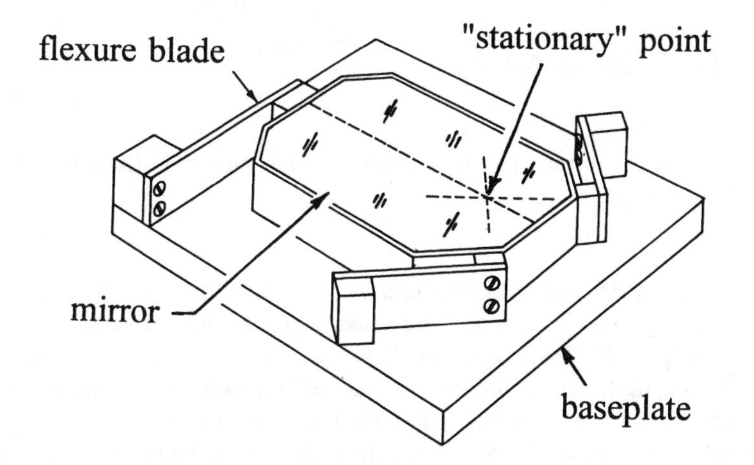

Fig. 7.20 Concept for a flexure mounting for a cell-mounted rectangular mirror. (From Yoder[37])

Figure 7.21 shows a concept for another flexure mount for a circular mirror. Here, the flexures are integral with the body of the ring-shaped mount. They typically would be machined by cutting slots in an electric discharge machining (EDM) operation. This mirror mount is an extension of a design concept advanced for lens mounting in the literature.[71] Once again, the blades are stiff in the axial direction and compliant in the radial direction as would be appropriate to negate decentrations due to temperature changes.

Modest-sized mirrors [say, those in the 15 to 24 in. (38.1 to 61 cm) diameter range] or smaller mirrors used in high precision, high performance applications may benefit from mounting in the manner shown in Fig. 7.22. Here, a circular mirror is held in a cell which, in turn, is supported by three tangent-arm flexures and three axial metering rod-type supports that also include flexures. Such a mount is insensitive radially

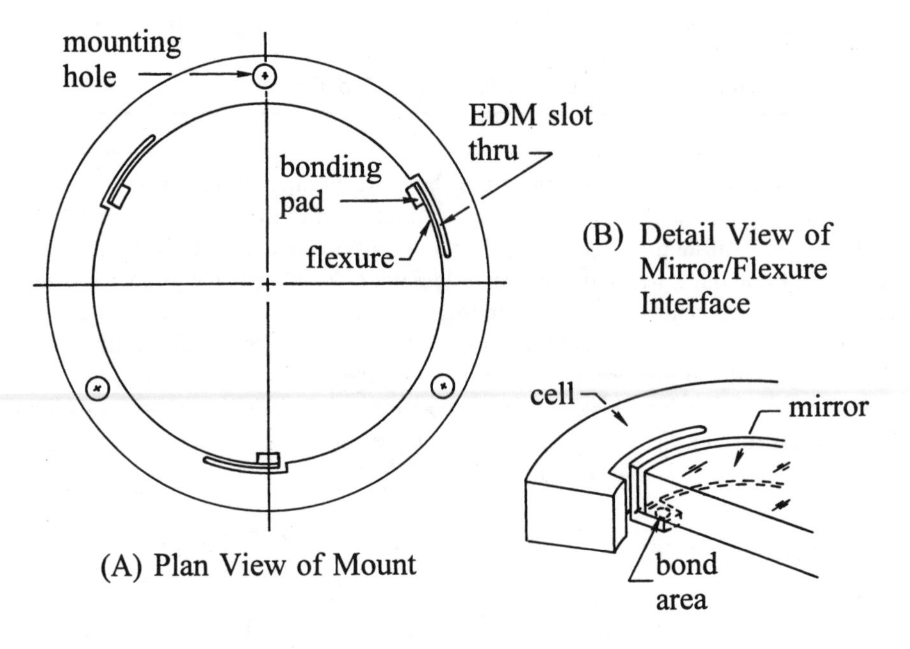

Fig. 7.21 Concept for a mirror mount with integral flexures. (Adapted from Bacich[71])

to temperature changes. The thermal compensation mechanism shown in the axial support makes the design less sensitive in that direction to temperature changes. The latter mechanism comprises selected lengths of dissimilar metals arranged in a re-entrant manner. In some manifestations of this concept, the cell has been omitted and the tangent-arms and metering rods bonded directly to the mirror substrate. As noted in Fig. 7.22, differential screws might be employed to make the lengths of the tangent-arms adjustable. "Turnbuckle" mechanisms as shown would facilitate axial adjustment. Such means would allow tilt adjustment of the mirror.

7.5 Center-mounts for circular-aperture mirrors

Some lightweighted mirrors in the size range of interest in this book are mounted on a hub that protrudes through a central perforation in the mirror substrate. An example is shown in Fig. 7.23. This is the rear end of a 150 in. (3.81 m) EFL, f/10 Cassegrainian telescope used in a photographic application for tracking missiles.[8] Both surfaces of the primary mirror are spherical; the first being the reflecting surface and the second shaped to reduce weight. Figure 6.7(d) applies. The first surface registers against a convex spherical seat on an integral shoulder of the hub. The seat radius is ground to match that of the mirror. A toroidal seat is provided on the hub. This surface is lapped to closely accommodate the perforation of the mirror. A threaded retaining ring bearing against the flat bevel at the center of the back surface of the mirror provides needed axial preload to constrain the mirror. The hub is inserted into the cast aluminum rear housing of the telescope and clamped in place with a threaded retaining ring.

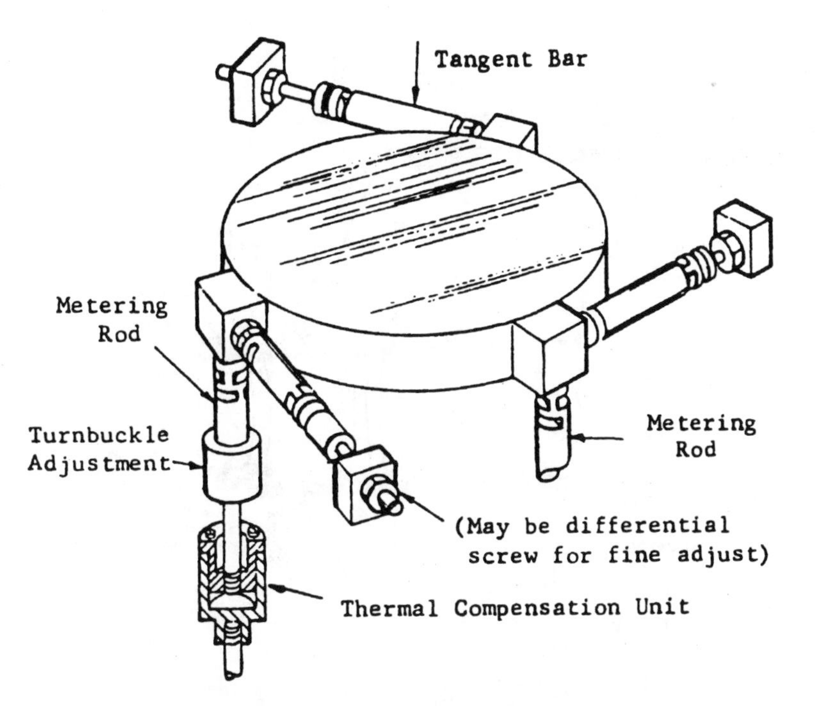

Fig. 7.22 Mirror mounting concept involving flexure-type tangent-arms and axial flexures

Small- to modest-sized single-arch lightweighted mirrors also are center-mounted on hubs since their rims typically are very thin so lack adequate strength to support the mirror. Designs usually follow the general lines of the mount shown in Fig. 7.23. More sophisticated, athermal designs, such as that described by Sarver et al.[72] and involving a conical interface, might be used with larger mirrors since the mirrors usually are more flexible and, hence, more susceptible to gravitational effects.

Double-arch lightweighted mirrors are typically mounted on three or more supports attached to the back surface of the substrate at its thickest point. Figure 7.24 due to Iraninjad et al.[73] shows such a design. This design was created to support a 20 in. (50.8 cm) diameter double-arch mirror with three equally spaced clamp and flexure assemblies oriented so the flexures were compliant in the radial direction, but stiff in all other directions. This allowed the aluminum mounting plate to contract differentially with respect to the fused silica mirror as the temperature was lowered to about 10 K. Each clamp was a "tee"-shaped Invar 36 piece that engaged a conical hole in the strong annulus of the mirror's back surface. These flexures were 91 mm (3.6 in.) long by 15 mm (0.6 in.) wide, twin parallel blades made of 0.04 in. (1.0 mm) thick 6Al4V ELI titanium. The blades were separated by 25 mm (1 in.).

This mount design was analyzed extensively and found to provide acceptable thermal performance and to withstand launch loads typical of the Space Shuttle as well as to survive (with damage) a crash landing of the Shuttle.

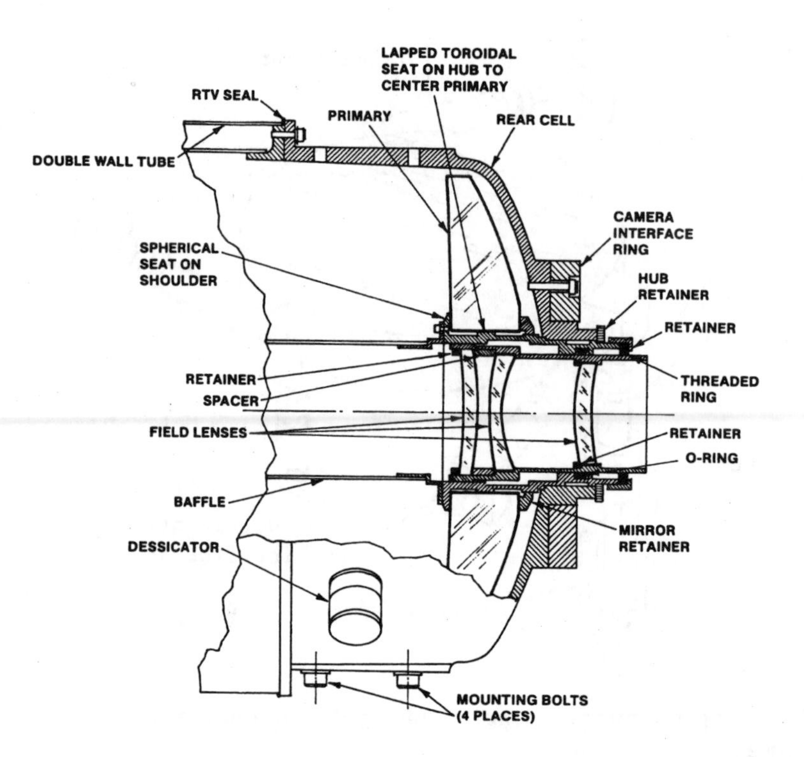

Fig. 7.23 Sectional view of the rear cell portion of a catadioptric lens featuring a hub-mounted 15 in. (38.1 cm) aperture mirror. (From Yoder[8] by courtesy of Marcel Dekker, Inc.)

7.6 Mounting metal mirrors

Small- and moderate-sized metal mirrors can often be mounted using the same techniques discussed above for non-metallic mirrors if there are no unusual requirements inherent in the application such as extreme temperatures (e.g., cryogenic applications), exposure to high-energy radiation (such as that from lasers or solar simulators), or extreme shock, or vibration. The prime differences between metallic and non-metallic mirror types have to do with differences in key mechanical properties such as density, Young's modulus, Poisson's ratio, thermal conductivity, CTE, and specific heat. We can take advantage of these differences by using metals whose unique properties allow significant improvements in performance, weight, environmental resistance, etc.

The preferred methods for supporting these metal mirrors involve mounting provisions built into the mirrors themselves. We illustrate such a feature in Fig. 7.25 which shows a section through a mirror with machined slots that isolate the mounting ears from the main part of the mirror so forces exerted when bolting the mirror to the mount shown at the bottom of the figure are not transmitted to the optical surface.[74] Figure 7.26 shows a close-up view of the back side of a rectangular metal mirror having the same feature. In this case, it is apparent that the mounting ears have been machined into the mirror by core-cutting parallel to the mirror face at three locations.[74]

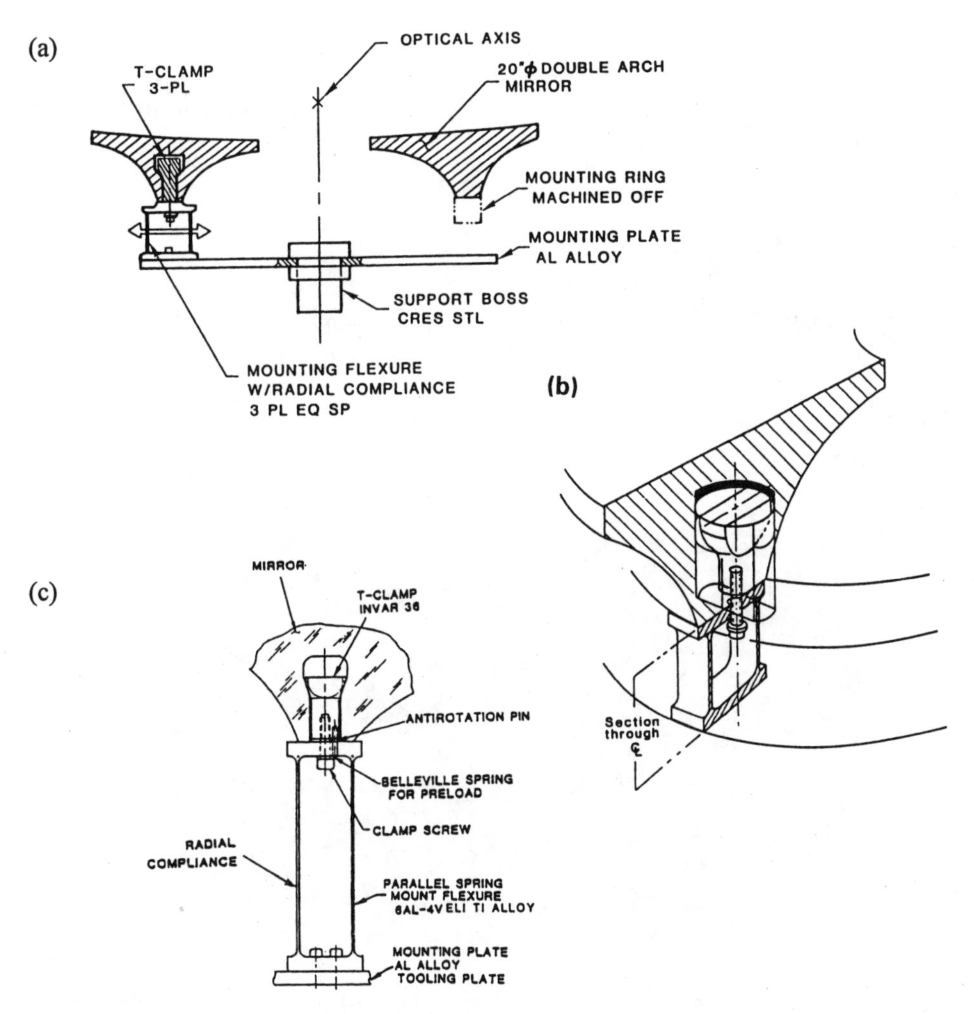

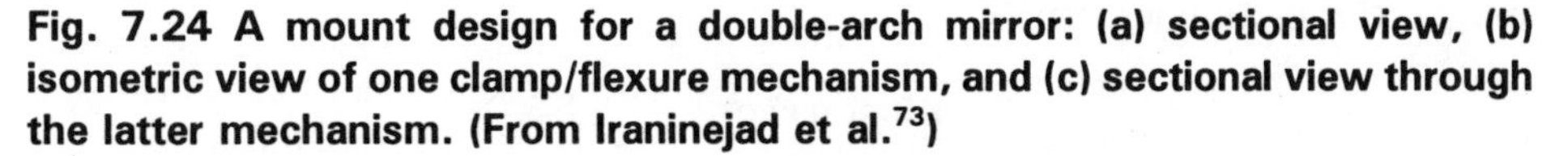

Fig. 7.24 A mount design for a double-arch mirror: (a) sectional view, (b) isometric view of one clamp/flexure mechanism, and (c) sectional view through the latter mechanism. (From Iraninejad et al.[73])

One major advantage of metallic mirrors is their compatibility with single point diamond machining which produces precision surfaces with minimal force exerted by the cutting tool on the surface being machined. This technique also results in accurate relationships between surfaces, especially when they can all be created without removing the part from the machine. When this is not feasible, key mounting surfaces can be machined first and then used as the references for turning the optical surfaces and other mounting interfaces. Figure 7.27 shows a case in point. Here, a circular mirror face is machined on the front side of the mirror blank, the slot for an O-ring seal and a pilot diameter for centration are cut, and, finally, the reference surface -A- is cut. All these operations are done without disturbing the alignment of the part to the spindle axis. Hence, they all have minimum relative errors.[77]

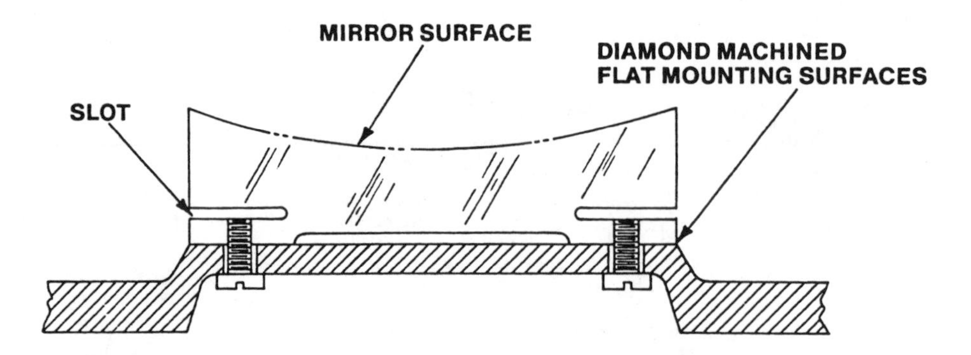

Fig. 7.25 Diagram of a strain-free mounting for a metal mirror. (From Zimmerman[74])

Fig. 7.26 Close-up view of a mounting ear (flexure) machined into a metal mirror. (From Zimmerman[74])

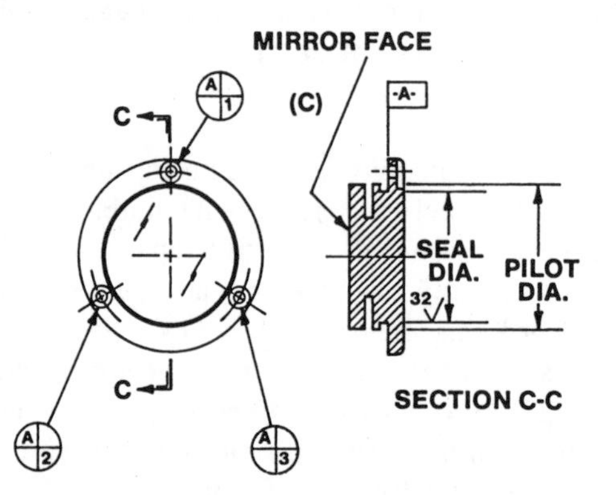

Fig. 7.27 Metal mirror with optical and interface surfaces cut without removing the part from the SPDT machine. (From Addis[75])

Another hardware example showing the advantage of metal mirrors for some applications is the multicomponent telescope shown in Fig. 7.28.[76] The primary diameter was 8 in. (20.3 cm); the entire assembly was made of 6061 aluminum. The individual components were SPDT machined with integral interface surfaces so they fit together to create an assembly with built-in precise alignment. The primary and secondary mirrors were designed to have spherical reference surfaces whose nominal optical centers coincided with the intersection of the axis and the system focal plane. The order of machining operations was chosen so as to allow alignment of the part on the turning machine using the interface and reference surfaces as the metrology features. This maximized accuracy of surface interrelationships even when the part had to be turned over and reattached to the spindle. Once all parts were machined, they all fit together without adjustment to produce the required optical performance. With a single material used in its construction, the system remained aligned at least to liquid nitrogen temperature.

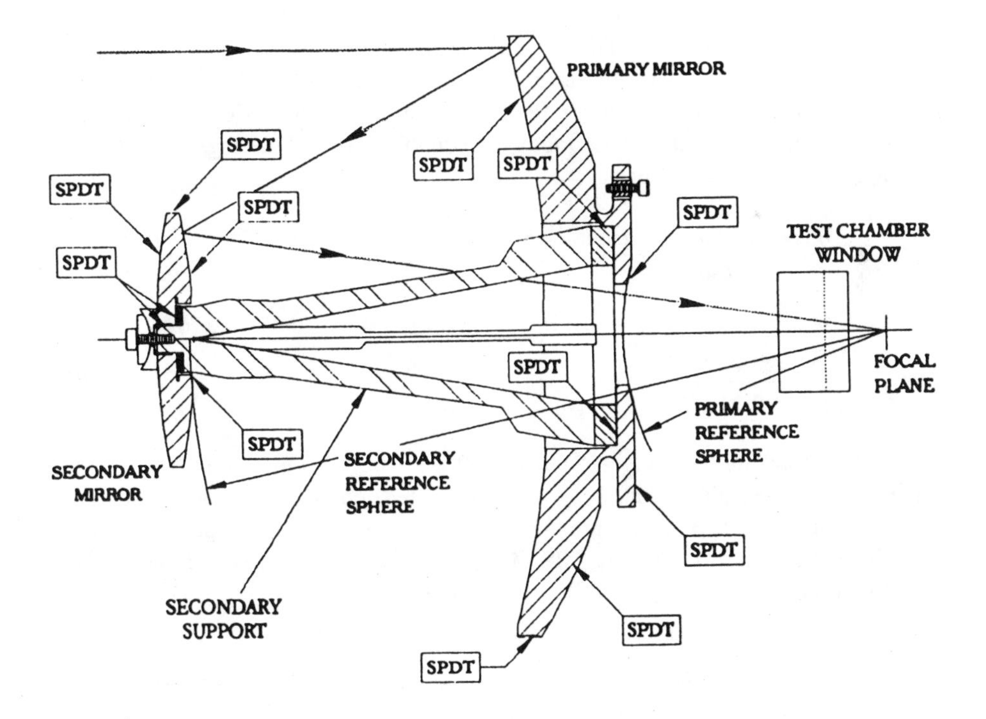

Fig. 7.28 Schematic of all-aluminum telescope system machined for easy and precise assembly by SPDT methods. (From Erickson et al[76])

In Sect. 6.3.3 we discussed a 7.3 in. (18.5 cm) diameter aluminum secondary mirror used in the Kuiper Airborne Observatory. That mirror, shown in Fig. 6.20, has a central mechanical interface with three coplanar pads and a central pilot diameter stud machined to facilitate mounting. Figure 6.21 shows the mirror, in section, attached to the scanning mechanism. Attachment is with three stainless steel bolts using Belleville washers to accommodate differential expansion. Minimal localized distortion of the optical surface that occurs due to clamping forces lies within the central obscuration of the system and thus causes no problem.

7.7 Gravitational effects on small mirrors

So far in this chapter we have ignored the effects of external forces such as gravity and operational accelerations on mirror surface figure. When the aperture is modest, the thickness and material choice conducive to stiffness, and the performance requirements not too high, the optic can be considered a rigid body and mounted semi-kinematically or even non-kinematically without performance problems. If, however, these attributes do not prevail, we must consider the effects of external forces. Gravity is the most prevalent so we will limit our discussion to that force. A special case is gravity release in space and the related problems of making and mounting the mirror so as not to be disturbed when gravity is missing.

The largest gravitational disturbances occur when the mirror axis is vertical. How the mirror is held then affects the magnitude of surface deformation and the resulting surface contour. Using Roark's theory for simple, unclamped plate flexure under uniform gravitational load,[45] we can use the following equations to find the sags, Δy_C and Δy_R of circular and rectangular mirrors:

$$\Delta y_C = \frac{3W(m-1)(5m+1)(a^2)}{16\pi E m^2 t^3}, \tag{7.5}$$

$$\Delta y_R = \frac{0.1422 w b^4}{E t^3 (1 + 2.21 \alpha^3)}, \tag{7.6}$$

where: W = total mirror weight, w = weight/area, m = 1/Poisson's ratio, a = semi-diameter or longest dimension, b = shortest dimension, E = Young's modulus, t = thickness, and α = b/a. The sags are measured at mirror center and represent changes in sag if the mirror is not flat. Figure 7.29 shows the geometry. Numerical Example No. 33 should be useful in assessing typical sag magnitudes.

Numerical Example No. 33: Gravity-induced sags of axis-vertical circular and rectangular mirrors.

(a) Assume a 20 in. diameter flat fused silica mirror with D/t = 6 is uniformly supported around its rim with axis vertical. Calculate its sag in waves of green light ($\lambda = 22\times10^{-6}$ in.).

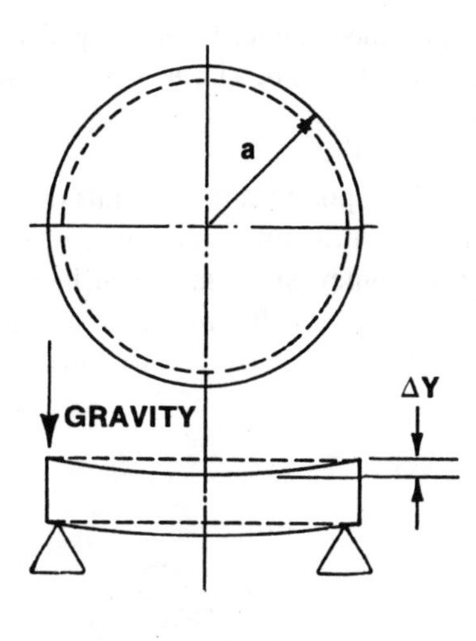

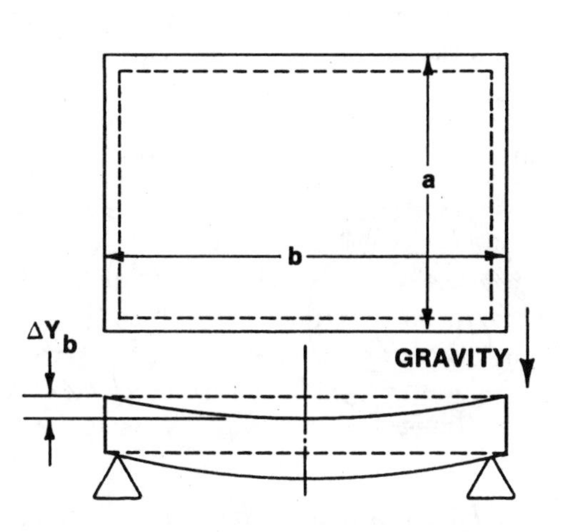

Fig. 7.29 Geometry of rim-supported circular and rectangular mirrors

From Table C5, $\rho = 0.0796$ lb/in.^{3}, $\upsilon = 0.167$, $E = 10.6 \times 10^6$ lb/in.^{2}

$$W = (\pi)(10^2)(20/6)(0.0796) = 83.3 \text{ lb}$$

From Eq. (7.5),

$$\Delta y_C = (3)(83.3)(5.988 - 1)((5)(5.988) + 1)(10^2) / (16\pi)(10.6 \times 10^6)(5.988^2)(3.333^3)$$
$$= 5.5 \times 10^{-6} \text{ in.} = 0.25\ \lambda_{GREEN}$$

(b) Change the above mirror to rectangular shape with a = 20 in. and b = 12.5 in.

$$\alpha = 12.5/20 = 0.625$$
$$W = (20)(12.5)(20/6)(0.0796) = 66.3 \text{ lb}, \ w = 66.3/(20)(12.5) = 0.265 \text{ lb/in.}^2$$

From Eq. (7.4),

$$\Delta y_R = (0.1422)(0.265)(12.5^4) / (10.6 \times 10^6)(20/6)^3(1 + (2.21)(0.625^3))$$
$$= 1.5 \times 10^{-6} = 0.07\ \lambda \text{ green}$$

We see that the circular mirror sags the most and that neither mirror mount can be considered for very high-performance applications.

If the mirrors are nominally flat, we would expect that contour lines of equal sag change on the deflected circular mirror's reflecting surface would be circles while those for the rectangular mirror would be generally rectangular.

If the same circular mirror considered above were supported at three points rather than all around the rim and if those points were located at different radial zones of the mirror aperture, the shapes of the surfaces would appear somewhat as indicated in Fig. 7.30. The support points are indicated by crosses. Although originally drawn for a large perforated mirror, the same patterns would be expected in smaller mirrors; only the scale would change.

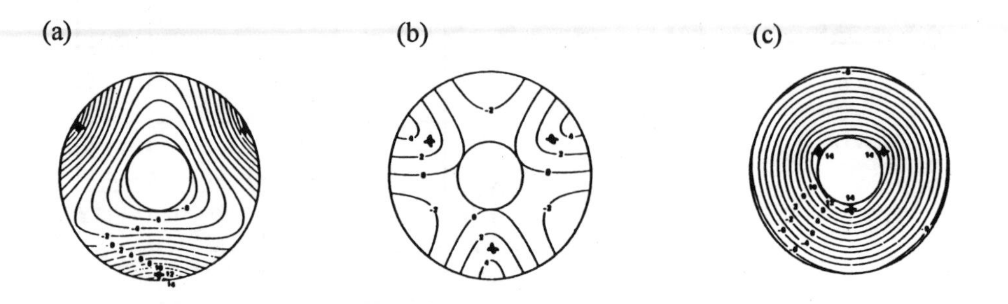

Fig. 7.30 Contour patterns for a circular mirror supported on three points at different zonal radii: (a) at 96%, (b) at 73%, and (c) at 38%. (From Malvick and Pearson[77])

The same general effect would be expected if the mirror were rectangular, but the changes in contour would modify the rectangular contour lines applicable when rim-supported. Adding axial support points would improve the symmetry and reduce the magnitude of the surface distortion for any mirror. Multipoint (Hindle) mounts[78,79] with six, nine, eighteen, twenty-seven, etc., support points acting through three or more "whiffle-trees" would typically be used, although more than nine points probably would be unlikely for the mirrors under 24 in. diameter considered here.

When a mirror is supported with its axis horizontal, the magnitude of gravity-induced distortion of the optical surface is reduced. The change in surface contour is, however, always non-symmetrical. This was explained by Schwesinger in 1954[80] using the geometry of Fig. 7.31. Radially directed forces varying in magnitude around the rim of the mirror and distributed approximately equally axially support the weight of each volume element of the substrate. If the mirror has one or both surfaces curved, elemental bending moments are produced. The resultant of all these moments is proportional to the mirror weight and the distance from the CG to the mirror mid-plane (dashed line). This resultant moment causes the surface to deform. Schwesinger expressed the rms figure error from a true sphere in the form of Eq. (7.5).

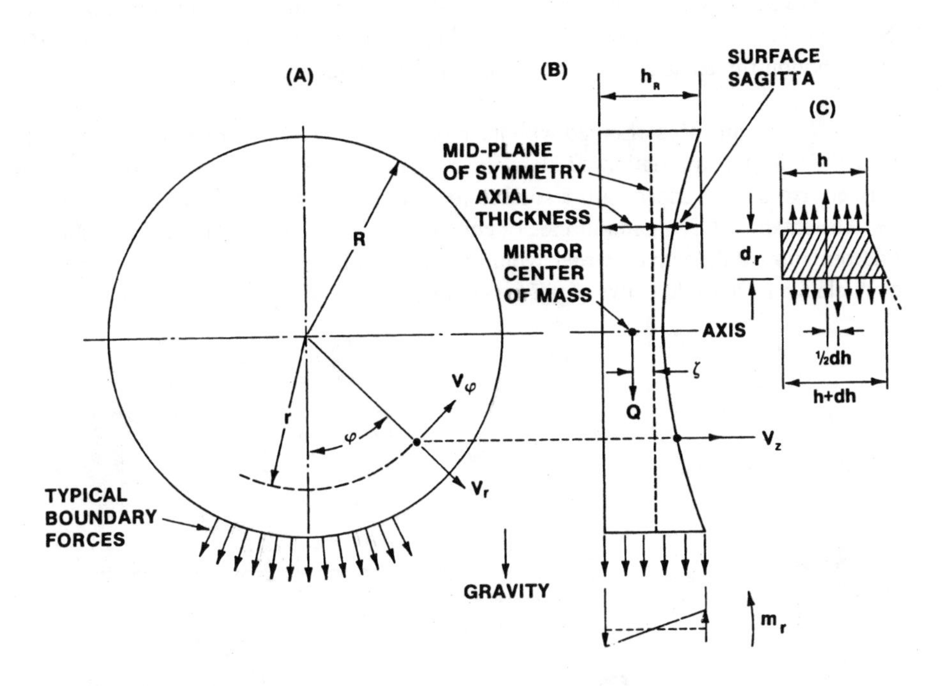

Fig. 7.31 Geometry of gravitational mirror bending with axis horizontal. (From Schwesinger[80])

$$\Delta_S = C_k(2\gamma a^2/E\lambda), \tag{7.5}$$

where C_k is a computed constant given where C_k is a computed constant given by Schwesinger for each of six types of mounts used to support a mirror radially, γ = weight/volume, a = semi-diameter, E = Young's modulus, and λ is wavelength.

If an axis-horizontal mirror is supported on two parallel horizontal posts as in the mount of Fig. 7.32, the value of C_k of Eq. (7.5) is 0.0548 for a flat mirror, 0.0832 for one with $K = sag/t_A = 0.1$, and 0.1152 for one with K = 0.2. The surface deformation thus grows with curvature of the mirror as well as with diameter. Equation (7.5) can be used to estimate deformation or, conversely, to find the maximum mirror diameter or curvature (i.e., EFL) that will suffer a given deformation.

When a mirror is supported on edge on two posts, the surface deforms in a particular manner depending upon the locations of the posts relative to the vertical centerline. Figure 7.33 shows typical contours for a mirror supported on posts at $\pm 30°$ and $\pm 45°$ from vertical.[83] In both cases, the mirror becomes astigmatic. Radial supports for the mirror are not shown in the figure.

The commercial mount shown in Fig. 7.32 can be used with mirrors in the 90 to 250 mm (3.5 to 9.8 in.) diameter range. Smaller mirrors can be mounted in mounts such as Newport Corporation's Model MM2-1A which has two Delrin rods for the mirror to rest on and a Nylon-tipped setscrew at top to apply radial preload.

When a mirror is oriented with its axis at an arbitrary elevation angle, the surface deformation is a combination of the axis-horizontal and axis-vertical conditions just discussed. No simple means are available to determine actual surface contours for this general orientation. FEA methods can, of course, be used. It may be sufficient to show that the deformation at the two extremes would be acceptable. It then would be assumed to be acceptable for an attitude between these extremes.

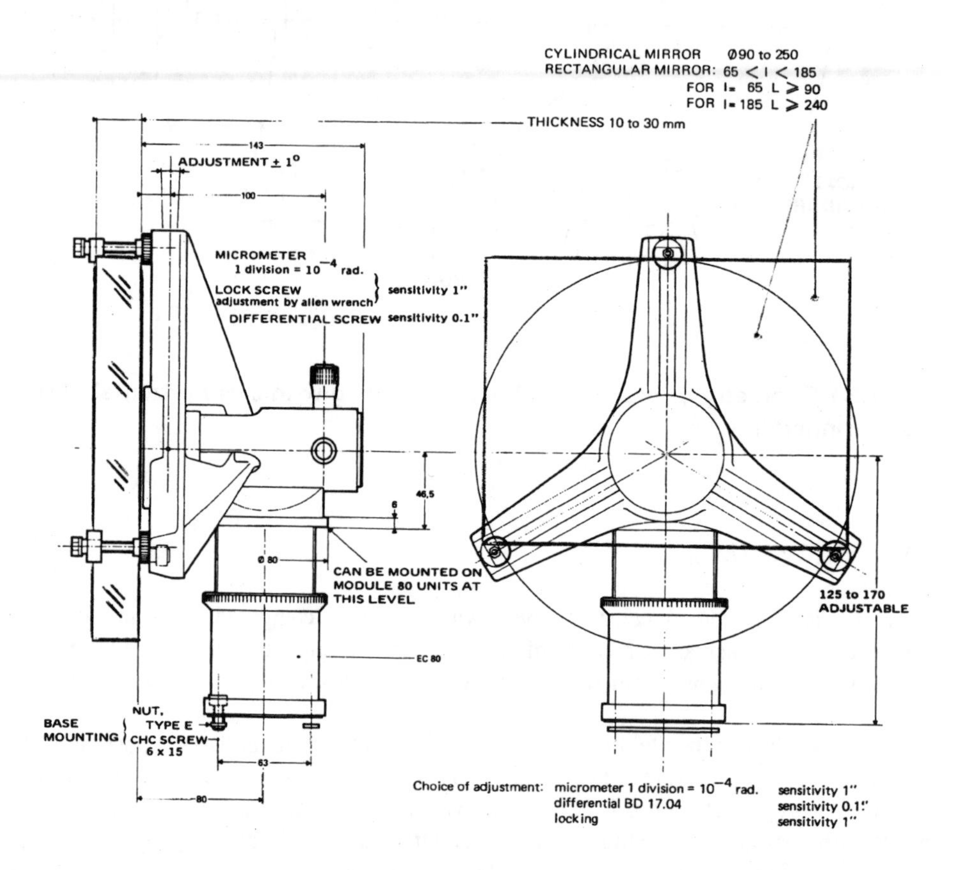

Fig. 7.32 Commercial mirror mount with two horizontal post support. (Courtesy of Newport Corporation)

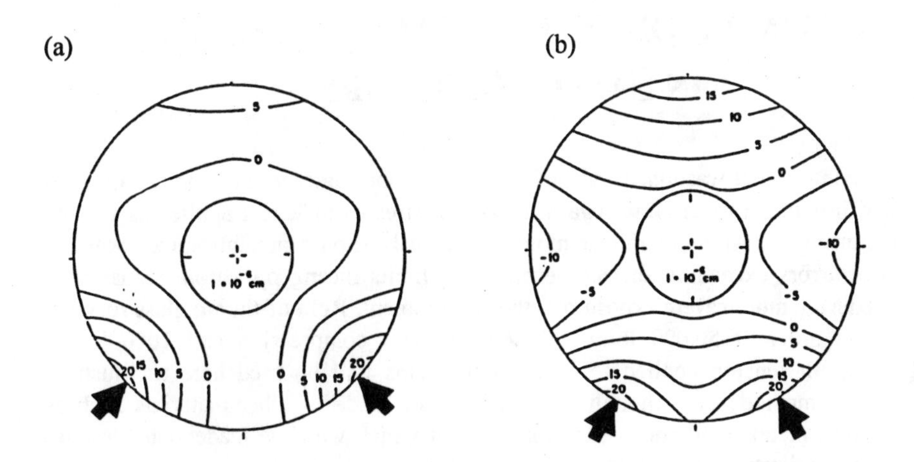

Fig. 7.33 Surface contours for a mirror with axis horizontal supported on horizontal posts at (a) $\pm 30°$ and (b) $\pm 45°$ from vertical. (From Malvick[81])

CHAPTER 8
ESTIMATION OF CONTACT STRESSES IN SMALL MIRRORS

Much of what was discussed in Ch. 5 of this book in regard to contact stress in prisms and in Ch. 5 of Ref. 1 in regard to contact stresses in lenses applies as well to mirror mountings. Designs of mirror mounts obviously should not introduce localized stress into mirrors exceeding surface deformation limits during operation nor damage thresholds during more severe exposure to the environment. Rule-of-thumb glass survival limits of the order of 50,000 lb/in.^{2} (3.4×10^8 Pa) for compression and 1000 lb/in.^{2} (6.9×10^6 Pa) for tension applied to lenses and prisms will be used here to illustrate computational methods even though most mirrors are made of other materials such as glass-ceramics, metals, and composites for some of which we have inadequate strength information. Similarly, we will consider stress build up in the mechanical parts that hold the mirrors in place.

In this chapter, we summarize the use of equations derived elsewhere[8] that convert mounting forces and geometrical relationships into material stresses. First, we consider compressive stresses in axially clamped mirrors, then radial stresses in mirrors that fit closely radially into their mounts, and then bending stresses due to improperly placed contact regions on mirrors. The final topic considered here (briefly) is the development of stresses in bonded mirrors.

8.1 Compressive stress in axially clamped mirrors

In Section 7.1 we discussed typical techniques for applying axial preload to a mirror to constrain it against mount features such as shoulders or locating hard points (pads or lugs). In these mounting concepts, preload is applied either all around the edge of the aperture or, preferably, at three or more localized areas distributed more or less symmetrically around that edge. Threaded retainers and continuous flanges (see Figs. 7.1 and 7.2) usually contact polished surfaces or bevels in lines which actually are narrow areas formed when the elastic materials deflect under compressive force. Axial contact stress is a maximum at the centers (laterally) of these areas. Assuming uniform distribution of preload along the contact "line" (i.e., around the aperture), the stress is uniform in this longitudinal direction. The stress at other locations within the mirror decreases with depth or lateral distance from the center of the contact area. Preload applied through cantilevered clips or some other form of springs (see Figs. 7.3 through 7.8) generates stress within the contact areas which may be lines or small areas. We first consider line contacts around the aperture, then local area contacts.

8.1.1 Line contact around the mirror edge

The axial contact stress S_A at a bevel or on the surface of a mirror preloaded at a height y_C from the axis can be estimated from the following equation adapted[8] from Roark[45]:

$$S_A = 0.798(K_1 p/K_2)^{1/2}, \tag{8.1}$$

where K_1 depends on the optomechanical interface design and the surface radius, K_2 depends on the elastic properties of the lens and mount materials (assumed here, for convenience, to be glass and metal), and p is the linear preload as determined from the total preload P by:

$$p = P/2\pi y_C. \tag{8.2}$$

The term K_1 will be discussed below in conjunction with various interface types. For all interface types, the term K_2 is given by Eq. (5.7) which is repeated here for convenience:

$$K_2 = [(1-\upsilon_G^2)/E_G] + [(1-\upsilon_M^2)/E_M], \tag{5.7}$$

where υ_G, E_G, υ_M, and E_M are Poisson's ratio and Young's modulus values for the glass and metal, respectively.

Figures 8.1 and 8.2 show geometric models that form the basis for Eq. (8.1) when the contact is at a spherical mirror surface or at a flat bevel on the edge of a mirror. The size of the narrow contact area in either case depends on the same parameters as the stress. Under light preload, the contact is essentially a "line" of length $2\pi y_C$. As the preload increases, the line contact widens due to elastic deformation and the resulting area is computed as $2\pi y_C \Delta y$, where Δy is the annular width of the deformed area and y_C is measured to the center of this area. The equation for Δy as adapted[8] from Roark[45] is

$$\Delta y = 1.6(K_2 p/K_1)^{1/2}, \tag{8.3}$$

where all terms except K_1 are as defined above.

The total preload P exerted upon a narrow "line" contact area divided by that area is the average contact stress, which we will here call S_{AVG}. The value of S_A calculated by Eq. (8.1) is the peak stress that occurs at the geometric center of the area. If we let S_A = (K)(S_{AVG}), it can easily be shown that K = (0.798)(1.6) and the peak axial contact stress is then always 1.277 times the average value.

Again applying equations from Roark[45], it can be shown that the value of K_1 in Eqs. (8.1) and (8.3) for any optomechanical interface is given by

$$K_1 = (D_1 \pm D_2)/D_1 D_2. \tag{8.4}$$

The "+" sign in this equation is used for convex lens surfaces and the "−" sign is used for concave lens surfaces. The term K_1 is always assigned a positive sign. Both D_1 and D_2 vary with design type. Contact on a convex or concave mirror surface makes D_1 = 2 (surface radius) whereas contact on a flat bevel makes D_1 = infinity. The term D_2

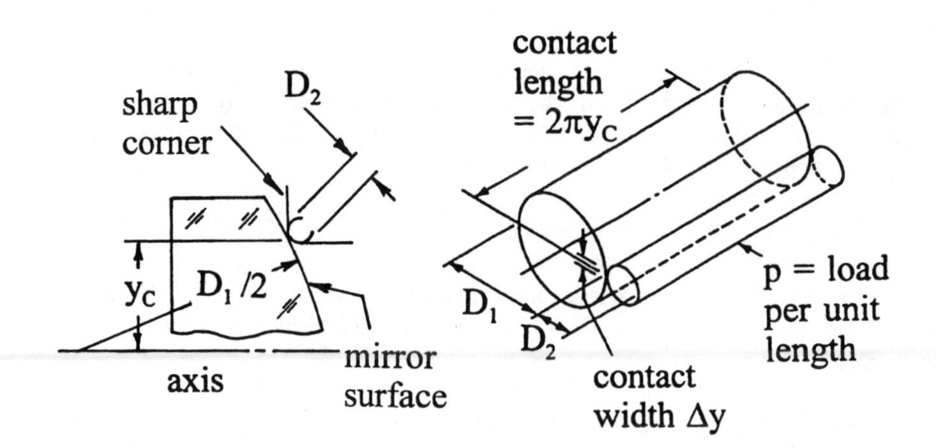

Fig. 8.1 Analytical model of the sharp corner interface on a convex spherical mirror surface

always is equal to twice the corner radius of the associated mechanical interface.

If the contacting part has a convex sharp corner burnished according to good machining practice during manufacture, we may assume $D_2 = 0.004$ in. (0.1 mm) per Delgado and Hallinan[82]. Then $K_{1\,SC} = (D_1 \pm 0.004)/0.004D_1$ (for US Customary units) or $(D_1 \pm 0.1)/0.1D_1$ (for SI units). For convex or concave surface radii larger than 0.2 in. (5.08 mm), the term D_2 can be ignored and the value of $K_{1\,SC}$ is constant[8] at 250/in. (10/mm). The error due to this approximation does not exceed 2%.

Numerical Example No. 34: Axial contact stress in a mirror with sharp corner interface on a convex reflecting surface.

Consider a convex fused silica mirror with the following dimensions: $D_G = 3.1$ in. (78.740 mm) and $R_1 = 18.0$ in. (1828.8 mm). The mirror lens is mounted in a 6061 aluminum cell with a sharp corner axial interface at $y_C = 1.5$ in. (38.1 mm) on R_1. What contact stress is developed at the interface if the total preload is 20 lb (88.964 N)? What is S_{AVG} and how does it compare to S_A?

From Table C5, $E_G = 1.06\times10^7$ lb/in.^{2} (7.3×10^4 MPa) and $\upsilon_G = 0.164$.
From Table C12, $E_M = 9.9\times10^6$ lb/in.^{2} (6.82×10^4 MPa) and $\upsilon_M = 0.332$.
From Eq. (8.2), $p = 20/(2\pi)(1.5) = 2.122$ lb/in. (371.617 N/m).

Using Eq. (5.7),

$$K_2 = [(1 - 0.164^2)/1.06\times10^7] + [(1 - 0.332^2)/9.9\times10^6] = 9.180\times10^{-8} + 8.988\times10^{-8}$$
$$= 1.817\times10^{-7} \text{ in.}^2/\text{lb } (2.635\times10^{-11} \text{ m}^2/\text{N}).$$

Since R_1 exceeds 0.2 in. (5.08 mm), we know that $K_{1\ SC}$ = 250/in. (10/mm).

From Eq. (8.1),

$$S_{A\ SC} = 0.798[(250)(2.122)/1.817\times10^{-7}]^{1/2} = 43{,}119 \text{ lb/in.}^2 \text{ (297.3MPa)}.$$

This stress level approaches the rule-of-thumb tolerance of 50,000 lb/in.^{2} (345 MPa), so the design is of questionable validity. A more precise estimate of the risk of breakage using statistical methods[36] is advised. Alternatively, another type of glass-to-metal interface with lower stress-generating characteristics could be employed.

From Eq. (8.3), $\Delta y = 1.6[(1.817\times10^{-7})(2.122)/250]^{1/2} = 6.283\times10^{-5}$ in. (1.596×10^{-3} mm),

Then, $S_{AVG} = 20 / (2\pi)(1.5)(6.283\times10^{-5}) = 33{,}775$ lb/in.^{2} (232.9 MPa).

The ratio of $S_{A\ SC}$ to S_{AVG} is 43,119/33,775 = 1.277 as predicted above.

When the mirror surface contacted by the mechanical constraint is convex, the stress for a given preload can be reduced by giving the mechanical interface a conical shape. The resulting tangential interface is depicted in the model of Fig. 8.2. The term D_2 of Eq. (8.4) is then infinite for the section through a conical interface. The value of K_1 therefore becomes:

$$K_{1\,TAN} = 1/D_1 = 0.5/R_1, \tag{8.5}$$

where R_1 is the mirror surface radius.

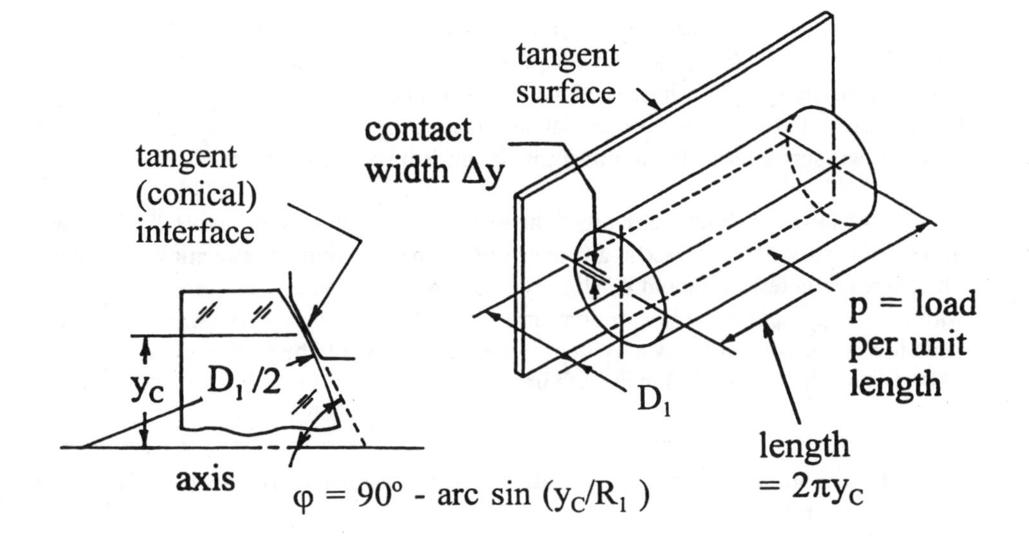

Fig. 8.2 Analytical model of the tangent interface on a convex spherical mirror surface

The axial contact stress developed in a mirror of given surface radius R_1 by a given preload with a tangential interface is smaller by a factor of about $(500R_1)^{1/2}$ than that with a sharp corner interface.

Numerical Example No. 35: Axial contact stress in a mirror with a tangent interface on a convex reflecting surface.
Repeat the last example to see what happens to S_A if the interface is changed to the tangential configuration.

Now, $K_{1\ TAN} = 0.5/18.0 = 0.0278/\text{in.}\ (0.0011/\text{mm})$.

By Eq. (8.1), $S_{A\ TAN} = 0.798[(0.0278)(2.122)/1.817\times10^{-7}]^{1/2} = 455\ \text{lb/in.}^2\ (3.1\ \text{MPa})$.

In comparison with the result of the stress calculation in Numerical Example No. 34, the stress with the tangential interface is found to be significantly reduced. We would expect the ratio $S_{A\ SC}/S_{A\ TAN}$ to be $[(500)(18)]^{1/2} = 94.9$. Not too surprisingly, this is what we obtain from 43,119/455.

Referring again to Fig. 8.1, we see that the corner of the retainer could be rounded with a larger radius than the sharp corner shown. We call this mechanical surface a toroidal (or donut-shaped) mechanical interface. As shown, it contacts a spherical mirror surface. A similar concept could be developed for contact on a concave surface. The term K_1 for interfaces on convex or concave surfaces is given by Eq. (8.4) with D_1 set equal to twice the mirror surface radius and D_2 set equal to twice the sectional radius, R_T, of the toroid. Remember to use the "+" sign with a convex mirror surface and the "-" sign with a concave surface.

Mechanical toroids contacting mirror surfaces are almost always convex, although a concave toroid *could* be used on a convex reflecting surface. The limiting case then would be matching radii, which is equivalent to a spherical interface. The infinite radius toroid case is the same as a tangential interface. Only a convex toroid can contact a concave lens surface, and the limiting case again is a spherical interface.

As has been shown for lenses,[8] the contact stress values using a toroidal interface instead of a tangential interface at a convex mirror surface would be essentially the same. The tangential interface should be slightly less expensive, hence is preferred. The toroid with $R_T = -R_{MIRROR}/2$ is the best option for a concave mirror surface; the stress is then the same low value as would prevail if a tangential interface could be used on such a surface. This is confirmed in the following example.

Numerical Example No. 36: Axial contact stress in a mirror with toroidal interface on a concave reflecting surface.
Repeat Numerical Example No. 34 changing R_1 to -18 (concave) and the interface to a toroid of radius $R_T = 9$ in. Note, the negative signs on the radii can be ignored.

Applying Eq. (8.4), $K_1 = (36 - 18) / (36)(18) = 0.0278$/in. (0.0011/mm).

By Eq. (8.1), $S_{A\ TOR} = 0.798[(0.0278)(2.122)/1.817\times10^{-7}]^{1/2} = 455$ lb/in.^{2} (3.1 MPa).

We note that the contact stress is reduced drastically from that obtained with a sharp corner interface and is equal to that with a tangential interface on a convex surface of the same absolute radius.

Figure 8.3 shows a toroidal mechanical interface on a flat bevel of a concave mirror. The contact again is at a height of y_C and is continuous around the edge of the mirror aperture. Equations (7.2) through (7.4) define the flange deflection that produces a particular total axial preload. The stress produced in the mirror can be estimated using Eqs. (8.1) through (8.4). Since D_1 is infinite, K_1 reduces to $1/D_2 = 0.5/R_T$. The value for R_T is chosen as an easily machined value with sufficient curvature that deflection of the thin section of the flange does not cause the edge of the curved pad to touch the bevel under the worst combination of mechanical tolerances. If that were to happen, a sharp corner interface could exist and the stress would go up significantly.

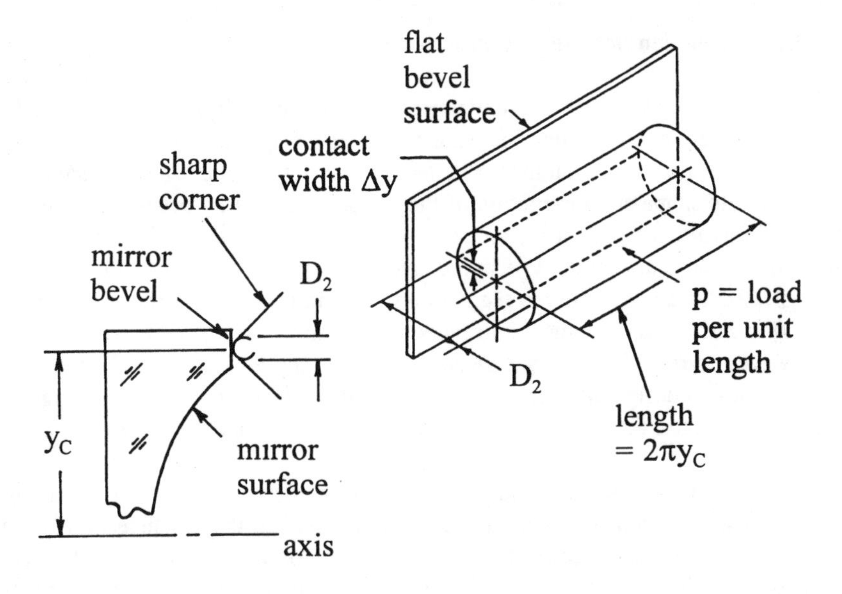

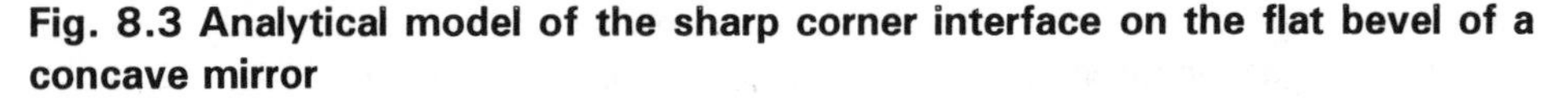

Fig. 8.3 Analytical model of the sharp corner interface on the flat bevel of a concave mirror

---

Numerical Example No. 37: Toroidal contact on a mirror bevel.
Assume that a mirror is to be held in place by a 6061 aluminum flange with toroidal interface on a flat bevel. (a) Calculate the stress in the mirror if $y_C = 7.5$ in., $R_T = 0.5$ in., $P = 380$ lb, and the mirror is made of Zerodur. (b) How does the stress increase if the contact is a sharp corner?

(a) By Eq. (8.2), $p = 380 / (2\pi)(7.5) = 8.064$ lb/in.

From Table C8a, $E_G = 13.6\times10^6$ lb/in.^{2} and $\upsilon_G = 0.24$
From Table C12, $E_M = 10.0\times10^6$ lb/in.^{2} and $\upsilon_M = 0.33$

By Eq. (5.7), $K_2 = [(1 - 0.24^2) / 13.6\times10^6] + [(1 - 0.33^2) / 10.0\times10^6] = 1.58\times10^{-7}$ in.^{2}/lb
From the above text, $K_1 = 0.5 / 0.5 = 1.0$

From Eq. (8.1), $S_{A\ TOR} = 0.798[(1.0)(8.064) / 1.58\times10^{-7}]^{1/2} = 5694$ lb/in.^{2}

This stress is well below the breaking stress for glass so presumably would be acceptable for this Zerodur mirror.

(b) From Eq. (8.4), $K_1 = (\infty + 0.004) / (\infty)(0.004) = 1 / 0.004 = 250$/in.

From Eq. (8.1), $S_{A\ TOR} = 0.798[(250)(8.064) / 1.58\times10^{-7}]^{1/2} = 90{,}140$ lb/in.^{2}

This stress is almost twice the damage threshold for glass so must be judged unacceptable for this application. The toroidal interface is much better.

8.1.2 Local area contacts on a mirror bevel

Figures 7.3 through 7.8 show mirrors constrained by springs with preload applied at several discrete points around the edge of the mirror aperture. Typically three contacts are used and they occur on flat mirror surfaces or on flat bevels. What was said in Section 5.1.1 about the shape and orientation of the surfaces contacting prisms also applies here. One should avoid sharp corner contact (see Fig. 5.1) if at all possible. Flat mechanical contacting surfaces also should be avoided since they can degenerate into sharp corners (see Fig. 5.2). This leaves curved, i.e., cylindrical and spherical, surfaces as the shapes of choice. Figure 5.3 shows such interfaces for prisms while Fig. 7.3 shows one such interface for a mirror. The design concept of Fig. 7.7 applies force to a flat mirror at three flat pads, but this design allows the pads to align themselves to the mirror surface thereby avoiding corner contact.

Spring clips such as are used in the design of Fig. 7.3 usually are designed as rectangular cantilevered beams of free length, L, equal to the distance from the nearest edge of the screw, washer, or post to the contact area on the mirror. A safety factor of at least 2 beyond the force needed to overcome worst-case anticipated dynamic forces is frequently employed to place an upper limit on preload so as not to overly stress the mirror. The equation given in Numerical Example No. 29 ($P_i = wa_Gf_S/N$) may be used to compute the preload force, P_i, required of any one of the N clips when a mirror of given weight is to be constrained against a given acceleration a_G with a safety factor f_S. Equation (5.1) for deflection, Δy, of one of the cantilevered springs to produce a specific load against a prism is repeated below for convenience in applying it to mirrors. Equation (5.2) for the tilt angle (in radians) of the spring also is repeated here.

$$\Delta y = (1 - \upsilon_M^2)(4L^3 P_i)/E_M b h^3, \tag{5.1}$$

where L is the free (cantilevered) length of the spring (in mm or in.)
P_i is the required force per spring (in N or lb)
E_M is Young's modulus for the spring material (in N/mm^2 or lb/in.^{2})
υ_M is Poisson's ratio for the spring material
b is the width of the spring (in mm or in.)
h is the thickness of the spring (in mm or in.).

$$\phi = (1 - \upsilon_M^2)(6L^2 P_i)/E_M b h^3. \tag{5.2}$$

The following example illustrates the use of these equations in mirror mount design.

Numerical Example No. 38: Cantilevered spring mirror mount.
Assume a mirror weighing 1.25 lb (0.567 kg) is to be constrained at a_G = 10 times gravity by three titanium springs touching the top surface of the mirror. Assume dimensions as follows: L = 0.625 in. (15.875 mm), b = 0.3185 in. (8.090 mm), and h = 0.035 in. (0.889 mm). Use the model of Fig. 7.3 (cylindrical pad touching the optical surface) and Eqs. (4.1) and (5.1) to find out how much each spring should be deflected from its relaxed condition in order to provide the proper constraining force. Applying Eq. (5.2), what is the angle of the spring end?

From Table C12, E_M = 16.5×10^6 lb/in.^{2} (1.14×10^5 Pa) and υ_M = 0.34
By Eq. (4.1): P_i = (1.25)(10) / 3 = 4.167 lb (18.535 N)
By Eq. (5.1): Δy = (1 - 0.34^2)(4)(0.625)^3(4.167) / (1.65×10^7)(0.3185)(0.035)^3
= 0.0160 in. (0.406 mm)
By Eq. (5.2): ϕ = (6)(0.625)^2(4.167) / (1.65×10^7)(0.3185)(0.035)^3
= 0.0434 radian =2.486°

The peak and average stresses generated at each short line contact on the top surface of the mirror in Fig. 8.3 are estimated using Eqs. (5.6) through (5.9). These equations are repeated below.

$$S_{Pcyl} = 0.564(p/K_2 R_{cyl})^{1/2}, \tag{5.6}$$

where: p is the linear preload = P_i/b (in N/mm or lb/in.)
b is the width of contact (in mm or in.)
R_{cyl} is the cylindrical radius of the surface contacting the prism (in mm or in.)
K_2 is given by Eq. (5.7).

where: υ_G and E_G are Poisson's ratio and Young's modulus for the glass (prism)
υ_M and E_M are Poisson's ratio and Young's modulus for the metal (pad).

$$K_2 = [(1 - \upsilon_G^2)/E_G] + [(1 - \upsilon_M^2)/E_M], \tag{5.7}$$

Since the mirror and the cylindrical pad are assumed to be elastic in this analysis, the force that presses them together causes each material to deform slightly along the line of contact. The deformed region is a rectangle of size b by Δx as shown in Fig. 5.5. Equation (5.8) is used to calculate Δx.

$$\Delta x = 2.263(K_2 p R_{cyl})^{1/2}. \tag{5.8}$$

If we divide the total contact preload by the contact area we obtain the average stress in the contact region as indicated in Eq. (5.9).

$$S_{Pavg} = P_i / b\Delta x. \tag{5.9}$$

The value of $S_{P\,cyl}$ derived from Eq. (5.6) is the peak stress along the line parallel to b at the center of the rectangular contact area. It can be shown analytically by dividing Eq. (5.6) by Eq. (5.9) that the peak stress is always 1.277 times the average stress.

Numerical Example No. 39: Contact stress in a flat mirror with spring-loaded cylindrical interfaces.

Assume that the spring force, P_i, = 4.167 lb (18.535 N) in Numerical Example No. 38 is applied through a cylindrical pad to a "line" contact of length 0.637 in. (15.875 mm) on the top of a ULE 7971 mirror. Let the spring and integral pad be made of beryllium copper. Assume, in turn, that the radius of the pad is 0.1, 1, 10, and 100 in. Calculate $S_{P\,cyl}$ and $S_{P\,avg}$ for each case. What conclusions can be drawn from these calculations?

R_{CYL}	given	(in.)	0.1	1.0	10	100
υ_G	Table C8a		0.176	0.176	0.176	0.176
E_G	Table C8a	(lb/in.^{2})	9.8×10^6	9.8×10^6	9.8×10^6	9.8×10^6
υ_M	Assumed		0.35	0.35	0.35	0.35
E_M	Table C12	(lb/in.^{2})	1.85×10^6	1.85×10^6	1.85×10^6	1.85×10^6
K_2	Eq. (5.7)	(in.^{2}/lb)	5.732×10^{-7}	5.732×10^{-7}	5.732×10^{-7}	5.732×10^{-7}
P_i	given	(lb)	4.167	4.167	4.167	4.167
p	P_i/b	(lb/in.)	6.542	6.542	6.542	6.542
$S_{P\,cyl}$	Eq. (5.6)	(lb/in.^{2})	6025	1905	603	190
Δx	Eq. (5.8)	(in.)	1.386×10^{-3}	4.382×10^{-3}	1.386×10^{-2}	4.382×10^{-2}
$S_{P\,avg}$	Eq. (5.9)	(lb/in.^{2})	4720	1492	472	149
$S_{P\,cyl} / S_{P\,avg}$			1.28	1.28	1.28	1.28

Conclusions: (1) longer R_{cyl} give lower stresses, (2) stresses shown here do not pose any damage problems, and (3) the ratios of $S_{P\,cyl}$ to $S_{P\,avg}$ are as indicated in the text.

Note that the spring deflection required to produce the specified preload is not affected by change in pad radius.

Examination of Eq. (5.6) shows that an increase in R_{cyl} from one value to another would reduce $S_{P\,cyl}$ by $[(R_{cyl})_2 / (R_{cyl})_1]^{1/2}$. The changes in Numerical Example No. 39 were by factors of 10 so we would expect the corresponding changes in $S_{P\,cyl}$ to be by factors of $10^{1/2} = 3.162$ and this is so. If we extrapolate those changes in the direction of reducing the cylindrical radius further, we find that for R_{cyl} = 0.001 in., S_{cyl} would increase to 19,053 lb/in.^{2} which would still provide a safety factor of 2.6 in comparison with the rule-of-thumb compression tolerance of 50,000 lb/in.^{2}. Hence, "line" contact such as is depicted in Figs. 5.1(a) and 5.1(b) would not be dangerous at the preload and geometry used here.

As in the case of prism mountings considered earlier, if the curved interface with the flat mirror surface were to be changed to a spherical contour and all other design features were to remain constant, the following equation would give the contact stress in the glass region surrounding the "point" contact.[45]

$$S_{Psph} = 0.578[P_i / (R_{sph}^2 K_2^2)]^{1/3}. \tag{5.10}$$

The radius, r_{sph}, of the elastically deformed circular contact region is given by Eq. (5.11) while the average stress in this region is as given by Eq. (5.12). The ratio of peak stress to average stress can be shown to be 1.5.

$$r_{sph} = 0.908(P_i R_{sph} K_2)^{1/3} \tag{5.11}$$

$$S_{Pavg} = P_i / \pi r_{sph}^2, \tag{5.12}$$

where all terms are as defined above.

Numerical Example No. 40: Contact stress in a mirror with spring-loaded spherical interface.

Repeat the last numerical example assuming spherical interfaces with R_{sph} = 0.1, 1.0, and 10.0 in. What conclusions can be drawn?

R_{sph}	given	(in.)	0.1	1.0	10.0
υ_G	Table C8a		0.176	0.176	0.176
E_G	Table C1	(lb/in.^{2})	9.8×10^6	9.8×10^6	9.8×10^6
υ_M	Assumed		0.35	0.35	0.35
E_M	Table C12	(lb/in.^{2})	1.85×10^6	1.85×10^6	1.85×10^6
K_2	Eq. (5.7)	(in.^{2}/lb)	5.732×10^{-7}	5.732×10^{-7}	5.732×10^{-7}
P_i	given	(lb)	4.167	4.167	4.167
$S_{P\,sph}$	Eq. (5.10)	(lb/in.^{2})	62,559	13,478	2904
r_{sph}	Eq. (5.11)	(in.)	0.0056	0.0121	0.0261
$S_{P\,avg}$	Eq. (5.12)	(lb/in.^{2})	42,283	9057	1946
$S_{P\,cyl} / S_{P\,avg}$			1.48	1.49	1.49

We see that (1) longer R_{sph} give lower stresses, (2) the contact stresses do not pose severe glass damage problems for R_{sph} greater than about 1.0 in., and (3) the ratios of $S_{P\,sph}$ to

$S_{P\ avg}$ are approximately as indicated in the text.

Note from Eq. (5.10) that changing R_{sph} by some factor, f_R, changes $S_{P\ sph}$ by $f_R^{-3/2}$ so, if the factor is 10 as in Numerical Example No. 40, we would expect the $S_{P\ sph}$ changes from column to column to be $10^{2/3}$ or 4.641. This is borne out by the calculations in the example.

8.1.3 Contact stress at mirror axial locating pads

In many mirror mounting designs, the width, d_P, of each metal locating pad touching the back of the mirror (see Fig. 7.3) is set equal the width, b, of the spring and each pad is circular. In such designs, the localized contact stress, S_{pad}, is estimated from Eq. (5.13). This stress should be compared to the rule-of-thumb glass compressive tolerance of 50,000 lb/in.^{2} to determine what safety factor, if any, exists for that aspect of the design.

$$S_{Pad} = 4P_i / \pi b^2. \tag{5.13}$$

This equation can easily be adjusted to accommodate other pad shapes. Equivalent stress is created in the metal pad, but the metal is more tolerable of compression so this is of little concern.

Numerical Example No. 41: Contact stress in a prism at a locating pad.
Assume that the spring forces in Numerical Example No. 40 are applied normally to three coplanar circular flat areas of diameters = 0.637 in. (16.180 mm) at the bottom of the prism. Assume uniform force distribution and, using data from that example, calculate the average stress, S_{Pad}, in the mirror back surface adjacent to one pad.

From Eq. (5.13), $S_{Pad} = (4)(4.167) / (\pi)(0.637)^2 = 13.1$ lb/in.^{2} (90,300 Pa).

As might be expected, this stress is negligible.

8.1.4 Stress in spring clips

Equation (5.3) defines the bending stress in the cantilevered beam forming the spring clips holding a mirror such as that in Fig. 7.3. This equation is:

$$S_B = 6LP_i / bh^2. \tag{5.3}$$

This stress should not exceed about 50% of the yield stress of the material used in the spring.

Numerical Example No. 42: Stress in a cantilevered spring.
Calculate the nominal bending stress in each of the springs used in Numerical Example No. 39 for the $R_{CYL} = 0.1$ in. case. Assume L = 0.625 in. and h = 0.040 in.

From the example: b = 0.637 in. and P_i = 4.167 lb.

From Eq. (5.3), $S_P = (18)(0.625)(4.167) / (0.637)(0.040^2) = 45{,}996$ lb/in.^{2}

From Table C12, the minimum yield strength of BeCu is 155,000 lb/in.^{2}. A very comfortable safety factor of 155,000 / 45,996 = 3.4 exists.

8.2 Radial stresses in rim-mounted mirrors

We know that changes in temperature cause differential expansion or contraction of mirrors and mount materials in both the axial and radial directions. In the following discussion of the effects of changes in radial dimensions, we assume:

- rotational symmetry of the optics and of the pertinent portions of the mount,
- that a small clearance between the mirror OD and the ID of the mount (or of hard radial locating pads, if used),
- that all optomechanical components are at a uniform temperature spatially, but not temporally.

The CTEs of the mirror materials (glass, ceramic, metal, or composite) and of the mount (usually metal) are again defined as α_G and α_M respectively while temperature changes are represented as $\pm\Delta T$.

The CTE of the mount usually exceeds that of a mirror mounted therein; the prime exception would be if Invar were used in the mount and a higher CTE material such as Pyrex were used in the mirror. In the usual case, a drop in temperature will cause the mount to contract radially towards the mirror's rim. Any radial clearance between these components will decrease in size and, if the temperature falls far enough, the ID of the mount will contact the OD of the mirror. Further temperature decreases will then cause radial force to be exerted upon the rim of the mirror. This force strains the mirror radially and creates radially directed stress. To the degree of approximation applied here, this strain and stress are axially symmetric. If the stress is large enough, the performance of the mirror will be adversely affected. Even larger stresses may cause failure of the optic and/or plastic deformation of the mount.

Temperature increases will, of course, cause the mount to expand away from the mirror thereby increasing any existing radial clearance or creating such a clearance. Significant increases in radial clearance may allow the optic to shift under external forces such as shock or vibration; alignment may then be affected.

8.2.1 Radial stress in a mirror at low temperature

The magnitude of the radial stress, S_R, for a given temperature drop, ΔT, can be estimated as:

$$S_R = -K_4 K_5 \Delta T, \qquad (8.6)$$

where

$$K_4 = (\alpha_M - \alpha_G)/[(1/E_G) + (D_G/2E_M t_C)], \tag{8.7}$$

$$K_5 = 1 + ((2\Delta r)/[D_G \Delta T(\alpha_M - \alpha_G)]), \tag{8.8}$$

and D_G = mirror OD
t_c = mount wall thickness outside the rim of the mirror
Δr = radial clearance.

Note that ΔT is negative for a temperature decrease. Also, $0 < K_5 < 1$. If Δr exceeds $D_G \Delta T(\alpha_M - \alpha_G)/2$, the mirror will not be constrained by the cell ID and radial stress will not develop within the temperature range ΔT due to rim contact.

Numerical Example No. 43: Estimation of radial stress in a radially-constrained mirror.

A mirror made of Ohara E6 glass with diameter of 20 in. is mounted in a 6061 aluminum cell machined to provide 0.0002 in. radial clearance for assembly at 68°F. The cell wall thickness is 0.25 in. at the mirror rim. What radial stress is developed in the mirror at T_{MIN} = -80° F?

From Table C8a, $E_G = 8.5\times10^6$ lb/in.^{2}
$\alpha_G = 1.5\times10^{-6}$ /°F

From Table C12, $E_M = 9.9\times10^6$ lb/in.^{2}
$\alpha_M = 13.1\times10^{-6}$ / °F

$\Delta T = -80 - 68 = -148$°F

From Eq. (8.7),
$K_4 = (13.1\times10^{-6} - 1.5\times10^{-6}) / [(1 / 8.5\times10^6) + (20 / (2)(9.9\times10^6)(0.25)]$
$= 2.79$ lb/in.^{2} °F

From Eq. (8.8),
$K_5 = 1 + [(2)(0.0002)/(20)(-148)(11.6\times10^{-6})] = 0.988$

From Eq. (8.6),
$S_R = -(2.79)(0.988)(-148) = 408$ lb/in^2

This stress would cause no damage threat to the mirror.

8.2.2 Tangential (hoop) stress in the mount wall

Another consequence of differential contraction of the mount relative to the mirror is that stress is built up within the mount in accordance with the following equation:

$$S_M = S_R D_G / 2 t_C, \tag{8.9}$$

where all terms are as defined above.

With this expression, one can determine if the cell is strong enough to withstand the force exerted upon the mirror without exceeding its elastic limit. If the yield strength of the mount material exceeds S_M, a safety factor exists.

Numerical Example No. 44: Hoop stress in a cell wall.
Estimate the stress in the cell wall for the preceding example.

By Eq. (8.9), $S_M = (408)(20) / (2)(0.25) = 16{,}320$ lb/in.^{2}

From Table C12, we find that the yield strength of the 6061 aluminum is 8000 to 40,000 lb/in.^{2}. The wall might fail (i.e., distort) at low temperature. To improve the design, the wall could be made thicker.

8.2.3 Growth of radial clearance at increased temperatures

The increase ΔGap_R in nominal radial clearance, Gap_R, between a mirror and its mount due to a temperature increase of ΔT from that at assembly can be estimated by the equation:

$$\Delta Gap_R = (\alpha_M - \alpha_G) D_G \Delta T / 2, \tag{8.10}$$

where all terms are as previously defined.

Neglecting axial constraint, whatever total radial clearance, Gap_R, exists between the mirror OD and mount ID allows the mirror to roll (i.e., tilt about a transverse axis) until its rim touches the mount ID at diametrically opposite points. This roll angle can be estimated by the equation:

$$Roll = \arctan(2 Gap_R / t_E), \tag{8.11}$$

where t_E is the mirror's edge thickness.

Numerical Example No. 45: Growth in radial clearance around a mirror at high temperature.
What increase in radial clearance exists in the mirror assembly described in Numerical Example No. 43 at $T_{MAX} = 160°$ F (71.1°C)?

The nominal radial clearance at assembly is 0.0002 in. (5.08×10^{-3} mm).

$\Delta T = 160 - 68 = 92°$ F (33.3° C).

By Eq. (8.10), $\Delta Gap_R = (13.1\times10^{-6} - 8.5\times10^{-6})(20)(92)/2 = 0.0042$ in. (0.107 mm).

The total nominal radial gap at T_{MAX} is then 0.0002 + 0.0042 = 0.0044 in. (0.112 mm).

Numerical Example No. 46: Roll (tilt) of a mirror within nominal expanded radial clearance at high temperature.

What is the maximum roll that the mirror of the preceding example can experience at maximum survival temperature, T_{MAX}? Assume that the mirror has an edge thickness t_E of 0.1.875 in. (47.625 mm).

The expanded radial gap around the lens is 0.0044 in.

By Eq. (8.11), Roll = arctan (2)(0.0044 / (1.875) = 0.005° = 18 arcsec.

8.3 Bending stresses in mirrors

8.3.1 Causes of bending

If the annular areas of contact between the mount and the mirror surfaces are not directly opposite (i.e., at the same height from the axis on both sides), a bending moment is created within the mirror whenever axial preload is applied. This moment tends to deform the mirror so one surface becomes more convex and the other surface becomes more concave as illustrated schematically in Fig. 8.4. Deformation of the reflecting surface may adversely affect the performance of the mirror.

The surface that becomes more convex is placed in tension while the other surface is compressed. Since glass-type materials break much more easily in tension than in compression, especially if the surface is damaged by scratches or has subsurface cracks, catastrophic failure may occur if the bending effect is large. The "rule of thumb" tolerance for tensile stress given earlier [1000 lb/in.^{2} (68.9 MPa)] applies.

8.3.2 Bending stress in the mirror

Bayar[83] indicates that an analytical model based upon a thin plane parallel plate (as in Fig. 8.4) that uses an equation from Roark[45] applies also to simple lenses. We here extend the analogy to include mirrors. The tensile stress in the surface made more convex due to bending of the mirror is given approximately by Eq. (8.12a).

$$S_T = \frac{3P}{2\pi m t_E^2}\left[0.5(m-1) + (m+1)\ln\frac{y_2}{y_1} - (m-1)\frac{y_1^2}{2y_2^2}\right], \qquad (8.12a)$$

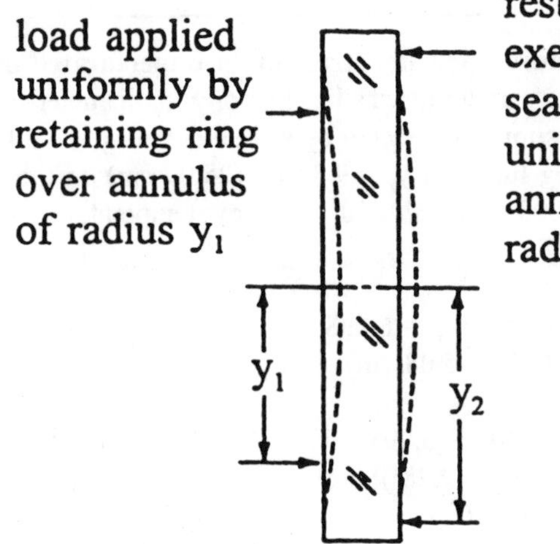

Fig. 8.4 Geometry allowing the estimation of bending moments from axial preloads applied at different heights on opposite sides of a plane-parallel mirror. (Adapted from Bayar[83])

where:

P = total applied axial preload
m = 1/Poisson's ratio for the mirror
t_E = mirror edge thickness
y_1 = smaller contact height
y_2 = larger contact height.

Equation (8.12a) applies to an unperforated plate. If the mirror has a central hole, the following equation should be used:

$$S_T = \frac{3P}{2\pi m t_E^2}[0.5(m-1)+(m+1)\ln\frac{y_2}{y_1}-(m-1)\frac{y_1^2}{2y_2^2}]-\frac{6M(y_2^2+b^2)}{(y_2^2-b^2)t_E^2}, \tag{8.12b}$$

where b is one-half the hole diameter and M is given by:

$$M = \frac{P}{8\pi m}[(m-1)+2(m+1)\ln\frac{y_2}{y_1}-(m-1)\frac{y_1^2}{y_2^2}]. \tag{8.13}$$

We should compare the stress from either equation to the above mentioned tension survival tolerance. To decrease the probability of breakage from this cause, the contact heights should be made equal within a few percent. Increasing the mirror's thickness also tends to reduce this danger.

Numerical Example No. 47: Bending (tensile) stress in an unperforated mirror.
A 20.0 in. (50.8 cm) diameter plane parallel fused silica mirror with diameter-to-thickness ratio of 10:1 and no central perforation is contacted by a shoulder at $y_1 = 9.50$ in. (24.13 cm) on one side and by a clamping flange at $y_2 = 9.88$ in. (25.09 cm) on the other side. Assume that the total preload is 5000 lb. (2.224×10^4 N) at low temperature. What tensile stress is created in the bent mirror?

From Table C8a, $\upsilon_G = 0.164$ so $m = 1/\upsilon_G = 6.098$

$t_E = 20 / 10 = 2.00$ in. (5.08 cm)

From Eq. (8.12a),

$S_T = [(3)(5000) / (2\pi)(6.098)(2.00)^2][(0.5)(6.098 - 1) + (6.098 + 1)(\ln(9.88 / 9.50)) - (6.098 - 1)(9.50^2 / (2)(9.88^2)]$

$= 95.0$ lb/in.^{2} (6.55×10^5 Pa).

This is well under the "rule of thumb" glass survival tolerance so the danger of damage at low temperature os low.

Numerical Example No. 48: Bending (tensile) stress in a perforated mirror.
Assume that the mirror in the preceding example has a central hole of 5.00 in. (12.70 cm) ID. If subjected to the same preload at low temperature, what is the tensile stress?

From Eq. (8.13):

$M = [5000 / (8\pi)(6.098)][(6.098 - 1) + (2)(6.098 + 1)(\ln(9.88 / 9.50)) - (6.098 - 1)(9.50^2 / 9.88^2)]$

$= 30.72$ lb

From Eq. (8.13):

$S_T = S_T$ (unperforated) $- [(6)(30.72)(9.88^2 + 2.50^2) / (9.88^2 - 2.50^2)]$

$= 94.97 - 52.38 = 42.58$ lb/in.^{2} (2.94×10^5 Pa)

This is even lower than the previous stress so should not be a problem.

8.3.3 Surface sag of a bent mirror

Roark[45] also gave the following equation for the change in sag at the center of the plate shown in Fig. 8.4 due to the bending moment exerted by the mount:

$$\Delta Sag_{max} = [\frac{3P(m^2-1)}{2\pi E_G m^2 t_E^3}][\frac{(3m+1)y_2^2-(m-1)y_1^2}{2(m+1)} - y_1^2(\ln\frac{y_2}{y_1}+1)], \tag{8.14}$$

where all terms are as defined earlier. This sag can be compared to some reasonable tolerance on surface deformation (such as $\lambda/2$ or $\lambda/50$) corresponding to the particular required system performance level and mirror location in that system.

Numerical Example No. 49: Change in sag of a reflecting surface due to bending moment.
Calculate the change in sag of the unperforated mirror described in Numerical Example No. 47.

From Table 8a, $E_G = 1.06\times10^7$ lb/in.^{2} (7.3×10^{10} Pa)

From Eq. (8.14),

$$\Delta Sag_{max} = [(3)(5000)(6.098^2 - 1) / (2\pi)(1.06\times10^7)(6.098^2)(2.50^3)][((3)(6.098 + 1)(9.88^2) - ((6.098 - 1)(9.50^2) / (2)(6.098 + 1)) - (9.50^2)(\ln(9.88 / 9.50) + 1)]$$

$$= 0.048 \text{ in. } (1.22 \text{ mm}) = 2233\ \lambda_{GREEN}$$

It is obvious that this mirror mounting design is unsatisfactory from the bending viewpoint even though the stress level is quite low. The design could be improved by making y_1 and y_2 more equal and/or by increasing the thickness of the mirror.

8.4 Stresses in bonded mirrors

In Sect. 7.2 we considered one technique for bonding the back of a mirror to a mount. There are three major sources of stress in the bonded joints between such mirrors and their mounts. These are shrinkage of the adhesive during cure, differential expansion and contraction at high and low temperatures, and the effect of acceleration that tends to pull the mirror away from the mount. We will consider each of these factors briefly.

Shrinkage during cure typically amounts to 2% to 10% of each dimension of the adhesive layer and may persist throughout the life of the device. Assuming that the material adheres well to both the mirror and mount surfaces throughout the contact area, the bonded surface of the optic is placed in tension. This force will tend to make the reflecting surface opposite become more convex. If the mirror is too thin, this effect may change the surface figure sufficently to degrade optical performance. Corrective actions include making sure that the mirror is as thick as reasonably possible, choosing an adhesive with minimal cure shrinkage, and minimizing the lateral dimensions of the bond (see Numerical Example No. 31). Using mirror substrate materials with high stiffness (i.e., large Young's modulus) also will help.

Thermal effects in bond joints due to mismatch of material CTEs have their greatest impacts at extreme temperatures. They tend to bend the mirror in much the same manner as cure shrinkage, but are temporary and usually reversible. In the common case with $\alpha_M > \alpha_G$, bending at higher temperatures would be expected to make the reflecting surface become more concave while the opposite change (to a more convex surface) would be expected at lower temperatures. Fracture of bonded optical parts has, on occasion, been attributed to high tensile forces exerted in the joint by low-temperature-induced shrinkage of the adhesive. The author knows of no simple analytical method to estimate mirror surface deformation or thermally induced stress due to this shrinkage. Finite element analysis methods can be used for this purpose, but they are beyond the scope of this book.

Acceleration directed normal to the bond joint and in the direction which places that joint in tension can cause sufficient force to break something. As was explained in Sect. 5.2 for prisms, the strength of the adhesive joint often is greater than the tensile strength of the optical material so fracture of the latter can occur under high tensile stress. The worst situation would be when this happens at low temperature so both differential contraction of the materials and the effect of acceleration act together.

Another temperature-related effect involving bonding of mirrors occurs in designs in which an elastomer such as epoxy or RTV sealant is placed between the mirror's rim and the ID of a mount. This material may form a continuous annular layer around the mirror or three or more localized pads symmetrically distributed around the rim. Since the CTEs of these elastomers exceed those of the optical and mount materials, temperature increase to the survival limit may cause the mirror to be placed in high compression radially. The resulting stress in the mirror may exceed the compression threshold for the material, thereby causing damage. Lesser effects occur at operational temperatures, but these may be of sufficient magnitude to degrade optical performance of the mirror. Athermal designs in which the thickness of the elastomer annular layer or pads is chosen so as to compensate nominally for the differential expansion would reduce this adverse effect.

CHAPTER 9

DESCRIPTIONS OF HARDWARE EXAMPLES

In this chapter are found descriptions and illustrations of eight examples of optical hardware involving mountings for prisms or small mirrors and using some of the concepts and designs described earlier. Some of these examples are described in more detail and in context with their applications in other publications. References are provided so the reader can explore these resources for items of particular interest.

9.1 Clamped penta prism

In Sect. 4.1, we briefly describe a concept for semi-kinematically mounting a penta prism by clamping it with three cantilevered spring clips against three coplanar pads on a baseplate and spring-loading it laterally with a single leaf spring against three locating pins. See Fig. 4.3. All six degrees of freedom are uniquely constrained. Contact areas are small; this minimizes bending moments that might distort the optical surfaces. Here, we show calculations used to establish the nominal mounting design.

The prism in this hardware example has an aperture, A, of 1.250 in. and is made of BK7 glass. Per Fig. 3.17, the volume of the prism is $1.5A^3 = 2.930$ in.^{3}. From Table C1, we find that the glass density is 0.091 lb/in.^{3} so the prism weight is 0.267 lb. When mounted, the prism is to remain in contact with all the constraints under 12 times gravity in any direction. The minimum clear aperture is 1.15 in. We will first consider details of the mount design in regard to constraint to motion perpendicular to the plane of reflection and then the design details for constraint in that plane.

9.1.1 Constraint perpendicular to the plane of reflection

Figure 9.1 shows one of three spring clips applied to the top of the prism. The preload to be provided by each spring is $P_i = wa_G/N = 1.066$ lb. Each spring has a rectangular cross section of b = 0.25 in. by h = 0.020 in. Its free length L is 0.375 in. These springs are made of beryllium copper. The linear preload per spring is $p = P_i/b = 4.266$ lb/in. Equation (5.1) says that the spring should be deflected by an amount equal

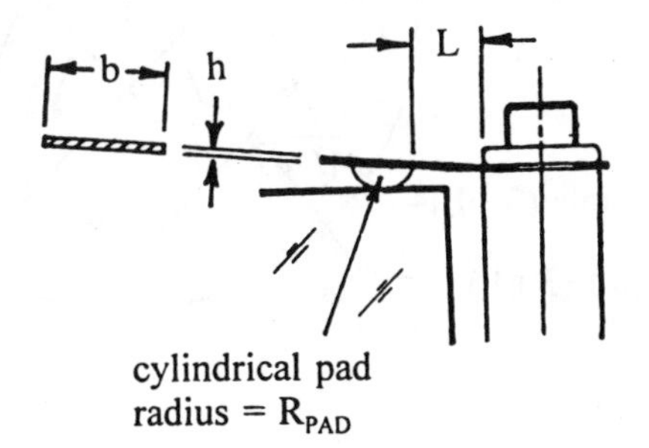

Fig. 9.1 Concept for one of three leaf springs with cylindrical pads applying preload to a prism.

to $\Delta y = (1 - \upsilon_M^2)(4L^3P_i/E_M bh^3)$. In Tables C1 and C12 we find the following parameters: $E_G = 1.17\times10^7$ lb/in.^{2}, $\upsilon_G = 0.208$, and $E_M = 18.5\times10^6$ lb/in.^{2}. The value for υ_M is not given, but can be assumed to be 0.35. Substituting, we obtain $\Delta y = 0.0053$ in. This deflection is attained at assembly by custom grinding the spacer under the spring. Care must be exercised to make sure the metal-to-metal interfaces are clean when assembled.

From Eq. (5.1), we calculate $K_2 = [(1 - \upsilon_G^2)/E_G] + [(1 - \upsilon_M^2)/E_M] = 1.289\times10^{-7}$ in.^{2}/lb. The spring clips each have integral convex cylindrical pads of radius $R_{cyl} = 0.25$ in. Equation (5.6) estimates the magnitude of stress in the prism at the pad interface as $S_{P\ cyl} = 0.564(p/K_2R_{cyl})^{1/2} = 6490$ lb/in.^{2}. Since glass can withstand compressive stress of about 50,000 lb/in.^{2}, this is not a problem. From Eq. (5.3), the bending stress in the spring is $S_B = 6LP_i/bh^2 = 23{,}985$ lb/in.^{2}. Comparing this to the yield stress of 155,000 lb/in.^{2} we derive find the safety factor to be 6.5. This is surely adequate.

9.1.2 Constraint in the plane of reflection

Figure 9.2 shows the straddling leaf spring holding the prism against the three locating pins. This BeCu spring measures 0.25 in. wide by 0.015 in. thick and has a total free length L of 1.040 in. Preload is applied to the prism through a convex cylindrical pad integral to the spring to hold the prism against the three cylindrical pins opposite. The pins are made of type 416 stainless steel and have diameters of 0.125 in. The total preload is again wa_G/N, but N=1 in this direction so $P_i = 3.204$ lb. Friction at all contacts is neglected. We may apply Eq. (5.4) to calculate the nominal spring deflection $\Delta x = (0.0625)(1 - \upsilon_M^2)(P_iL^3)/E_M bh^3 = 0.0127$ in. This deflection is obtained by grinding the two spacers at assembly.

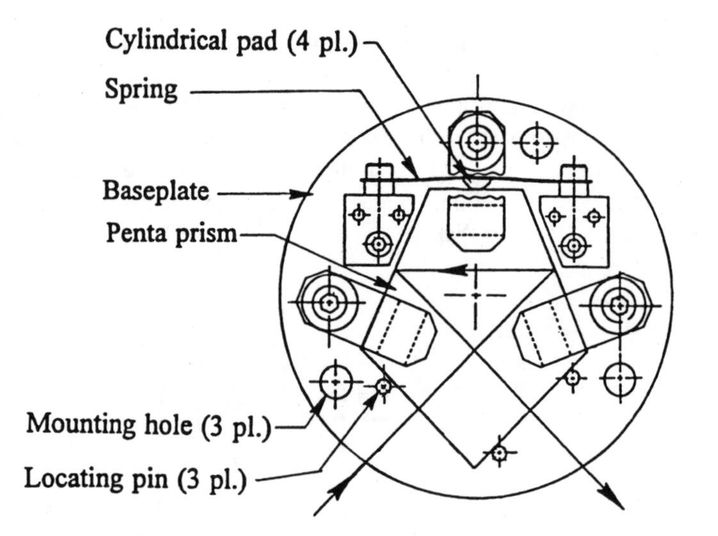

Fig. 9.2 Concept for one straddling leaf spring with a cylindrical pad that applies transverse preload to a prism.

Figure 9.3(a) shows the forces acting on the prism. P_i is directed along the prism centerline. It has equal components in the X and Y directions of P_i cos 45°. For static equilibrium, the sum of forces in the X and Y directions must be zero. Hence,

$$P_1 = P_i \cos 45^o \tag{9.1}$$

and

$$P_2 + P_3 = P_i \cos 45^o. \tag{9.2}$$

The sum of moments about any point must also be zero. Choosing the 90° corner of the prism as the reference, we write:

$$P_1 d_1 - P_2 d_2 - P_3 (d_2 + d_3) = 0 \tag{9.3}$$

where d_1, d_2, and d_3 are the distances from the prism corner to the pin contacts.

Figure 9.3(b) shows the locations of the pins corresponding to P_1 and P_2. The latter is located just outside the circle inscribed into the square prism face (not shown) while the former is located just outside the clear aperture (dashed line). Note the assymetry of the pin locations. This is important in that it ensures that some positive force is exerted against all three pins. The third pin is located 0.442 in. from the prism centerline as indicated in Fig. 9.3(a).

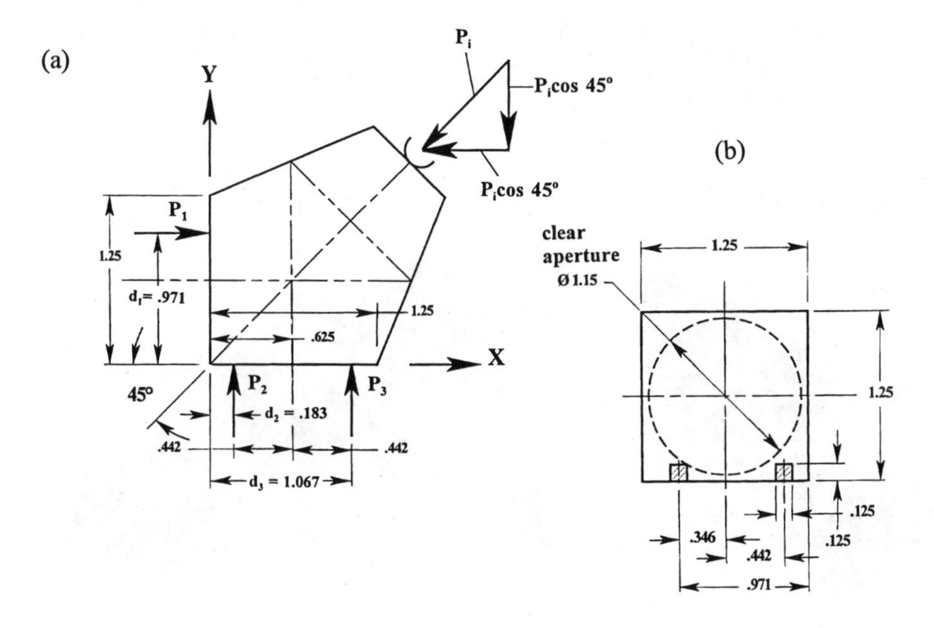

Fig. 9.3 View (a) shows the forces acting on the prism in the plane of reflection while view (b) shows the locations of two of the cylindrical pins. Dimensions are inches.

In this design, $d_1 = 0.625 + 0.346 = 0.971$ in., $d_2 = 0.625 - 0.442 = 0.183$ in., and $d_3 = 0.625 + 0.442 = 1.067$ in. Substituting into Eqs. (9.1) through (9.3) we obtain: $P_1 = 2.265$ lb, $P_2 = 0.592$ lb, and $P_3 = 1.673$ lb.

The worst case for stress is at the single pin on the entrance face of the prism. There, K_1 is $0.5/R_{pin} = 8$ since $R_{pin} = 0.0625$ in. In Table C12, we find E_M to be 29×10^6 lb/in.^{2} and $\upsilon_M = 0.3$. K_2 is then 1.128×10^{-7}. The line contact on each pin is 0.125 in. long so the linear preload normal to the single pin is $p_1 = (2.265)/0.125 = 18.120$ lb/in. The stress in the prism is determined by applying Eq. (8.1) as $0.798(K_1p/K_2)^{1/2}$ or 28,607 lb/in.^{2} This stress level is a little higher than we would wish. It could be reduced by lengthening the pin contact on the prism face or increasing the pin diameter. The stress in the BeCu leaf spring is found from Eq. (5.5) as $1.5P_iL/bh^2 = 88{,}858$ lb/in.^{2} The safety factor here is 1.7; this is slightly smaller than would be desired.

9.2 Bonded Porro prism erecting system for a binocular

The U.S. Army's M19 7×50 binocular, developed in the 1950s as a replacement for the World War II vintage modified commercial binoculars is shown in Fig. 9.4. This binocular was a totally new design featuring significantly reduced weight and size, improved optical performance, large-quantity producibility, and improved reliability and maintainability as compared to all prior designs. These advantages were achieved by making the device modular, with only five optomechanical parts, all of which were interchangeable in any reasonably clean location without adjustment and without special tools.[25,84] Because the success of modular construction depends largely on the detailed optomechanical design, the availability of optically based tooling, and special care exercised during manufacture, we describe how the prealigned Porro prism erector was assembled and incorporated into the body housing.

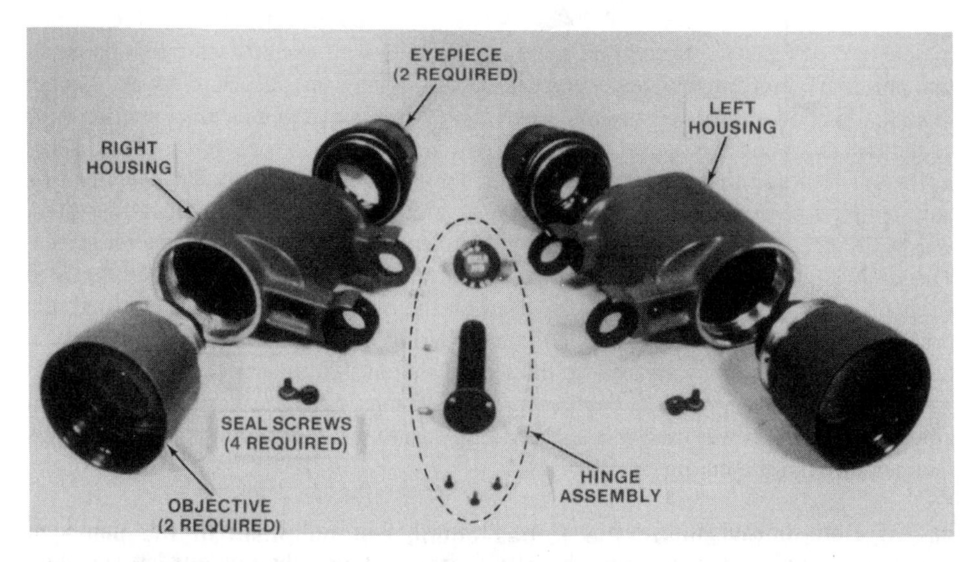

Fig. 9.4 Exploded view of the Binocular M19 showing its modular construction. (From Yoder[8] by courtesy of Marcel Dekker, Inc.)

Photographs of the Porro prism cluster are shown in Fig. 9.5. The prisms were made of high index (649338) glass to ensure TIR at the reflecting surfaces and were tapered to have minimal volume without vignetting.

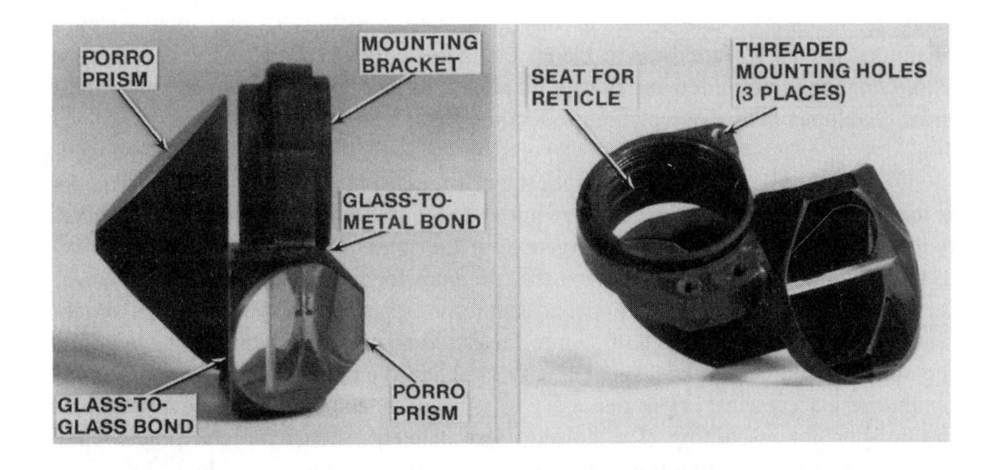

Fig. 9.5 Photographs of the M19 bonded Porro prism subassembly. (From Yoder[8] by courtesy of Marcel Dekker, Inc.)

The first step in assembly was to bond one Porro prism to a die cast aluminum bracket with adhesive per MIL-A-4866 (such as Summers Milbond) in a fixture built to exacting tolerances and carefully maintained throughout use. After that bond was cured, the prism/bracket subassembly was mounted in a second precision fixture. Ultraviolet-curing optical cement (Norland 61) was applied to the appropriate portion of the prism's hypotenuse surface. The second prism was positioned with respect to the first prism so that the input and output axes were parallel (pointing) and displaced by the proper distance. In addition, that prism was rotated in the interface plane to correct rotation (tilt) of the image around the optical axis. A video camera and monitor were used to display to the operator pointing and tilt relationships for the prism assembly. The operator first adjusted the free prism laterally until the image of a reticle projected through the system was positioned within a prescribed rectangular tolerance envelope on the monitor screen. Then, while maintaining the image inside this rectangle, the prism was rotated slightly to align tilt reference indicators also displayed on the monitor screen. Once adjusted, the prism was clamped in position in the fixture. Curing of the cement took place under a bank of ultraviolet lamps adjacent to the setting station. Multiple setting/curing fixtures were necessary to support the required production rate. After curing, the same optical alignment apparatus was used as a test device to ensure that the desired prism setting was retained through the curing process.

Both housings started out as identical thin-walled, vinyl-clad aluminum investment castings; they were machined differently to form their unique left and right shapes. The wall thickness was nominally 1.524 mm. Over this, 0.381 mm of soft vinyl was applied prior to machining of the critical mounting seats for the eyepiece, the prism assembly, and the objective. The locations of the eyepiece and prism assembly seats were established mechanically during the machining process. Normally, with a rigid, stable part, these would not have presented unusual problems despite the very demanding tolerances required. However, the structural flexibility of the thin-walled housing was a serious handicap. In addition, due to the soft vinyl, it was not possible to locate reliably from any of the vinyl-clad surfaces nor to clamp on them without cosmetic damage. Elaborate fixtures relying upon a few previously machined surfaces which were not vinyl clad had to be developed before acceptable production yields and rates were attained.

Machining of the housing with the prism assembly installed was the critical step in obtaining the module precision required to permit interchangeability. Collimation requirements (divergence and dipvergence) of the monocular's optical axis with respect to the hinge pin centerline were such that the bore for the objective had to be properly located within 0.0127 mm and the requirement for perpendicularity between the objective seat and the optical axis was 0.0051 mm measured across the objective seat. In addition, the objective seat had to be located axially to obtain the proper flange focal distance. To obtain these accuracies, it was necessary to use optical alignment techniques to position the housing for machining. The housing was held in a transferable setting/machining fixture. The housing initially was positioned by optical alignment and locked in place on the fixture at an off-line setting station. Then the fixture was transferred to the spindle of a numerically controlled multi-tool lathe for final machining. Multiple transferable fixtures were provided so setting and machining could proceed in parallel.

The fixture and the optical alignment technique used at the setting station are shown schematically in Fig. 9.6. The fixture base was designed to mate precisely with the lathe spindle such that the fixture centerline was coincident with the rotational axis of the

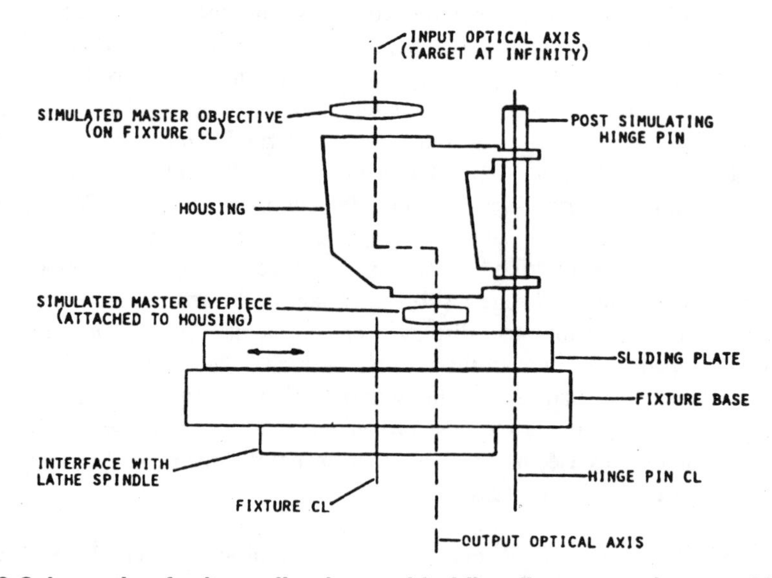

Fig. 9.6 Schematic of prism-adjusting and holding fixture used to machine the M19 body housing with prealigned prism installed. (From Trsar et al.[84])

spindle during machining. Therefore, the mounting seat for the objective would be machined concentric with the fixture centerline. Atop the fixture base was a sliding plate which could be translated laterally. This plate carried a post simulating a binocular hinge pin. The post centerline was always parallel to the fixture centerline.

An optical system in the setting station (not shown in Fig. 9.6) provided an image of a target at infinity along the input optical axis which was coincident with the fixture axis. A master objective was mounted at a fixed location in the setting station and centered on this axis. This objective formed an image of the target at an image plane inside the housing. This image was then viewed through a master eyepiece (temporarily attached to the housing) by a video camera, with the output being displayed on a video monitor. The proper flange focus position for machining the objective lens seat in the housing was obtained by moving the housing vertically along the hinge post until best focus was obtained on the video monitor. The housing was then clamped to the post and sliding plate. Axial positioning of the housing on the fixture now was completed, but lateral adjustment to obtain collimation was still needed.

After focus adjustment, the housing and sliding plate assembly were adjusted laterally (in two directions) with respect to the fixture base and the master objective until the required collimation conditions were achieved. This was indicated by a predetermined positioning of the target image on the video monitor. The sliding plate was then locked to the fixture base and the assembly transferred to the CNC lathe for machining of the objective mounting seat.

A similar procedure was used to orient the housing for machining the eyepiece interface. The result was a body housing with a prealigned prism assembly that would mate properly with any objective module and any eyepiece module to form one half of the binocular instrument. The left and right housing assemblies also were properly aligned to fit together at the hinge without adjustment.

9.3 Large flexure-mounted mirror assembly for a microlithography mask projection system

Large prisms may pose difficult mounting problems, not only because of their weight and size, but also because large separations of mounting supports may cause significant thermal expansion problems due to the use of dissimilar materials. For example, the three-component prism shown in Figs. 9.7 and 9.8 is 6 in. wide and 7.3 in. long. The prisms are made of Zerodur; they are mounted on a cast aluminum base structure. In order to accommodate the different CTEs, flexures are used to attach the prism to the structure. The basic principle of the flexure mount is explained in Section 4.4 in conjunction with Fig. 4.21. We here describe the prism and the mounting interface design for the specific device of Figs. 9.7 and 9.8.

The prism actually comprises a triplet of first-surface mirrors fashioned as a right angle reflector and a mirror version of an Amici prism. Each mirror is inclined at 45° to

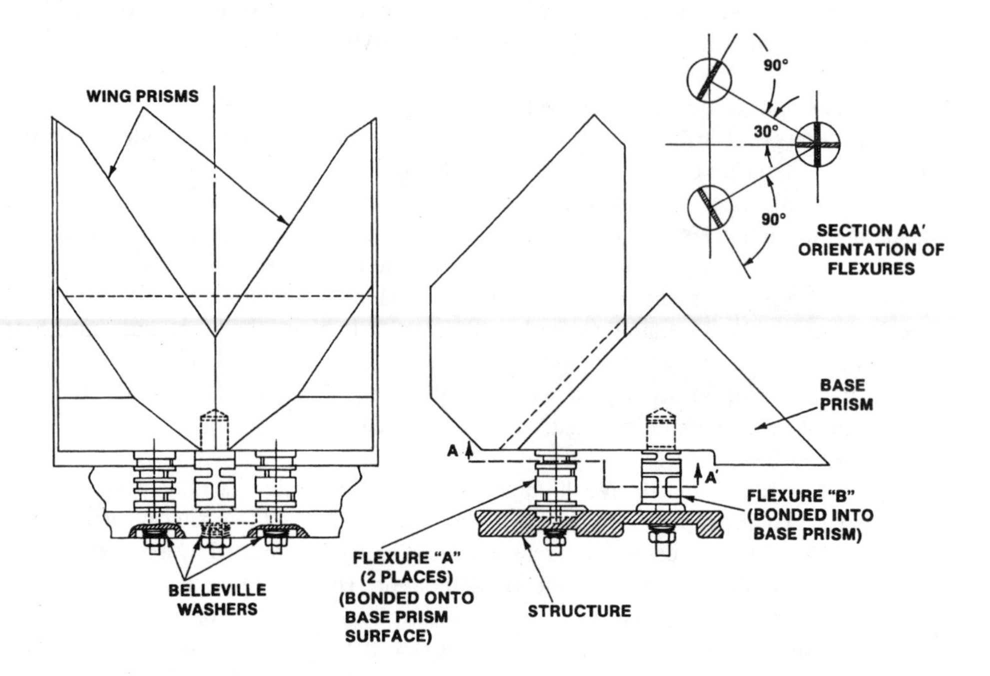

Fig. 9.7 Schematic of a large prism assembly attached to structure through flexure posts. (Courtesy of SVG Lithography Systems, Wilton, CT)

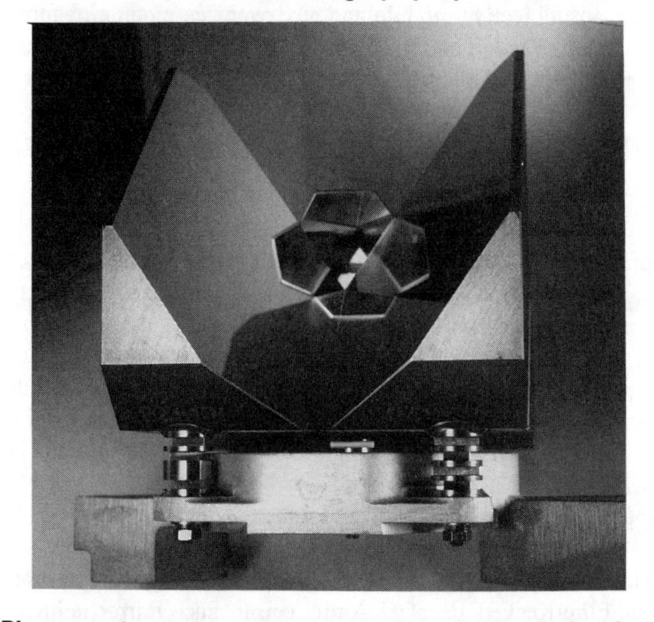

Fig. 9.8 Photograph of the prism assembly of Fig. 9.7. (Courtesy of SVG Lithography Systems, Wilton, CT)

the optical axis of a microlithography mask projection system as shown in Fig. 9.9. The beam from a light condenser (not shown) reflects from one face (R_1) of the base prism to a 1:1 magnification imaging system. It then reflects from both surfaces (R_2 and R_3) of the wing prisms to form the image on the silicon wafer.

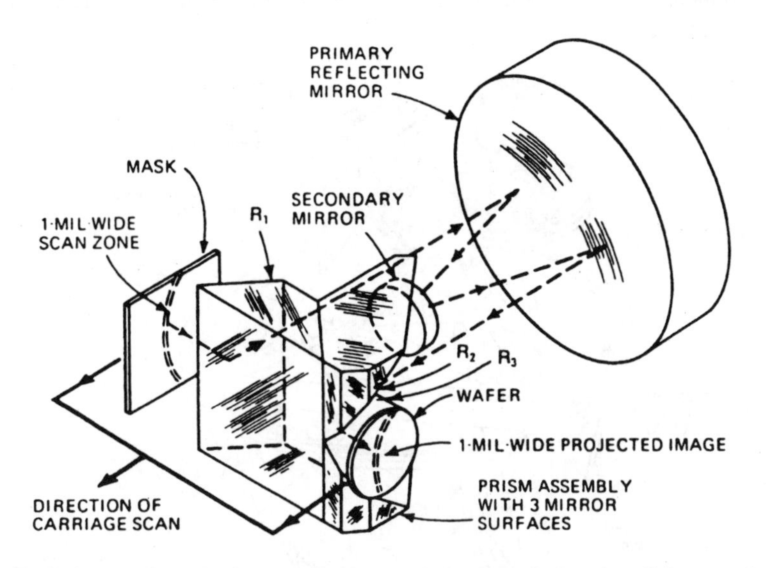

Fig. 9.9 Schematic of the optical system for the microlithography mask projection system. (Courtesy of SVG Lithography Systems, Wilton, CT)

The wing prisms are optically contacted to one face of the right angle base prism that is, in turn, mounted to the structure by the three flexures attached to the hypotenuse of the base prism. The multiple flexure blades are 0.020 in. thick and 0.120 in. long; they are machined into 0.750 in. diameter Invar posts. Post "B" has a torsion flexure and a universal joint; it is bonded with 3M 2216 epoxy into a hole in the base prism. The other two posts are bonded to the lower surface of the base prism. One of these posts has one universal joint and a single flexure blade which is compliant in the direction towards Post "B". The third post has two universal joints. The section view AA of Fig. 9.7 shows the locations of the posts at corners of a nearly equilateral triangle and the orientations of some flexure blades. Each post has a threaded section at the bottom that passes through holes in the base plate and are secured with nuts acting through stacks of Belleville washers to preload the threads and provide required axial compliance for anticipated temperature changes.

9.4 Mounting for large prisms in the Keck II Echellette Spectrograph/Imager (ESI)

The Echellette Spectrograph and Imager (ESI), being developed for use at the Cassegrain focus of the Keck II 10-m telescope, employs two large (approx. 25 kg each) prisms for cross dispersion. In order to maintain optical stability in the operational modes, these prisms must maintain a fixed angle relative to the nominal spectrograph optical axis under a variety of flexural and thermal loads. Gravity or thermally induced motions of the optical elements, stress induced deformations of the optical surfaces, and thermally-induced changes in the refractive index of the materials making up the components. The major components of the ESI are shown in Fig. 9.10. The ESI has three scientific modes:

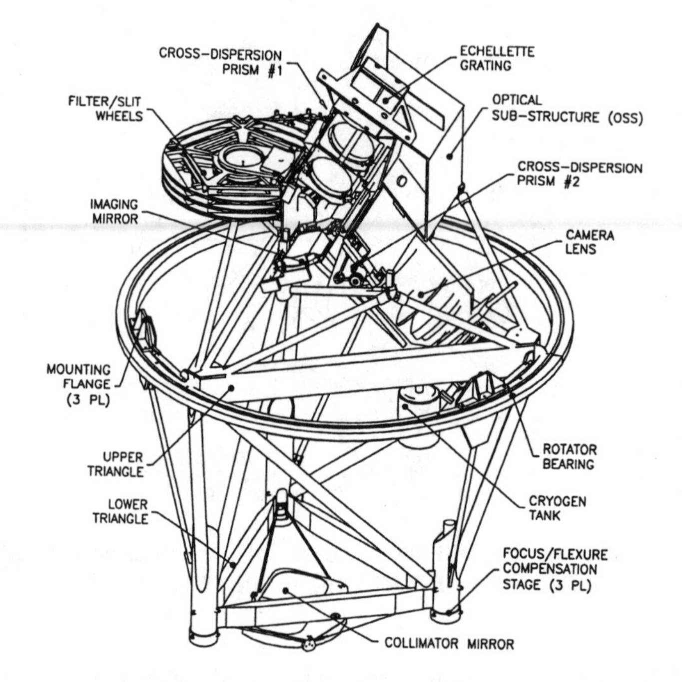

Fig. 9.10 Major components of the Echellette Spectrograph and Imager (ESI) developed for use with the Keck II telescope. (From Sheinis et al.[85])

medium resolution Echellette mode, low resolution prismatic mode, and imaging mode. To switch from one mode to another, one prism must move out of the beam as shown in Fig. 9.11. That prism is mounted on a single-axis stage. A mirror is moved into the beam to switch to the direct imaging mode. The spectrograph optical system is described[86,87] by Epps and Miller and by Sutin. In this section, we describe a novel concept for mounting the large prisms described by Shein, Nelson and Radovan[85] that has been developed to provide the required stability.

The design philosophy for the ESI is characterized by the use of determinate structures or space frames wherever possible. A determinate structure is one that constrains the six degrees of freedom of a solid body by six structural elements or struts connected to the outside world at six points or nodes. Up to three pairs of the nodes may be degenerate. Struts are used in compression and tension only. Thus, deflections of the struts are linear with length, as opposed to struts or plates used in bending where the deflections are proportional to the third power of the part length. Other examples can be found in the description[88] by Radovan et al. of the active collimator used for tilt correction

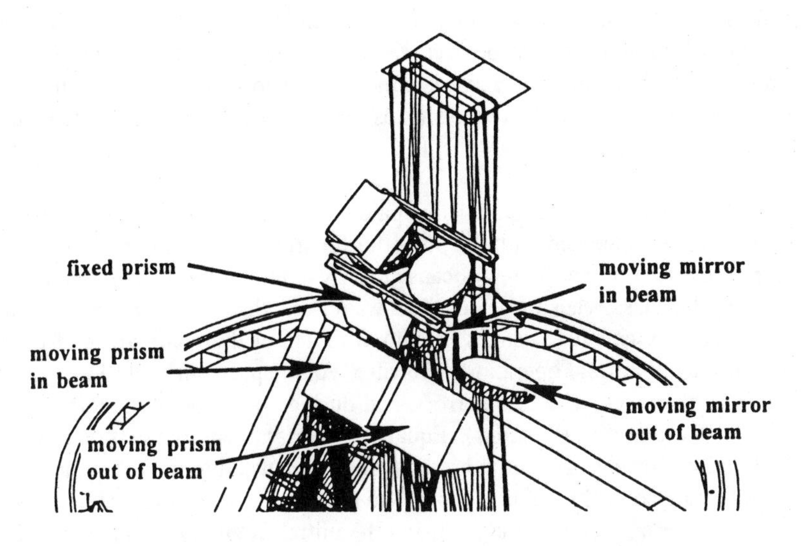

Fig. 9.11 Another view of the ESI showing the dispersing prisms. (Adapted from Sutin[87])

of the ESI and in the description[89] by Bigelow and Nelson of the space frame which provides the backbone for the entire instrument. The desirable features of this type of mounting arise because no moments can be imparted at the strut connections. This has the advantage that distortions of one structural member will introduce displacements without inducing stresses in a second member (i.e., an optical component) mounted on the first member. The challenge to the designer is to create a design that allows only motions of each component and of combined groups of optics such that the combined set of errors are smaller than the tolerances prescribed by the error budget. To do this the mechanical designer must work very closely with the optical designer to establish the predicted performance of the system prior to actual construction.

In the ESI, the cross dispersing prisms are in collimated light; therefore, to first order, small translations of the prisms will produce pupil motion only and no corresponding image motion. Tilts of the prisms will produce combinations of the following: image motion, change of the cross dispersion direction, change in the amount of cross dispersion, change in the anamorphic magnification factor, and increased distortion. Therefore, the most important stability criteria for the prisms is control of tip and tilt with almost no requirement on displacement stability. The sensitivities for ESI, are ±0.013, ±0.0045, and ±0.014 arcsec of image motion for ±l arcsec of prism tilt about the X, Y, and Z axes, respectively. The desired spectrograph performance is ±0.06 arcsec of image motion without flexure control and ±0.03 arcsec of image motion with flexure control for a two hour integration. For a reasonable choice of the allowable percentage of the total error allotted to the prism motion, these sensitivities give a requirement of less than ±1.0, ±2.0 and ±1.0 arcsec rotation about the X, Y and Z axes. The normal operating temperature range for the Keck instruments is 2 ± 4°C, and the total range seen at the summit of Mauna Kea is -15 to 20°C. The instrument must maintain all the above translational and rotational specifications over the entire working temperature range. Therefore the prism mounts need to be designed to be athermal with respect to tilts over this working temperature range and must keep stresses below the acceptable limits over the complete temperature range of the site as well as extremes experienced during shipping.

In addition, careful attention must be paid to the stresses induced into the prisms. Not only must one be concerned with the potential for fracture of the bond joint or of the glass, but stress induced in the glass will cause a corresponding local change in the index of refraction of the glass, causing a possible wavefront distortion. In order to insure against glass breakage, calculated stress is limited to 2% of the yield stress of the glass at the bond joint under normal operation. This also would apply during to loads occuring in shipping, in earthquakes, with drive errors, and during collisions of the telescope with other objects in the dome (e.g. cranes). Equally important requirements of the mount design are: 1) minimization of measurable hysteresis because that limits the accuracy of the open loop flexure control system; 2) ability to make a one time alignment adjustment of the prism tilts over a 30 arcmin range during the initial assembly; and 3) removability of the prisms for recoating, with a repeatable alignment position upon reassembly.

While, in general, the ESI design approach was to hold optics with determinate space frames and to interconnect optical assemblies with space frames, it was found to be easier and adequate to build the central structure upon which all of the optical elements and assemblies (except the collimator mirror) were mounted as an optical substructure (OSS) comprising a bolted plate structure. The prisms are attached to the OSS through 6 struts; the actual attachment to each prism consists of two parts: a pad which is permanently bonded to the prism and a mating steel coupling plate that is attached to the ends of the struts. A similar plate couples the other ends of the struts to the structure. Bolted and pinned joints at each coupling plate allow the prism to be readily and repeatably removed from its determinate support system and reinstalled accurately.

The fixed and moveable prism mounting designs are shown in Figs. 9.12 and 9.13, respectively. Joined pairs of struts are connected to each prism in one point on each of the three nonilluminated faces via a tantalum pad bonded to the prism. The CTE of tantalum ($6.5 \times 10^{-6}/°C$) closely matches that of the BK7 prisms ($7.1 \times 10^{-6}/°C$). The struts attach either directly to the OSS (in the case of the fixed prism) or to the translation stage (in the case of the moveable prism) which, in turn, is bolted to the OSS. The largest refracting faces of the fixed and moveable prisms measure 36.0 by 22.8 cm and 30.6 by 28.9 cm, respectively. The glass path is long (>80 cm) so it was necessary for the refractive index to be unusually uniform throughout the prisms. They were made of Ohara BSL7Y glass having measured index homogeneity better than $\pm 2 \times 10^{-6}$. This achievement is believed to represent the state-of-the-art for prisms of this size.

Each pair of struts was milled from a single piece of ground steel stock. Since each strut should constrain only one degree of freedom of the prism, crossed flexures were cut into each end of each strut to remove four degrees of freedom (one rotational and one translational per flexure pair). The fifth degree of freedom, axial rotation, is removed by the low torsional stiffness of the strut/flexure combination.

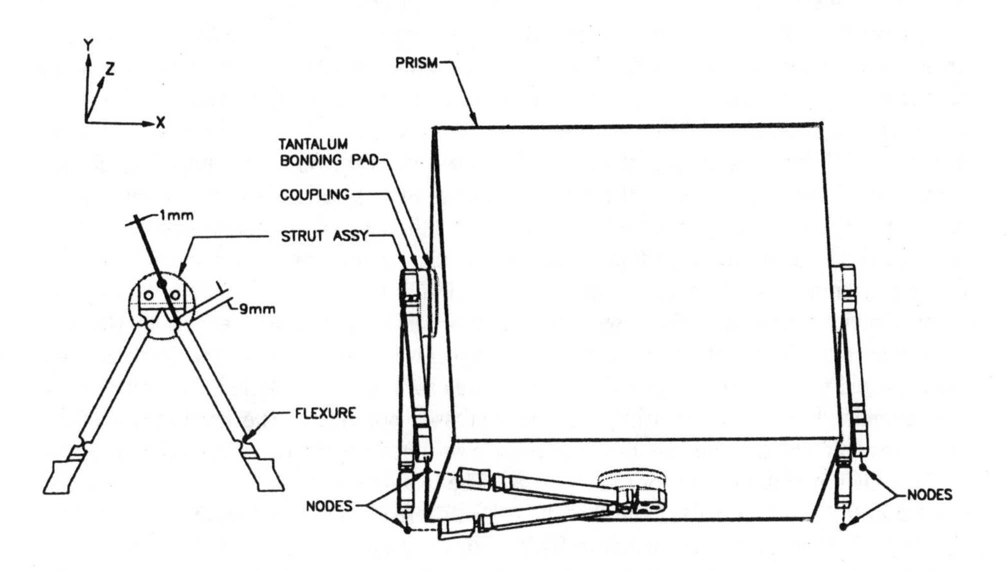

Fig. 9.12 Sketch of the fixed prism assembly showing its 6-strut mounting configuration. (From Sheinis et al.[85])

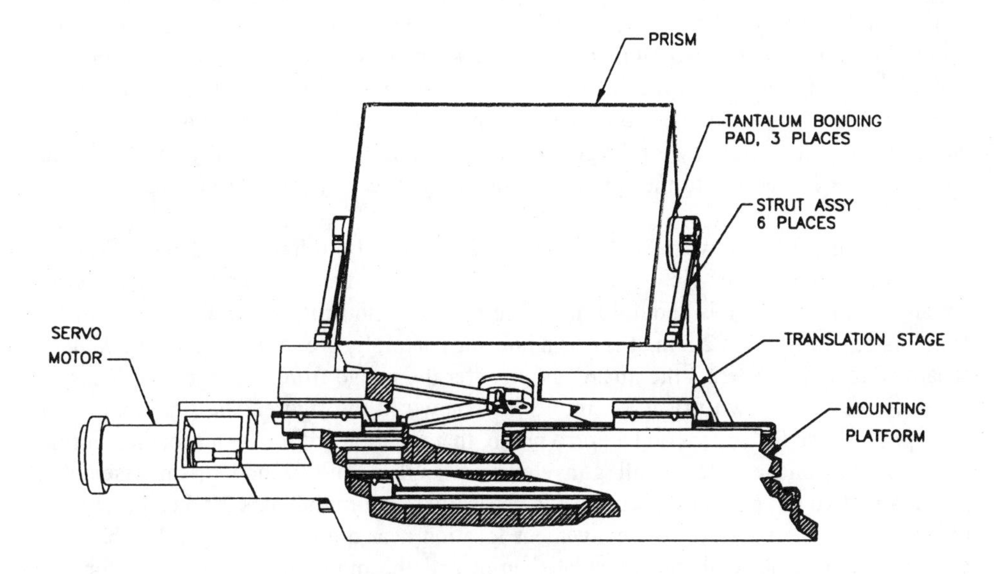

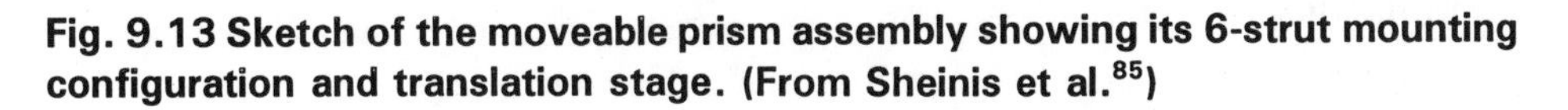
Fig. 9.13 Sketch of the moveable prism assembly showing its 6-strut mounting configuration and translation stage. (From Sheinis et al.[85])

Flexure thicknesses and lengths were designed to impart less stress than the self weight loading of the prism into the prism pad connection and to be below the elastic limit over the full range of adjustment, while keeping the strut as stiff as possible. Pad areas were chosen to give a self weight induced stress of 125 KPa. If we consider the tensile strength of glass to be 7 MPa[90], this gives a safety factor of 50. The glass-to-metal bond adhesive was Hysol 9313 and the thickness was chosen to be the same (0.25 mm) as developed[91] by Iraninejad et al. during development of the bonded connections for the Keck primary segments. To confirm these choices, extensive stress testing over various temperature ranges was performed for BK7-to-tantalum bonds. Several samples of BK7 were fabricated with the same type of surface finish specified on the prisms. These were bonded to tantalum and steel pads mechanically similar to the actual bonding pads for the prism mounts. These assemblies were subjected to tensile and shear loads up to 10 times the expected loading in the instrument. The test jigs were then cycled 20 to 30 times over the expected temperature range of the Mauna Kea summit. None failed. The joints were then examined for stress birefringence under crossed polarizers. The level of wavefront error was calculated to be less than the limit prescribed by the error budget in the case of the tantalum pad, but not for the steel pad. The tantalum material was then designated for use in the bonding pads. Note that the CTE difference between tantalum and BK7 is $0.6\times10^{-6}/^{o}C$ whereas the best match to BK7 reported elsewhere[8] is $1.7\times10^{-6}/^{o}C$ using 6Al-4-V titanium.

9.5 Mountings for prisms in an articulated telescope

Conventionally, the main weapon of an armored vehicle (tank) is operated by a gunner who has two optical intruments that can be used to acquire and fire at hostile targets. The primary fire control sight is usually a periscope protruding through the turret ceiling while the secondary sight usually is a telescope protruding through the front of the turret alongside and attached by linkages to the weapon. Key design features of a typical embodiment of the latter type instrument are discussed here. The specific telescope considered is of the articulated form, i.e., it is hinged towards its midsection so the front end can swing in elevation with the gun while the rear section is essentially fixed in place so the gunner has access to the eyepiece without significantly moving his head.

Figure 9.14 shows the optical system schematically. It has a fixed magnification of 8-power and a field of view of about 8°. The exit pupil diameter is about 5 mm so the entrance pupil diameter is about 40 mm. The telescope housing diameter throughout its length is generally about 2.5 in.; the prism housings naturally are somewhat larger. Widely separated relay lenses erect the image and transfer the image from the objective focus to the eyepiece focus. Two prism assemblies are shown in the figure. The first comprises three prisms, two 90° prisms and a Porro prism, that function within the mechanical hinge and keep the image erect at all gun elevation angles. The second prism assembly, comprising two 90° prisms, offset the axis vertically and turn that axis 20° in a horizontal plane to bring the eyepiece to a convenient location near the gunner's eye. The balance of this section deals with the articulated joint and the mounting arrangement for the prisms therein.

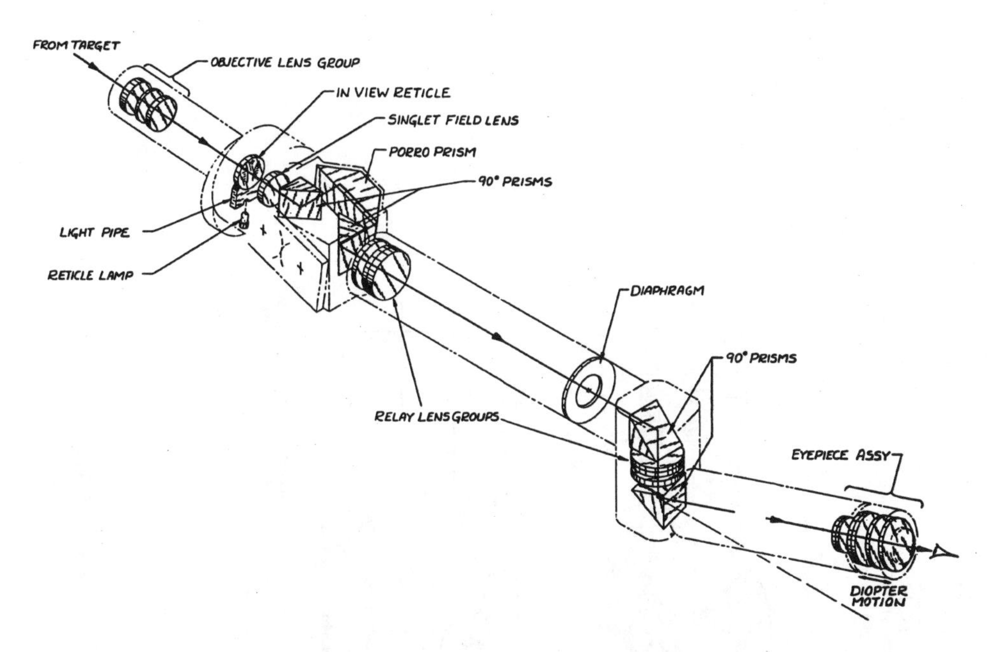

Fig. 9.14 Optical schematic of the articulated telescope. (Courtesy of the U.S. Army.)

The articulated joint mechanism is shown in Fig. 9.15. The first right angle prism of the articulated joint is mounted in a "Housing, 90° Prism" (see Fig. 9.16). The prism is bonded to a bracket that is attached with two screws and two pins to a plate that is, in turn, attached with four screws to the housing. After assembly and alignment, a cover is installed over the screws and sealed in place. Surface "W" of that housing attaches to the exit end of the reticle housing.

The second right angle prism is mounted in a "Housing, Erector" as indicated in Fig. 9.17. It also is bonded to a bracket that is attached with two screws and two pins to a plate that is, in turn, screwed fast to the housing. The note in this figure indicates the alignment requirements for the prism. Surface "W" mentioned there is shown in Fig. 9.10. After alignment, a cover is sealed over the screws.

As shown in Fig. 9.18, the Porro prism is contained within a separate housing and, together with a gear housing on the opposite side of the telescope, forms the mechanical link between the telescope's front and rear portions. The action of the gear train keeps the prism oriented angularly midway between the front and rear portions of the telescope. This angular relationship maintains the erect image. The housing for the Porro prism is made of hardened stainless steel since it acts as a bearing for the angular motion. The rotary joints in the assembly are sealed with lubricated O-rings that seat in grooves in the mating parts. The prism is bonded to a bracket that is attached to a cover by two screws riding in two slots. After installation of the bonded prism assembly to the housing, the prism is slid in the slots to adjust the optical path through the assembly; the screws are then secured and the plate pinned in place. A protective cover is then installed.

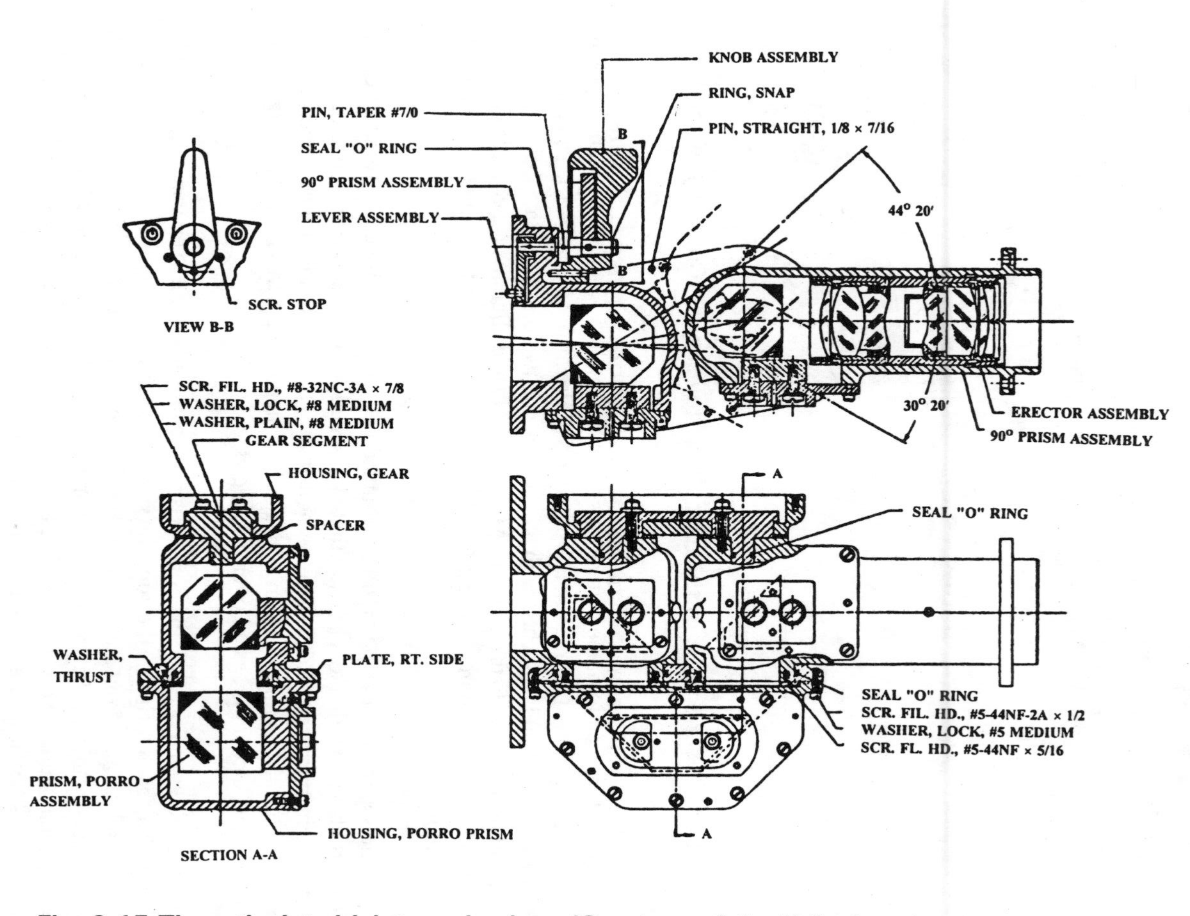

Fig. 9.15 The articulated joint mechanism. (Courtesy of the U.S. Army)

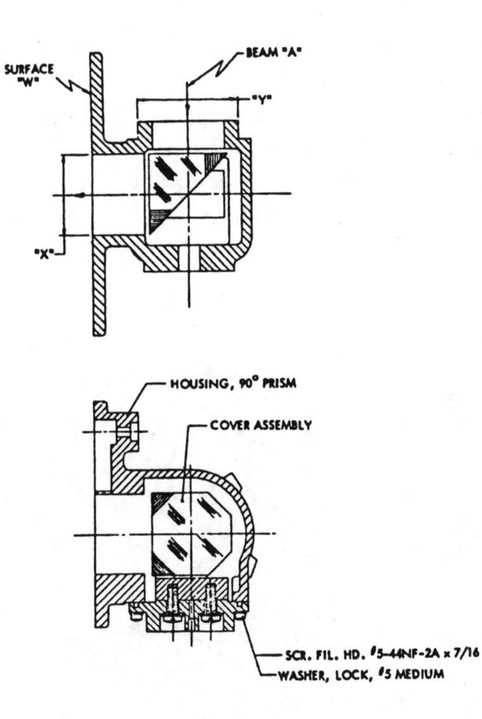

Fig. 9.16 First right angle prism assembly. (Courtesy of the U.S. Army)

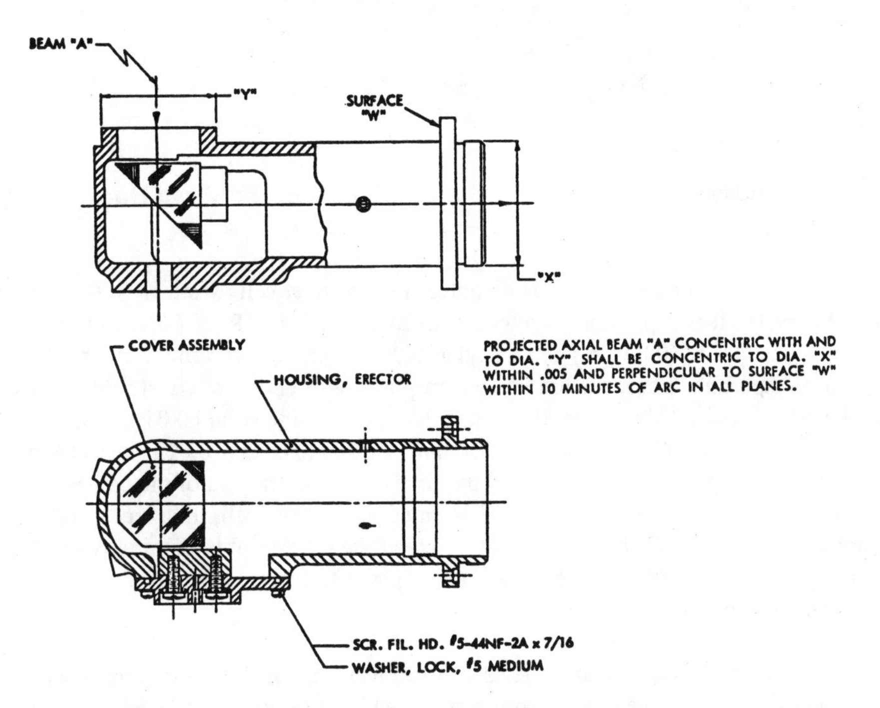

Fig. 9.17 Second right angle prism assembly. (Courtesy of the U.S.Army)

9.6 Mounting for the GOES telescope secondary mirror

The configuration of the Cassegrainian telescope used in NASA's Geostationary Operational Environmental Satellite (GOES) is shown in Fig. 9.18, The aperture of the primary mirror is 12.25 in. (31.1 cm) while that of the secondary mirror is 1.53 in. (3.9 cm). Particular attention is paid here to the mounting of the latter mirror.[92]

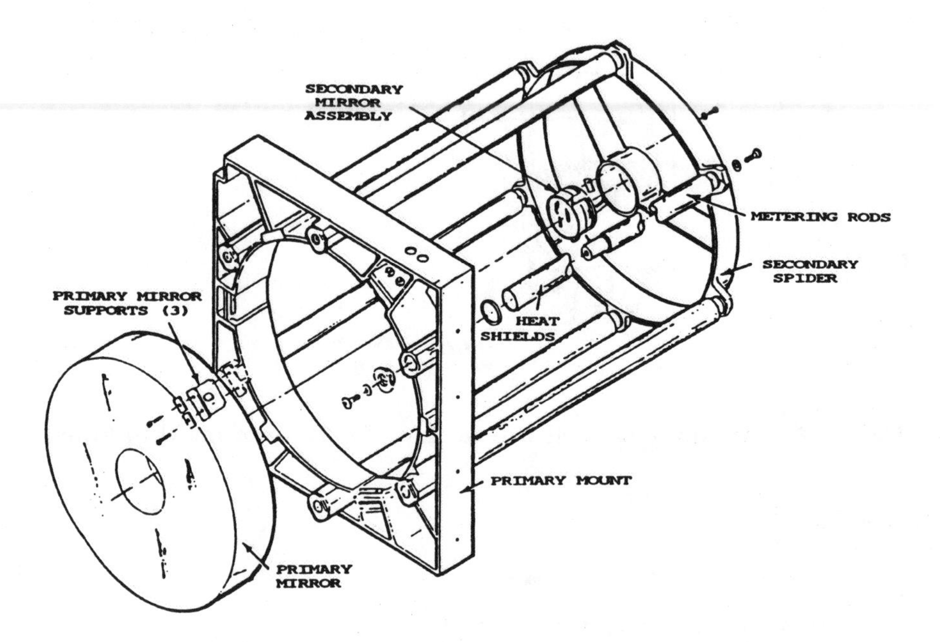

Fig. 9.18 Partially exploded view of the GOES telescope. (From Hookman[92])

The ULE secondary mirror is mounted into an Invar cell as shown in the section view of Fig. 9.19. It is supported radially and axially by pads of RTV-566 and registered against three 0.002 in. (0.05 μm) thick Mylar pads equally spaced around the periphery of the mirror aperture. The pads are bonded in place with epoxy to ensure that they do not move. The radial RTV pads are 0.20 in. (5.1 mm) in diameter and 0.01 in. (0.25 mm) thick while the axial pads are 0.20 in. (5.1 mm) in diameter and 0.025 in. (0.64 mm) thick. The Invar retaining ring is clamped by three screws to the end of the cell as shown in the exploded view of Fig. 9.20. When bottomed against the cell, the cured axial RTV pads are compressed 0.002 in. (0.05 mm) to preload the subassembly nominally by about 2.15 lb (9.6 N). The radial pads are located at the neutral plane of the mirror to minimize bending due to radial forces.

To minimize temperature effects due to mismatch of the Invar mirror cell and the aluminum spider, the cell is supported at the ends of three flexure blades machined integrally into the mount shown at right in Fig. 9.20. The mount is 6061-T6 aluminum.

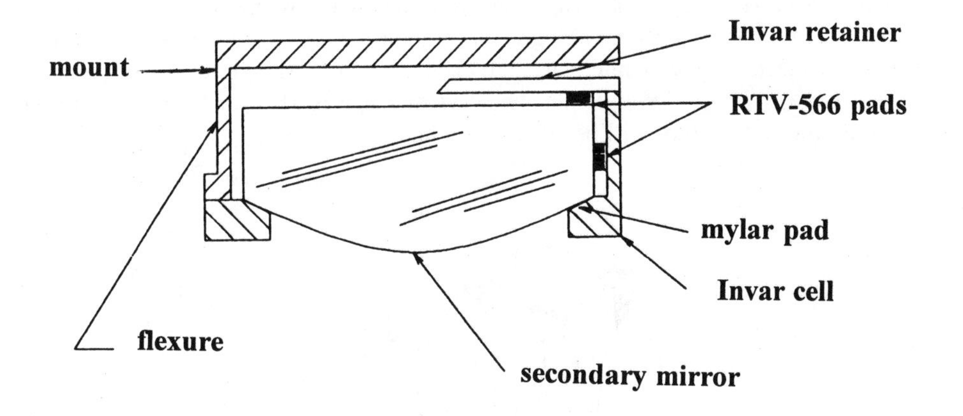

Fig. 9.19 Sectional view of the secondary mirror mount. (From Hookman[92])

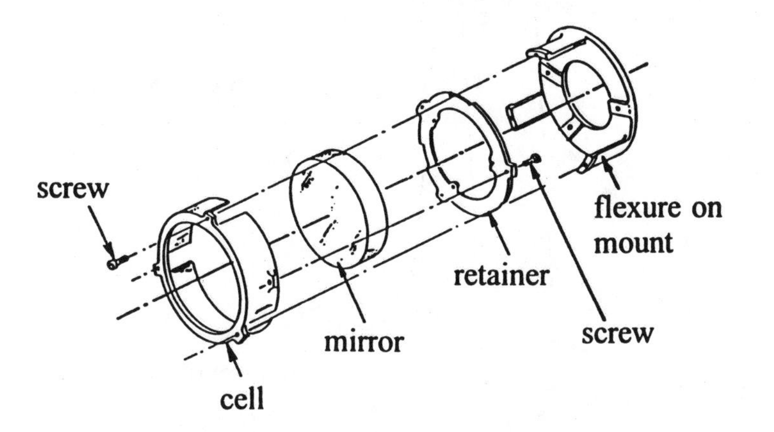

Fig. 9.20 Exploded view of the secondary mirror mount. (From Hookman[92])

The flexure blades are 0.5 in (12.7 mm) long, 0.32 in. (8.1 mm) wide, and 0.020 in. (0.5 mm) thick. Temperature changes do not disturb the location or orientation of the secondary mirror with this design. Tests of the spider-mounted mirror indicated that the natural frequency of the assembly was 830 Hz. This safely avoided strong driving frequencies of the application.

9.7 Mounting for the FUSE spectrograph gratings

The Far Ultraviolet Spectroscopic Explorer (FUSE) is a NASA sponsored, low earth orbiting astrophysical observatory designed to provide high spectral resolution observations across a 905 to 1195 Å spectral band. Figure 9.21 shows schematically the optical arrangement of the spectrograph.[93] Light is collected by four off-axis parabolic telescope mirrors (not shown) and focused onto four slit mirrors which act both as moveable entrance slits for the spectrometer and as mirrors that direct the visible starfield to fine error sensors (also not shown). The diverging light passing through the slits is diffracted and re-imaged by four holographic grating mount assemblies (GMAs). The spectra are collected on two microchannel plate detectors. The on-orbit temperature operating range is 15° to 25° C while the survival range is -10° to 40° C. During an observation, the temperature will be stabilized within 1° C.

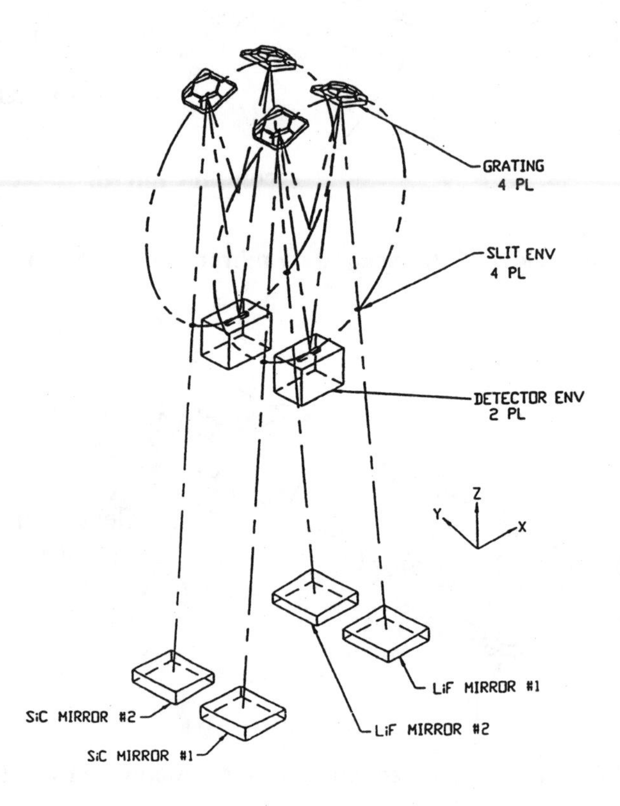

Fig. 9.21 Optical arrangement of the FUSE spectrograph. (From Shipley et al.[93])

The four gratings are identical in size (266 by 275 by 68.1 mm). They are made of Corning 7940 fused silica, class 0, grade F. This material was chosen to accommodate the process of adding holographic gratings, for its low CTE and availability. The weight of each blank is reduced by machining the rib pattern shown in Fig. 9.22 in its back surface. Two corners are removed to accommodate an anticipated envelope interference. Strength and fracture control requirements dictate that the blank be acid etched after machining. The optical surface is polished to a spherical radius with maximum $\lambda/10$ figure error.

The mounting interface is at three ribs in the central hexagon. Three U-shaped brackets are slipped over and bonded to the ribs as indicated in Fig. 9.23. The holographically recorded grating spacings are customized for two UV spectral bands; the

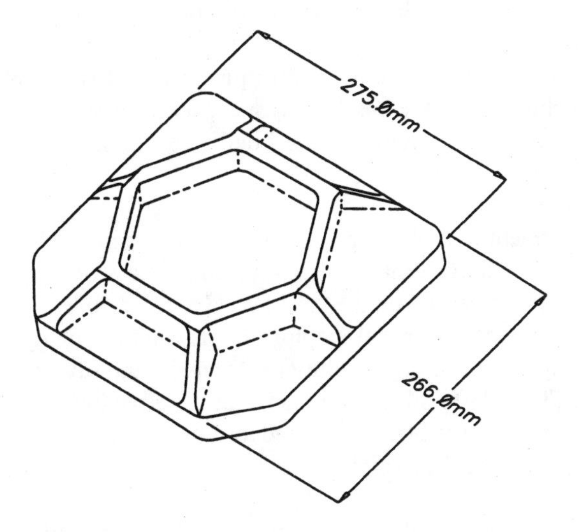

Fig. 9.22 View of the back side of the grating blank showing its machined rib structure. (From Shipley et al.[93])

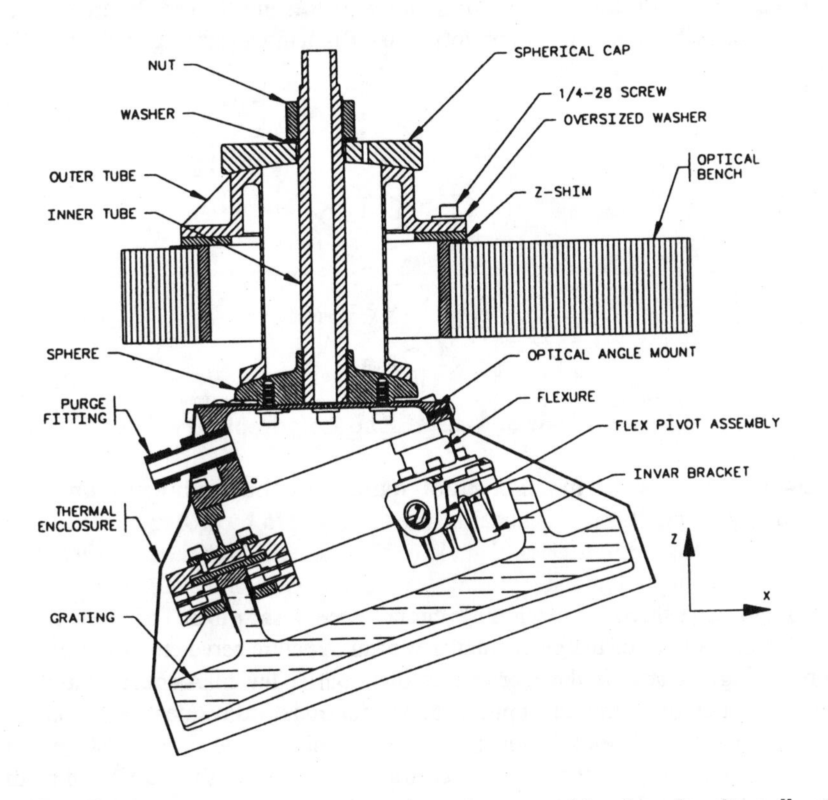

Fig. 9.23 Section view of the grating mount assembly showing its adjustment provisions. (From Shipley et al.[94])

ruled surfaces are coated with LiF or SiC to optimize performance in these bands.

The Invar mounting brackets are bonded in place with Hysol EA9396 epoxy. Tests of bonded samples showed that the bond strength consistently exceeded 4000 lb/in.^{2} with some samples exceeding 5000 lb/in.^{2}. This bond strength is judged more than adequate for the application.

Calculations of fracture probability using Gaussian and Weibull statistical methods were inconclusive.[94] The fact that the non-optical surfaces of the gratings were not polished contributes significantly to the problem. In order to ensure success, a conservative mechanical interface design was employed; this was thoroughly evaluated by finite element means throughout the design evolution. One major improvement was to add the flex pivots shown in Fig. 9.23 to allow compliance in the directions perpendicular to the radial flexures. The radial flexure blades were reduced in length to accommodate this addition while maintaining mechanism height.

Figure 9.24 shows details of the flex pivot installation. Each pivot consists of outer and inner pivot housings, two 0.625 in. diameter welded cantilever flex pivots, and eight cone point setscrews. Each pivot's location is maintained by the setscrews which are driven into shallow conical divots machined at two places in each cantilevered end of the flexure. Rigorous vibration testing of prototype and flight model grating mounts confirmed the design.

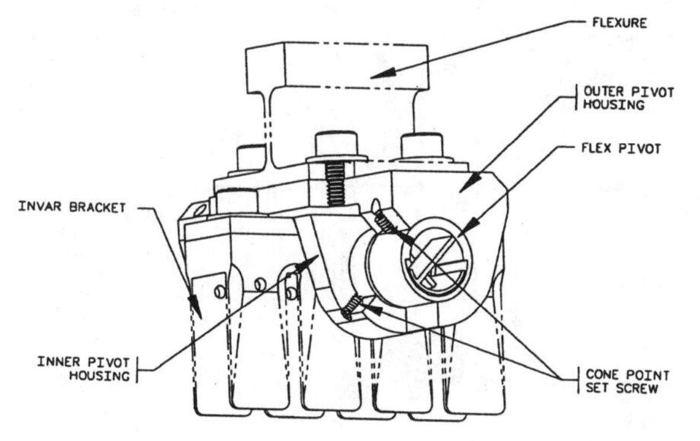

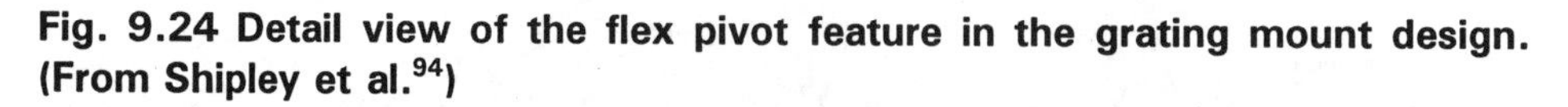
Fig. 9.24 Detail view of the flex pivot feature in the grating mount design. (From Shipley et al.[94])

The wedge-shaped optical angle mount seen in Fig. 9.23 between the radial flexures and the bottom of the outer tubular central structure serves to orient the grating at the proper angle relative to the coordinate system of the device. A spacer (shim) is used between the outer tube and the optical bench for axial adjustment. The outer tube interfaces with spherical seats at top and bottom. This allows fine adjustment of the angular orientation by external motorized screws in an alignment fixture. Clamping of this adjustment is accomplished by torquing the nut at the top of the assembly. Optical cubes attached to the backs of the gratings are used with theodolites as the metrology means during alignment.

Titanium was used extensively in the grating mount because of its high strength and relatively low CTE. All titanium parts except the flexures are Tiodized[95] to reduce friction between mating surfaces during alignment and to facilitate cleaning at assembly. The convex sphere and the spherical washer of Fig. 9.23 are made of Type 17-4 PH stainless steel, the nut is Type 303 stainless steel, and the Z-shim is a Type 400 stainless steel.

9.8 Mounting for the IRAS cryogenic beryllium mirror

The Infrared Astronomical Satellite (IRAS) was developed for NASA and was launched to a ~900 km altitude earth orbit in January 1983. It is shown in Fig. 9.25. The primary and secondary mirrors, as well as the main structure of the telescope were made of beryllium.[96] Figure 9.26 shows the optomechanical layout of the telescope.[97] The f/2 primary had a 24 in. (61 cm) diameter and was mounted on flexures to minimize temperature-induced misalignments and surface distortions when assembled at room temperature, tested (at 40 K), and operated at cryogenic temperatures of 2 K. The design and mounting for this mirror is the subject of this section.

Fig. 9.25 Photographs of the IRAS telescope. (From Schreibman and Young[96])

The primary was made of optical grade beryllium I-70A made by Kawecki-Berylco. This material was known to have superior CTE homogeneity as compared to other types of that material. Homogeneity was specified as <76 ppm/°C. The mirror was machined from a cylindrical billet to form a lightweighted structure generally as indicated in Fig. 9.26. The radial ribs were located at 20° intervals; the circumferential ribs were spaced progressively closer to each other as their radial distance from the axis increased. The mirror was cantilevered at three points at 120° intervals on flexures attached to pads machined into the back surface of the mirror at the 9.2 in. (23.4 cm) radial zone containing the radial center of mass.

Since these machined pads could not be absolutely parallel or coplanar, minute residual errors would be expected to translate into moments that would distort the mirror's

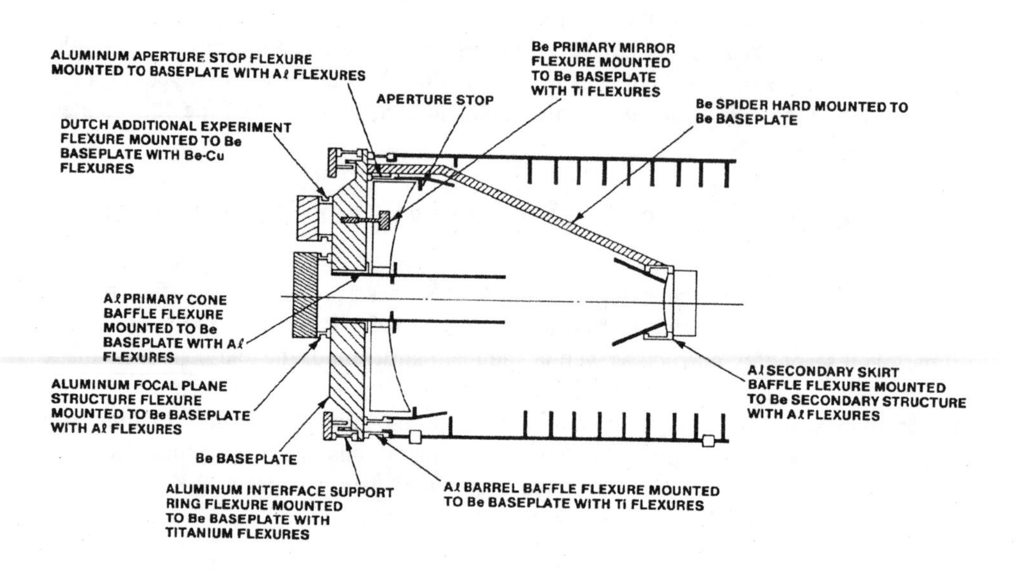

Fig. 9.26 Optomechanical layout of the IRAS telescope. (Adapted from Young and Schreibman[97])

optical surface when the attachment bolts were tightened or when the temperature changed significantly. Figure 9.27 illustrates the changes in reflecting surface contours resulting from three types of possible errors in location or orientation of the mounting pads. Tolerances of $\pm$ 25 μin. ($\pm$0.63 μm) were imposed at the points labled "T".

Mounting-imposed and gravity-release mirror distortions due to "within tolerance" errors were minimized by the design of the "T"-shaped flexures shown in Fig. 9.28. The diagrams indicate the stiff and compliant axes. These flexures were made of 5Al-2.5Sn titanium alloy with a CTE closely matching that of beryllium. The alloy was further designated as extra low interstitial (ELI). Vukobratovich et al. subsequently pointed out[98] that Ti-6Al-4V ELI would be a better material for such flexures. The flexures actually used were designed to support the mirror with axis horizontal during testing since that was necessary for the mirror to fit into an available cryogenic test chamber. NASTRAN finite element analysis indicated that the surface distortions due to gravity would not exceed 0.2 wave rms at $\lambda = 633$ nm after removal of defocus, decentration, and tilt. The system error budget allowed 0.1 wave rms for this error.

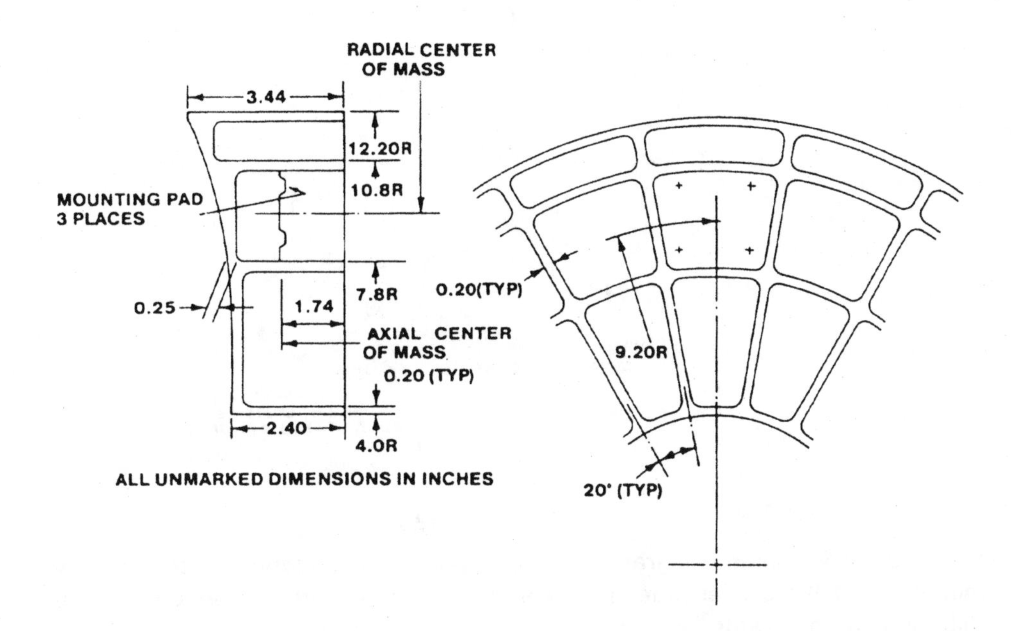

Fig. 9.27 Detail views of the lightweighted beryllium primary used in IRAS. (From Young and Schreibman[97])

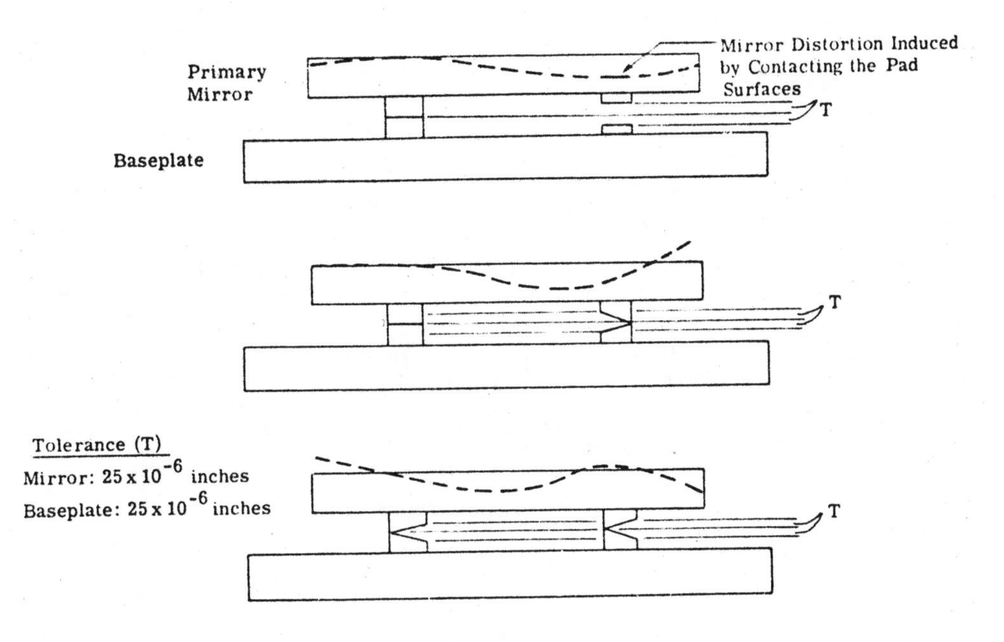

Fig. 9.28 Possible effects of machining errors in mounting pads on mirror surface figure. (From Schreibman and Young[96])

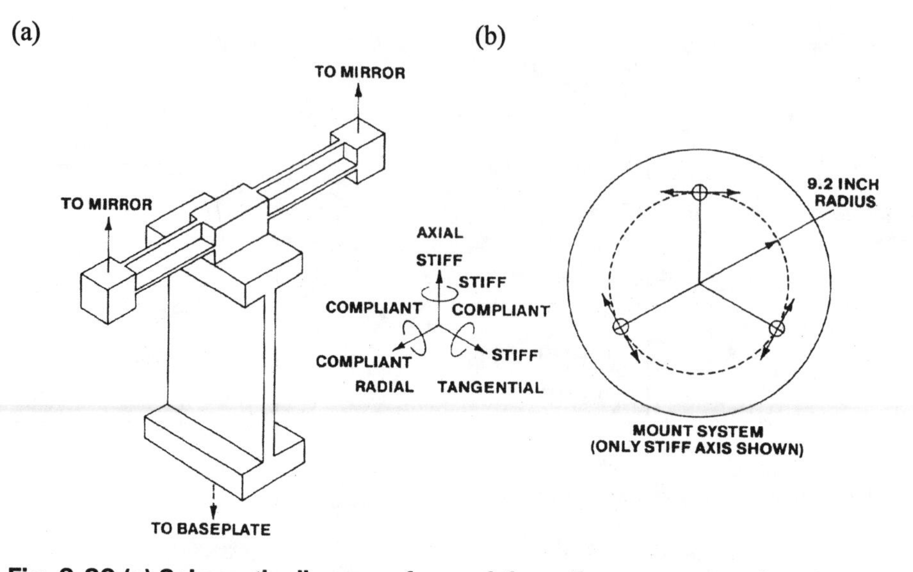

Fig. 9.29 (a) Schematic diagram of one of three flexures supporting the primary mirror in the IRAS telescope, (b) Arrangement of flexures on mount. (From Schreibman and Young[96])

APPENDIX A

PERTINENT UNIT CONVERSION FACTORS

To facilitate conversion from the U.S. Customary (formerly English) System to the metric (SI) System, we tabulate below the standard factors for changing the units commonly used in measuring selected physical parameters mentioned in this text. In most cases, this involves multiplying the value in U.S. Customary units by the factor listed. Conversion in the reverse direction requires division by the same factor.

To change **length** in
- feet (ft) to inches (in.), multiply by 0.3048.
- inches (in.) to millimeters (mm), multiply by 25.4.

To change **weight** in
- pounds (lb) to kilograms (kg), multiply by 0.4536.
- ounces (oz) to grams (g), multiply by 28.3495.

To change **force** in
- pounds (lb) to newtons (N), multiply by 4.4482.
- kilograms (kg) to newtons, multiply by 9.8066.

To change **linear preload or spring constant (stiffness)** in
- lb/in. to N/m, multiply by 175.126.

To change **pressure, stress or units for Young's modulus** in
- lb/in.^{2} (psi) to N/m^2, multiply by 6894.757.
- lb/in.^{2} (psi) to pascals (Pa), multiply by 6894.757.
- lb/in.^{2} (psi) to N/mm^2, multiply by 6.895×10^{-3}.
- atmospheres to Pa, multiply by 1.103×10^{5}.
- atmospheres to lb/in.^{2}, multiply by 14.7.
- pascals (Pa) to megapascals (MPa), divide by 10^6.

To change **torque** or **bending moment** in
- lb-in. to N-m, multiply by 0.11298.
- oz-in. to N-m, multiply by 7.0615×10^{-3}.
- lb-ft to N-m, multiply by 1.35582.

To change **density** in
- lb/in.^{3} to g/cm^2, multiply by 27.6799.

To change **acceleration** in
- gravitational units (G) to m/sec^2, multiply by 9.80665.
- ft/sec^2 to m/sec^2, multiply by 0.3048.

To change **temperature** in degrees F to degrees C, subtract 32 and multiply by 5/9.
- degrees C to degrees F, multiply by 9/5 and add 32.
- degrees C to degrees K and add 273.1.

Appendix B1
Extreme Service Environments

Environment	Normal	Severe	Extreme	Example of Extreme
Low temperature	293 K	222 K	2.4 K	Cryogenic satellite telescope
High temperature	300 K	344 K	423 K	White cell for combustion studies
Low pressure	88 KPa	57 KPa	0	Satellite telescope
High pressure	108 KPa	1 MPa	138 MPa	Submersible window
Humidity	25-75% RH	100% RH	(under water)	Submersible window
Acceleration (here a_G)	2	12	11,000	Gun-launched projectile
Vibration	200×10^{-6} m/s rms, $f \geq 8$ Hz	0.04 g^2/Hz $20 \leq f \geq 100$ Hz	0.13 g^2/Hz $30 \leq f \geq 1500$ Hz	Satellite launch

From Vukobratovich[3]

Appendix B2

Vibration Power Spectral Densities

Environment	Frequency, f (Hz) (Hz)	Power Spectral Density, PSD
Navy warships	1 - 50	0.001 g^2/Hz
Minimum integrity test (MIL-STD -810E)	20 - 1000	0.04 g^2/Hz
	1000 - 2000	-6 dB/octave
Typical aircraft	15 - 100	0.03 g^2/Hz
	100 - 300	+4 dB/octave
	300 - 1000	0.17 g^2/Hz
	≥1000	-3 dB/octave
Thor-delta launch vehicle	20 - 200	0.07 g^2/Hz
Titan launch vehicle	13 - 30	+6 dB/octave
	30 - 1500	0.13 g^2/Hz
	1500 - 2000	-6 dB/octave

Environment	Frequency, f (Hz)	Power Spectral Density (PSD)
Ariane launch vehicle	5 - 150	+6 dB/octave
	150 - 700	0.04 g^2/Hz
	700 - 2000	-3 dB/octave
Space shuttle (orbiter keel location)	15 - 100	+6 dB/octave
	100 - 400	0.10 g^2/Hz
	400 - 2000	-6 dB/octave

From Vukobratovich[3]

APPENDIX C

SELECTED MECHANICAL PROPERTIES OF MATERIALS

On the following pages we reproduce tables of material properties derived from various sources as noted. Included are:

Table C1 - Selected mechanical properties of 68 selected Schott glasses ranked in order of increasing value of refractive index, n_d (Page 1 of 4)

Rank	Glass name	Glass type	Young's modulus (E_G)		Poisson's ratio (υ_G)	$K_G=(1-\upsilon_G{}^2)/E_G$		Thermal expansion coefficient (α_G)		Density (ρ)	
			USC units (lb/in^2)	SI units (N/m^2)		USC units (in^2/lb)	SI units (m^2/N)	USC units (1/°F)	SI units (1/°C)	USC units (lb/in^3)	SI units (g/cm^3)
1	K10	501564	9.57e+06	6.60e+10	0.192 a	1.006e-07	1.459e-11	3.6e-06	6.5e-06	0.091	2.52
2	ZKN7	508612	1.03e+07	7.10e+10	0.214	9.266e-08	1.344e-11	2.5e-06	4.5e-06	0.090	2.49
3	BK1	510635	1.07e+07	7.40e+10	0.210	8.906e-08	1.292e-11	4.3e-06	7.7e-06	0.089 a	2.46 a
4	K7	511604	1.00e+07	6.90e+10	0.218	9.518e-08	1.380e-11	4.7e-06	8.4e-06	0.091	2.53
5	KF3	515547	9.57e+06	6.60e+10	0.216	9.959e-08	1.444e-11	4.5e-06	8.1e-06	0.092	2.56
6	BK7	517642	1.17e+07	8.10e+10	0.208	8.144e-08	1.181e-11	3.9e-06	7.1e-06	0.091	2.51
7	UBK7	517643	1.17e+07	8.10e+10	0.212	8.129e-08	1.179e-11	3.9e-06	7.0e-06	0.091	2.51
8	BaLKN3	518602	1.04e+07	7.20e+10	0.212	9.146e-08	1.326e-11	4.4e-06	7.9e-06	0.094	2.61
9	PK2	518651	1.22e+07	8.40e+10	0.209	7.850e-08	1.138e-11	3.8e-06	6.9e-06	0.091	2.51
10	PK50	521697	9.57e+06	6.60e+10	0.235	9.870e-08	1.431e-11	4.9e-06	8.8e-06	0.094	2.59
11	K5	522595	1.03e+07	7.10e+10	0.227	9.211e-08	1.336e-11	4.6e-06	8.2e-06	0.094	2.59
12	KF9	523515	9.72e+06	6.70e+10	0.202	9.871e-08	1.432e-11	3.8e-06	6.8e-06	0.098	2.71
13	LLF6	532488	9.14e+06	6.30e+10	0.205	1.048e-07	1.521e-11	4.2e-06	7.5e-06	0.102	2.81
14	BaK2	540597	1.03e+07	7.10e+10	0.233	9.184e-08	1.332e-11	4.4e-06	8.0e-06	0.103	2.86
15	LLF1	548458	8.70e+06	6.00e+10	0.210	1.098e-07	1.593e-11	4.5e-06	8.1e-06	0.106	2.94
16	KzFN1	551496	8.85e+06	6.10e+10	0.224	1.074e-07	1.557e-11	3.9e-06	7.1e-06	0.098	2.71
17	PSK3	552635	1.22e+07	8.40e+10	0.226	7.789e-08	1.130e-11	3.4e-06	6.2e-06	0.105	2.91
18	BaK5	557587	1.04e+07	7.20e+10	0.241	9.020e-08	1.308e-11	4.3e-06	7.8e-06	0.109	3.02
19	SK11	564608	1.15e+07	7.90e+10	0.239	8.229e-08	1.194e-11	3.6e-06	6.5e-06	0.111	3.08
20	BaK50	568580	1.17e+07	8.10e+10	0.259	7.941e-08	1.152e-11	2.1e-06 a	3.7e-06 a	0.106	2.93
21	BaK1	573575	1.07e+07	7.40e+10	0.253	8.721e-08	1.265e-11	4.2e-06	7.6e-06	0.115	3.19

Table C1 - Selected mechanical properties of 68 selected Schott glasses ranked in order of increasing value of refractive index, n_d (Page 2 of 4)

Rank	Glass name	Glass type	Young's modulus (E_G)		Poisson's ratio (υ_G)	$K_G=(1-\upsilon_G^2)/E_G$		Thermal expansion coefficient (α_G)		Density (ρ)	
			USC units (lb/in^2)	SI units (N/m^2)		USC units (in^2/lb)	SI units (m^2/N)	USC units (1/°F)	SI units (1/°C)	USC units (lb/in^3)	SI units (g/cm^3)
22	LF7	575415	8.56e+06	5.90e+10	0.217	1.114e-07	1.615e-11	4.4e-06	7.9e-06	0.116	3.20
23	LF4	578416	8.70e+06	6.00e+10	0.219	1.094e-07	1.587e-11	4.5e-06	8.1e-06	0.116	3.21
24	BaLF4	580537	1.10e+07	7.60e+10	0.244	8.532e-08	1.237e-11	3.6e-06	6.4e-06	0.115	3.17
25	LF5	581409	8.56e+06	5.90e+10	0.226	1.109e-07	1.608e-11	5.1e-06	9.1e-06	0.116	3.22
26	BaF3	583465	9.28e+06	6.40e+10	0.261	1.004e-07	1.456e-11	4.3e-06	7.8e-06	0.118	3.28
27	SK5	589613	1.22e+07	8.40e+10	0.256	7.670e-08	1.112e-11	3.1e-06	5.5e-06	0.119	3.30
28	F3	596392	8.41e+06	5.80e+10	0.224	1.129e-07	1.638e-11	4.4e-06	8.0e-06	0.128	3.55
29	F14	601382	8.41e+06	5.80e+10	0.218	1.132e-07	1.642e-11	4.4e-06	7.9e-06	0.124	3.44
30	F5	603380	8.41e+06	5.80e+10	0.220	1.131e-07	1.641e-11	4.4e-06	8.0e-06	0.125	3.47
31	BaF4	606439	9.57e+06	6.60e+10	0.247	9.809e-08	1.423e-11	4.4e-06	7.9e-06	0.126	3.50
32	SK2	607567	1.13e+07	7.80e+10	0.263	8.228e-08	1.193e-11	3.3e-06	6.0e-06	0.128	3.55
33	F4	617366	7.98e+06 a	5.50e+10 a	0.225	1.190e-07 a	1.726e-11 a	4.6e-06	8.3e-06	0.129	3.58
34	SSKN8	618498	1.20e+07	8.30e+10	0.256	7.763e-08	1.126e-11	3.9e-06	7.1e-06	0.120	3.33
35	F2	620364	8.41e+06	5.80e+10	0.225	1.129e-07	1.637e-11	4.6e-06	8.2e-06	0.130	3.61
36	FN11	621362	1.22e+07	8.40e+10	0.230	7.774e-08	1.128e-11	4.2e-06	7.5e-06	0.096	2.66
37	BaF8	624470	1.06e+07	7.30e+10	0.260	8.806e-08	1.277e-11	3.9e-06	7.0e-06	0.133	3.67
38	F7	625356	7.98e+06 a	5.50e+10 a	0.239	1.182e-07	1.714e-11	5.4e-06 a	9.8e-06 a	0.131	3.62
39	F1	626357	8.12e+06	5.60e+10	0.229	1.167e-07	1.692e-11	4.8e-06	8.7e-06	0.132	3.65
40	BaSF1	626390	8.99e+06	6.20e+10	0.242	1.047e-07	1.518e-11	4.7e-06	8.5e-06	0.132	3.66
41	F6	636353	8.27e+06	5.70e+10	0.231	1.145e-07	1.661e-11	4.7e-06	8.5e-06	0.136	3.76
42	SF12	648338	8.70e+06	6.00e+10	0.228	1.089e-07	1.580e-11	4.3e-06	7.8e-06	0.135	3.74

Table C1 - Selected mechanical properties of 68 selected Schott glasses ranked in order of increasing value of refractive index, n_d (Page 3 of 4)

Rank	Glass name	Glass type	Young's modulus (E_G) USC units (lb/in^2)	Young's modulus (E_G) SI units (N/m^2)	Poisson's ratio (υ_G)	$K_G=(1-\upsilon_G^2)/E_G$ USC units (in^2/lb)	$K_G=(1-\upsilon_G^2)/E_G$ SI units (m^2/N)	Thermal expansion coefficient (α_G) USC units (1/°F)	Thermal expansion coefficient (α_G) SI units (1/°C)	Density (ρ) USC units (lb/in^3)	Density (ρ) SI units (g/cm^3)
43	SF2	648339	7.98e+06 a	5.50e+10 a	0.231	1.187e-07	1.721e-11	4.7e-06	8.4e-06	0.139	3.86
44	LaKN22	651559	1.31e+07	9.00e+10	0.268	7.111e-08	1.031e-11	3.7e-06	6.6e-06	0.135	3.73
45	BaSF2	664358	9.57e+06	6.60e+10	0.245	9.820e-08	1.424e-11	4.6e-06	8.2e-06	0.141	3.90
46	SF19	667330	8.41e+06	5.80e+10	0.228	1.127e-07	1.635e-11	4.3e-06	7.7e-06	0.145	4.02
47	SF5	673322	8.12e+06	5.60e+10	0.233	1.164e-07	1.689e-11	4.6e-06	8.2e-06	0.147	4.07
48	LaF20	682482	1.36e+07	9.40e+10	0.273	6.788e-08	9.845e-12	4.1e-06	7.4e-06	0.140	3.87
49	BaF50	683445	1.35e+07	9.30e+10	0.266	6.889e-08	9.992e-12	4.6e-06	8.3e-06	0.137	3.80
50	SF8	689312	8.12e+06	5.60e+10	0.233	1.164e-07	1.689e-11	4.6e-06	8.2e-06	0.152	4.22
51	SF15	699301	8.70e+06	6.00e+10	0.235	1.086e-07	1.575e-11	4.4e-06	7.9e-06	0.147	4.06
52	SFN64	706303	1.35e+07	9.30e+10	0.250	6.950e-08	1.008e-11	4.7e-06	8.5e-06	0.108	3.00
53	SF1	717295	8.27e+06	5.70e+10	0.234	1.143e-07	1.658e-11	4.5e-06	8.1e-06	0.161	4.46
54	SF18	722293	8.12e+06	5.60e+10	0.238	1.161e-07	1.685e-11	4.5e-06	8.1e-06	0.162	4.49
55	SF10	728284	9.28e+06	6.40e+10	0.232	1.019e-07	1.478e-11	4.2e-06	7.5e-06	0.155	4.28
56	SF53	728287	8.41e+06	5.80e+10	0.238	1.121e-07	1.626e-11	4.6e-06	8.2e-06	0.161	4.45
57	SF13	741276	9.28e+06	6.40e+10	0.233	1.019e-07	1.478e-11	3.9e-06	7.1e-06	0.158	4.36
58	SF54	741281	8.41e+06	5.80e+10	0.237	1.122e-07	1.627e-11	4.3e-06	7.7e-06	0.165	4.56
59	SF14	762265	9.43e+06	6.50e+10	0.235	1.002e-07	1.454e-11	3.7e-06	6.6e-06	0.164	4.54
60	SF55	762270	8.12e+06	5.60e+10	0.247	1.156e-07	1.677e-11	4.6e-06	8.2e-06	0.171	4.72
61	SF11	785258	9.57e+06	6.60e+10	0.237	9.860e-08	1.430e-11	3.4e-06	6.1e-06	0.171	4.74
62	SFL56	785261	1.32e+07	9.10e+10	0.255	7.084e-08	1.027e-11	4.8e-06	8.7e-06	0.118 b	3.28 b
63	SF56	785261	8.41e+06	5.80e+10	0.242	1.119e-07	1.623e-11	4.4e-06	7.9e-06	0.178	4.92

Table C1 - Selected mechanical properties of 68 selected Schott glasses ranked in order of increasing value of refractive index, n_d (Page 4 of 4)

Rank	Glass name	Glass type	Young's modulus (E_G)		Poisson's ratio (υ_G)	$K_G=(1-\upsilon_G{}^2)/E_G$		Thermal expansion coefficient (α_G)		Density (ρ)	
			USC units (lb/in^2)	SI units (N/m^2)		USC units (in^2/lb)	SI units (m^2/N)	USC units (1/°F)	SI units (1/°C)	USC units (lb/in^3)	SI units (g/cm^3)
64	LaSF32	803304	1.64e+07	1.13e+11	0.256	5.702e-08	8.270e-12	4.4e-06	7.9e-06	0.127	3.52
65	LaSFN30	803464	1.80e+07 a	1.24e+11 a	0.293 a	5.083e-08 a	7.372e-12 a	3.4e-06	6.2e-06	0.161	4.46
66	SFL6	805254	1.35e+07	9.30e+10	0.260	6.913e-08	1.003e-11	5.0e-06	9.0e-06	0.122 b	3.37 b
67	SF6	805254	8.12e+06	5.60e+10	0.248	1.155e-07	1.676e-11	4.5e-06	8.1e-06	0.187 a	5.18 a
68	LaSFN9	850322	1.58e+07	1.09e+11	0.286	5.808e-08	8.424e-12	4.1e-06	7.4e-06	0.160	4.44
Ratio (max./min.)			2.25		1.53	2.34		2.57		2.11	

Glass selection from Walker.[11] Data (except for K_G) from Schott.[10]
Note: "a" indicates extreme low or high value, "b" indicates lightweight version.

Table C2 - Comparison of 11 lightweight optical glasses (L) with the nearest standard glass types (S)

Glass name	n_d	v_d	Density (g/cm^3)	Weight reduction (%)	Resistance to: acid "SR"	alkali "AR"	Knoop hardness "HK"
(L) FN11	1.62096	36.18	2.66	26	1	1.0	510
(S) F2	1.62004	36.37	3.61		1	2.3	370
(L) SFL4	1.75520	27.21	3.37	30	1.0	1.3	500
(S) SF4	1.75520	27.58	4.79		51	2.3	330
(L) SFL6	1.80518	25.39	3.37	35	2	1	500
(S) SF6	1.80518	25.43	5.18		52	2.3	310
(L) SFL56	1.78470	26.08	3.28	33	2	1.3	530
(S) SF56A	1.78470	26.08	4.92		3	2.2	330
(L) SFL57	1.84666	23.62	3.55	36	1.3	1	510
(S) SF57	1.84666	23.83	5.51		52	2.3	300
(L) SFN64	1.70585	30.30	3.00	26	1-2	1.2	500
(S) SF15	1.69895	30.07	4.06		1	1.2	370
(L) BaSF64A	1.70400	39.38	3.20	19	2	1.2	540
(S) BaSF13	1.69761	38.57	3.97		51	1.2	450
(L) LaSF32	1.80349	30.40	3.52	28	1-2	1.0	560
(S) LaSF8	1.80741	31.61	4.87		52	1.2	420
(L) LaSF36A	1.79712	35.08	3.60	20	3	1.0	570
(S) LaSF33	1.80596	34.24	4.48		51	1.2	440
(L) LaKL12	1.67790	54.93	3.32	19	52	2.2	600
(S) LaKN12	1.67790	55.20	4.10		53	4.2	470
(L) LaKL21	1.64048	59.75	2.97	21	53	4.2	590
(S) LaK21	1.64050	60.10	3.74		53	4.2	460

From Schott[10]

Table C3 - Selected optical and mechanical characteristics of commonly used optical plastics (Page 1 of 2)

Properties*	ASTM method	Units	Methyl methacrylate (acrylic)	Polystyrene	Polycarbonate	Methyl methacrylate styrene co-polymer (NAS)	Styrene acrylonitrile (SAN)	Allyl diglycol carbonate (CR39)
Refractive index (n_D)	D542		1.491(7)	1.590(3)	1.586(0)	1.564	1.567-1.571	1.504
Abbe value (v_D)	D542		57.2	30.8	34.0	35	38	56
$dn/dT (\times 10^{-5})$		(/° C)	-8.5	-12.0	-14.3	-14.0		-14.3
Haze	D1003	(%)	<2	<3	<3	<3	3	3
Luminous transmittance (0.125 in. thickness)	D1003	(%)	92	87-92	85-91	90	88	93
Deflection temperature ($\times 10^{-5}$)	D648-56	(/° F)						
3.6° F/min., 264 psi			198	180	280		99-104	
3.6° F/min., 66 psi			214	230	270	212	100	
CTE ($\times 10^{-5}$)	D696-44	(/° F)	3.6	3.5	3.8	3.6	3.6-3.7	6.3 @25-75°C

Table C3 - Selected optical and mechanical characteristics of commonly used optical plastics (Page 2 of 2)

Properties*	ASTM method	Units	Methyl methacrylate (acrylic)	Polystyrene	Polycarbonate	Methyl methacrylate styrene co-polymer (NAS)	Styrene acrylonitrile (SAN)	Allyl diglycol carbonate (CR39)
Recommended maximum continuous service temperature		(° F)	175	175	240	175	212	
Water absorption (immersed 24 hr @ 73°F)	D570-63	(%)	0.3	0.2	0.15	0.15	0.2	
Specific gravity (ρ)	D792		1.19	1.06	1.20	1.09		
Hardness (0.25 in. sample)	D785-62		M97	M90	M70	M75		
Thermal conductivity (k)		(cal/sec-cm-° C)	4.96	2.4-3.3	4.65	4.5	2.9	5

* Material formulation data should be confirmed prior to design and specification.
Source: From manufacturer's publications, U.S. Precision Lens,[99] and Wolpert.[100]

Table C4 - Optomechanical properties of selected alkali halides and alkaline earth halides (Page 1 of 2)

Material name and symbol	Refractive index @ λ in μm	dN/dT @ λ in μm ($\times 10^{-6}$ /°C)	CTE ($\times 10^{-6}$/° C)	Young's modulus ($\times 10^{10}$ N/m^2)	Poisson's ratio	Density (g/cm^3)	Knoop hardness (kg/mm^2)
Barium fluoride (BaF_2)	1.463 @ 0.63 1.458 @ 3.8 1.449 @ 5.3 1.396 @ 10.6	-16.0 @0.63 -15.9 @3.4 -14.5 @10.6	6.7 @75K 18.4 @300K	5.32	0.343	4.89	82 (500 g)
Calcium fluoride (CaF_2)	1.431 @ 0.7 1.420 @ 2.7 1.411 @ 3.8 1.395 @ 5.3	-10.4 @0.66 - 8.1 @3.4	18.9 @300K	9.6	0.29	3.18	160-178
Potassium bromide (KBr)	1.555 @ 0.6 1.537 @ 2.7 1.529 @ 8.7 1.515 @ 14	 -41.9 @1.15 -41.1 @10.6	25.0 @ 75K	2.69	0.203	2.75	7 (200 g)
Potassium chloride (KCl)	1.474 @ 2.7 1.472 @ 3.8 1.469 @ 5.3 1.454 @ 10.6	-36.2 @1.15 -34.8 @10.6	36.5	2.97	0.216	1.98	7.2 (200 g)
Lithium fluoride (LiF)	1.394 @ 0.5 1.367 @ 3.0 1.327 @ 5.0	-16.0 @0.46 -16.9 @1.15 -14.5 @3.39	5.5	6.48	0.225	2.63	102-113 (600 g)

Table C4 - Optomechanical properties of selected alkali halides and alkaline earth halides (Page 2 of 2)

Material name and symbol	Refractive index @ λ in μm	dN/dT @ λ in μm ($\times 10^{-6}$ /°C)	CTE ($\times 10^{-6}$/° C)	Young's modulus ($\times 10^{10}$ N/m^2)	Poisson's ratio	Density (g/cm^3)	Knoop hardness (kg/mm^2)
Magnesium fluoride (MgF_2)	1.384 @ 0.4o* 1.356 @ 3.8o* 1.333 @ 5.3o*	+0.88@1.15 +1.19@3.39	14.0(∥) 8.9 (⊥)	16.9	0.269	3.18	415
Sodium chloride (NaCl)	1.525 @ 2.7 1.522 @ 3.8 1.517 @ 5.3	-36.3 @3.39	39.6	4.01	0.28	2.16	15.2 (200 g)
Thallium bromo-iodide (KRS5)	2.602 @ 0.6 2.446 @ 1.0 2.369 @ 10.6 2.289 @ 30	-254 @0.6 -240 @1.1 -233 @10.6 -152 @40	58	1.58	0.369	7.37	40.2 (200 g)

* Birefringent material: o = ordinary axis

From Yoder[8] and Tropf et al[101]

Table C5 - Mechanical properties of selected IR-transmitting glass and other oxides (Page 1 of 3)

Material name and symbol	Refractive index @ λ in μm	dN/dT @ λ in μm ($\times10^{-6}$ /°C)	CTE ($\times10^{-6}$/° C)	Young's modulus ($\times10^{10}$ N/m^2)	Poisson's ratio	Density (g/cm^3)	Knoop hardness (kg/mm^2)
Aluminum oxynitride (ALON)	1.793 @ 0.6 1.66 @ 4.0		5.8	32.2	0.24	3.71	1970
Calcium alumino-silicate (Schott IRG11)	1.684 @ 0.55 1.608 @ 4.6 1.635 @ 3.3		8.2 @ 293-573 K	10.8	0.284	3.12	608
Calcium alumino-silicate (Corning 9753)	1.61 @ 0.5 1.57 @ 2.5		6.0 @ 293-573 K	9.86	0.28	2.798	600 (500 g)
Calcium alumino-silicate (Schott IRGN6)	1.592 @ 0.55 1.562 @ 2.3 1.521 @ 4.3		6.3 @ 293-573 K	10.8	0.284	3.12	608
Fluoride glass (Ohara HTF1)	1.51 @ 1.0 1.49 @ 3.0	-8.19	16.1	6.42	0.28	3.88	311

Table C5 - Mechanical properties of selected IR-transmitting glass and other oxides (Page 2 of 3)

Material name and symbol	Refractive index @ λ in μm	dN/dT @ λ in μm ($\times10^{-6}$ /°C)	CTE ($\times10^{-6}$/° C)	Young's modulus ($\times10^{10}$ N/m^2)	Poisson's ratio	Density (g/cm^3)	Knoop hardness (kg/mm^2)
Fluoro-phosphate glass (Schott IRG9)	1.488 @ 0.55 1.469 @ 2.3 1.458 @ 3.3		16.1 @ 293-573 K	7.7	0.288	3.63	346 (200 g)
Germanate (Corning 9754)	1.67 @ 0.5 1.63 @ 2.5 1.61 @ 4.0		6.2 @ 293-573 K	8.41	0.290	3.581	560 (100 g)
Germanate (Schott IRG2)	1.899 @ 0.55 1.841 @ 2.3		8.8 @ 293-573 K	9.59	0.282	5.00	481 (200 g)
Lanthanum dense flint (Schott IRG3)	1.851 @ 0.55 1.796 @ 2.3 1.776 @ 3.3		8.1 @ 293-573 K	9.99	0.287	4.47	541 (200 g)
Lead silicate (Schott IRG7)	1.573 @ 0.55 1.534 @ 2.3		9.6 @ 293-573 K	5.97	0.216	3.06	379

Table C5 - Mechanical properties of selected IR-transmitting glass and other oxides (Page 3 of 3)

Material name and symbol	Refractive index @ λ in μm	dN/dT @ λ in μm ($\times 10^{-6}$/°C)	CTE ($\times 10^{-6}$/° C)	Young's modulus ($\times 10^{10}$ N/m^2)	Poisson's ratio	Density (g/cm^3)	Knoop hardness (kg/mm^2)
Sapphire* (Al_2O_3)	1.684 @ 3.8 1.586 @ 5.8	13.7	5.6 (∥) 5.0 (⊥)	40.0	0.27	3.97	1370 (1000 g)
Fused silica (Corning 7940)	1.566 @ 0.19 1.460 @ 0.55 1.433 @ 2.3 1.412 @ 3.3	9.5-11.2 @ 0.5-2.5 μm	-0.6 @ 73K 0.52 @ 278-308 K 0.57 @ 273-473 K	7.3	0.164	2.202	500

* Birefringent material.
From Yoder[8] and Tropf et al.[101]

Table C6 - Mechanical properties of diamond and selected IR-transmitting semiconductor materials

Material name and symbol	Refractive index @ λ in μm	dN/dT @ λ in μm ($\times 10^{-6}$ /°C)	CTE ($\times 10^{-6}$/° C)	Young's modulus ($\times 10^{10}$ N/m^2)	Poisson's ratio	Density (g/cm^3)	Knoop hardness (kg/mm^2)
Diamond (C)	2.382 @2.5 2.381 @5.0 2.381 @10.6		-0.1 @ 25K 0.8 @ 293K 5.8 @ 1600K	105	0.1	3.51	9000
Indium antimonide (InSb)	3.99 @8.0	4.7	4.9	4.3		5.78	225
Gallium arsenide (GaAs)	3.1 @10.6	1.5	5.7	8.29	0.31	5.32	721
Germanium (Ge)	4.055 @2.7 4.026 @3.8 4.015 @5.3 4.00 @10.6	4.0 @ 250-350K	2.3 @100K 5.0 @200K 6.0 @300K	10.37	0.278	5.323	800
Silicon (Si)	3.436 @2.7 3.427 @3.8 3.422 @5.3 3.148 @10.6	1.3	2.7-3.1	13.1	0.279	2.329	1150

From Yoder,[8] Amirtharaj and Seiler,[102] and manufacturer's data sheets.

Table C7 - Mechanical properties of selected IR-transmitting chalcogenide materials

Material name and symbol	Refractive index @ λ in μm	dN/dT @ λ in μm ($\times 10^{-6}$/°C)	CTE ($\times 10^{-6}$/° C)	Young's modulus ($\times 10^{10}$ N/m^2)	Poisson's ratio	Density (g/cm^3)	Knoop hardness (kg/mm^2)
Arsenic trisulfide (AsS_3)	2.521 @0.8 2.412 @3.8 2.407 @5.0	85 @0.6 17 @ 1.0	26.1	1.58	0.295	3.43	180
$Ge_{33}As_{12}Se_{55}$ (AMTIR-1)	2.605 @1.0 2.503 @8.0	101 @1.0 72 @10.0	12.0	2.2	0.266	4.4	170
IRG 100 (Schott)	2.723 @1 2.620 @4 2.607 @8	103 @2.5 56 @10.6	15.0	2.1	0.261	4.67	150
Zinc sulfide (ZnS)	2.36 @0.6 2.257 @ 3.0 2.246 @ 5.0 2.192 @10.6	63.5 @0.63 49.8 @1.15 46.3 @10.6	4.6	7.45	0.29	4.08	230
Zinc selenide (ZnSe)	2.61 @0.6 2.438 @3.0 2.429 @5.0 2.403 @10.6	91.1 @0.63 59.7 @1.15 52.0 @10.6	5.6 @163K 7.1 @273K 8.3 @473K	7.03	0.28	5.27	105

From Yoder[8] and Tropf et al.[102]

Table C8a - Mechanical properties of selected nonmetallic mirror substrate materials

Material	Source	CTE. α, ppm/°C (ppm/°F)	Young's modulus, E 10^{10} N/m² (10^6 lb/in.²)	Poisson's ratio υ	Density ρ, g/cm³ (lb/in.³)	Specific heat C_P, J/kg K (Btu/lb °F)	Thermal conductivity k, W/m K (Btu/hr ft °F)	Knoop hardness	Surface smooth-, ness, Å (rms)
Duran 50	Schott	3.2 (1.8)	6.17 (8.9)	0.20	2.23 (0.0806)	835 (0.20)	1.02 (0.59)		5
Borosilicate crown E6	Ohara	2.8 (1.5)	5.86 (8.5)	0.195	2.18 (0.0788)				
Fused silica	Corning or Heraeus	0.56 (0.31)	7.32 (10.6)	0.167	2.202 (0.0796)	741 (0.177)	1.37 (0.8)	500	5
ULE 7971	Corning	0.015 (0.008)	6.77 (9.8)	0.176	2.20 (0.0795)	766 (0.183)	1.31 (0.76)	460	5
Zerodur	Schott	0 ± 0.05 (0 ± 0.03)	9.06 (13.6)	0.24	2.53 (0.0914)	821 (0.196)	1.64 (0.95)	630	5
Zerodur-M	Schott	0 ± 0.05 (0 ± 0.03)	8.9 (12.9)	0.25	2.57 (0.0928)	810 (0.194)	1.6 (0.92)	540	5
Cer-Vit C-101*	Owens-Illinois	0 ± 0.03 (0 ± 0.02)	9.18 (13.3)	0.252	2.50 (0.0903)	840 (0.20)	1.70 (1.0)	540	5

Adapted from Yoder[8] * Obsolete material

Table C8b - Mechanical properties of selected metallic and composite mirror substrate materials

Material	CTE. α, ppm/°C (ppm/°F)	Young's modulus, E 10^{10} N/m^2 (10^6 lb/in.^{2})	Poisson's ratio υ	Density ρ, g/cm^3 (lb/in.^{3})	Specific heat C_P, J/kg K (Btu/lb °F)	Thermal conductivity k, W/m K (Btu/hr ft °F)	Hard-ness	Surface smooth-, ness, Å (rms)
Beryllium I-70A	11.3 (6.3)	28.9 (42)	0.08	1.85 (0.067)	1820 (0.436)	194 (112)		60-80[a]
Aluminum 6061-T6	23.0 (12.8)	6.90 (10.0)	0.33	2.71 (0.098)	960 (0.23)	171 (99)	95 Brinell	200?
Copper OFHC	16.7 (9.3)	11.7 (17)	0.35	8.94 (0.323)	385 (0.092)	392 (226)	40 Rockw. F	40
Molybdenum TZM	5.0 (2.8)	31.8 (46)	0.32	10.2 (0.368)	272 (0.065)	146 (84.5)	200 Vickers	10
Silicon carbide RB-30% Si	2.64 (1.47)	31.0 (45)		2.92 (0.106)	660 (0.16)	158 (91)		
Silicon carbide RB-12% Si	2.68 (1.49)	37.3 (54.1)		3.11 (0.112)	680 (0.68)	147 (85)		
Silicon carbide CVD	2.4 (1.3)	46.6 (67.6)	0.21	3.21 (0.116)	700 (0.17)	146 (84)	2540 Knoop (500 g)	
SXA metal matrix 30% SiC_p in 2024 Al[b]	12.4 (6.9)	11.7 (17)		2.90 (0.105)	770 (0.18)	130 (75)		

From Yoder[8]   [a] Sputtered surface   [b] Contains SiC particles of mean size 3.5 μm (0.0014 in.) per Advanced Composite Materials Corp.

Table C9 - Comparison of material figures of merit especially pertinent to mirror design (Page 1 of 2)

	Weight and self-weight deflection proportionality factors				Thermal distortion coefficients	
Preferred	$(E/\rho)^{1/2}$ Resonant freqency for same geometry (arbitrary units) large	ρ/E Mass or deflection for same geometry (arbitrary units) small	ρ^3/E Deflection for same mass (arbitrary units) small	$(\rho^3/E)^{1/2}$ Mass for same deflection (arbitrary units) small	α/k Steady state (μm/W) small	α/D Transient (sec/m^2 K) small
Pyrex	5.3	3.53	1.76	0.420	2.92	5.08
Fused silica	5.7	3.04	1.46	0.382	0.36	0.59
ULE fused silica	5.5	3.30	1.61	0.401	0.02	0.04
Zerodur	6.0	2.78	1.78	0.422	0.03	0.07
Aluminum: 6061	5.0	3.97	2.90	0.538	0.13	0.33
MMC: 30% SiCAl	6.3	2.49	2.11	0.459	0.10	0.22
Beryllium: I-70-H	12.5	0.64	0.22	0.149	0.05	0.20
Beryllium: I-220-H	12.5	0.64	0.22	0.149	0.05	0.20
Copper: OFC	3.6	7.64	61.1	2.471	0.53	0.14
Invar 36	4.2	5.71	37.0	1.924	0.10	0.38

Table C9 - Comparison of material figures of merit especially pertinent to mirror design (Page 2 of 2)

	Weight and self-weight deflection proportionality factors				Thermal distortion coefficients	
Preferred	$(E/\rho)^{1/2}$ Resonant freqency for same geometry (arbitrary units) large	ρ/E Mass or deflection for same geometry (arbitrary units) small	ρ^3/E Deflection for same mass (arbitrary units) small	$(\rho^3/E)^{1/2}$ Mass for same deflection (arbitrary units) small	α/k Steady state (μm/W) small	α/D transient (sec/m^2 K) small
Super Invar	4.3	5.49	36.3	1.906	0.03	0.12
Molybdenum	5.6	3.15	32.8	1.812	0.04	0.09
Silicon	7.5	1.78	0.97	0.311	0.02	0.03
SiC: HP alpha	11.9	0.70	0.72	0.268	0.02	0.03
SiC: CVD	12.0	0.69	0.71	0.267	0.02	0.03
SiC: RB-30% Si	10.7	0.88	0.73	0.270	0.01	0.03
Stainless steel: 304	4.9	4.15	26.5	1.629	0.91	3.68
Stainless steel: 416	5.2	3.63	22.1	1.486	0.34	1.23
Titanium: 6Al4V	5.1	3.89	7.63	0.873	1.21	3.03

From Paquin[14]

Table C10a - Characteristics of aluminum alloys used in mirrors

Alloy No.	Form	Hardenable	Remarks
1100	Wrought	No	Relatively pure; low strength; can be diamond turned
2014/2024	Wrought	Yes	High strength and ductility; multiphase; must be plated
5086/5456	Wrought	No	Moderate strength when annealed; weldable; available in large plate
6061	Wrought	Yes	Low alloy; all purpose; reasonably high strength; weldable; can be diamond turned and/or plated; all forms readily available
7075	Wrought	Yes	Highest strength; usually plated; strength more temperature sensitive than others
B201	Cast	Yes	Sand or permanent mold cast; high strength; can be diamond turned
A356/357	Cast	Yes	Sand or permanent mold cast; moderate strength; most common; extensive processing for dimensional stability
713/Tenzalloy	Cast	Yes	Sand or permanent mold cast; moderate strength
771/Precedent 71A	Cast	Yes	Sand cast; moderate strength; very stable; expensive casting procedures required; easiest to machine

From Paquin[9]

Table C10b - Characteristics of aluminum matrix composites

Property	Instrument grade	Optical grade	Structural grade
Matrix alloy	6061-T6	2124-T6	2021-T6
Volume % SiC	40%	30%	20%
SiC form	Particulate	Particulate	Whisker
CTE ($\times 10^{-6}$/K)	10.7	12.4	14.8
Thermal conductivity (W/m K)	127	123	n/a
Young's modulus (MPa)	145	117	127
Density (g/cm^3)	2.91	2.91	2.86

From Mohn and Vukobratovich[103]

Table C10c - Beryllium grades and some of their properties

Property	O-50	I-70-H	I-220-H	I-250	S-200-FH
Max. beryllium oxide content (%)	0.5	0.7	2.2	2.5	1.5
Grain size (μm)	15	10	8	2.5	10
2% offset yield strength (MPa)	172	207	345	544	296
Microyield strength (MPa)	10	21	41	97	34
Elongation (%)	3.0	3.0	2.0	3.0	3.0

From Paquin[97]

Table C10d - Characteristics of major silicon carbide types

SiC Type	Structure/ composition	Density	Fabrication process	Properties*	Remarks
Hot pressed	>98% alpha plus others	>98%	Powder pressed in heated dies	High E, ρ, K_{ic}, MOR; lower k	Simple shapes only; size limited
Hot isostatic pressed	>98% alpha/beta plus others	>99%	Hot gas pressure on encapsulated preform	High E, ρ, K_{ic}, MOR; lower k	Complex shapes possible; size limited
Chemically vapor deposited	100% beta	100%	Deposition on hot mandrel	High E, ρ, k; lower K_{ic}, MOR	Thin shape or plate forms; built-up shapes
Reaction bonded	50-92% alpha plus silicon	100%	Cast, prefired, porous preform fired with silicon infiltration	Lower E, ρ, MOR, k; lowest K_{ic}	Complex shapes readily formed; large sizes; properties are silicon content dependent

* K_{ic} is fracture toughness, MOR is modulus of rupture

From Paquin[14]

Table C11 - Techniques for machining, finishing, and coating materials for optical applications

Material	Figure control method*	Surface finish method*	Coatings
Al alloys	SPDT, Cs, CM, EDM, ECM, IM, Po	Po	MgF_2, SiO, SiO_2, Au, EN, and most others
Al matrix	HIP, CS, SDT, EDM, ECM, IM, Po, CM	EN	MgF_2, SiO, SiO_2, Au, EN, An, and most others
Al castings			
A-201	Cs, EDM, ECM, IM, SPDT	Po or coated EN	Same as Al
A-356.0			Coat Ni
520			Same as Al
Al silicon hypereutectic 393.2	Cs, EDM, CE, IM, SPDT, Gr, CM	Coated EN, Po	Coat Ni
Beryllium alloys	CM, EDM, ECM, EM, Gr, HIP, not SPDT	Bare or EN	None or coat EN
Magnesium alloys	SPDT, Cs, CM, EDM, ECM, IM	EN	Same as Al
SiC	HIP, CVD, RB	Po, EN	Vacuum processes
Steels	CM, EDM, ECM, Gr, not SPDT	Po, EN	EN
Titanium	CM, HIP, ECM, EDM, Gr, not SPDT	Po	EN, Tiodize
Glass, quartz, ULE, Zerodur	Cs, Gr, IM, Po	Po	Most types of metals and non-metals

* Legend: An = anodize, CE = chemical etching, CM = conventional machining, Cs = cast, CVD = chemical vapor deposition, ECM = electrochemical etching, EDM = electrical discharge machining, EN = electroless nickel, Gl = glazing, Gr = grinding, HIP = hot isostatic pressing, IM = ion milling, Po = polishing, RB = reaction bonding, SPDT = single point diamond turning

Adapted from Engelhaupt[4]

Table C12 - Mechanical properties of selected metals used for mechanical parts in optical instruments (Pg. 1 of 4)

Material	CTE (α) $\times10^{-6}$/°C ($\times10^{-6}$/°F)	Young's modulus* (E_M) $\times10^{10}$ Pa ($\times10^{6}$ lb/in.^{2})	Yield strength* $\times10^{7}$ Pa (10^{3}lb/in.^{2})	Poisson's ratio (υ_M)	Density (ρ) g/cm^3 (lb/in.^{3})	Thermal Conductivity* (k) W/m-K (Btu/hr-ft-°F)	Hardness[a]
Aluminum 1100	23.6 (13.1)	6.89 (10.0)	3.4-15.2 (5-22)		2.71 (0.098)	218-221 (126-128)	23-44 Brinell
Aluminum 2024	22.9 (12.7)	7.31 (10.6)	7.6-39.3 (11-57)	0.33	2.77 (0.100)	119-190 (69-110)	47-130 Brinell
Aluminum 6061	23.6 (13.1)	6.82 (9.9)	5.5-27.6 (8-40)	0.332	2.68 (0.097)	154-180 (89-104)	30-95 Brinell
Aluminum 7075	23.4 (13.0)	7.17 (10.4)	10.3-50.3 (15-73)		2.79 (0.101)	142-176 (82-102)	60-150 Brinell
Aluminum 356	21.4 (11.9)	7.17 (10.4)	17.2-20.7 (25-30)		2.68 (0.097)	150-168 (87-97)	60-70 Brinell
Beryllium S-200	11.5 (6.4)	27.6-30.3 (40-44)	20.7 (30)		1.85 (0.067)	220 (127)	80-90 Rockwell-B
Beryllium I-400	11.5 (6.4)	27.6-30.3 (40-44)	34.5 (50)		1.85 (0.067)	220 (127)	100 Rockwell-B

Table C12 - Mechanical properties of selected metals used for mechanical parts in optical instruments (Pg. 2 of 4)

Material	CTE (α) $\times 10^{-6}$/°C ($\times 10^{-6}$/°F)	Young's modulus* (E_M) $\times 10^{10}$ Pa ($\times 10^{6}$ lb/in.^{2})	Yield strength* $\times 10^{7}$ Pa (10^{3}lb/in.^{2})	Poisson's ratio (υ_M)	Density (ρ) g/cm^3 (lb/in.^{3})	Thermal Conductivity* (k) W/m-K (Btu/hr-ft-°F)	Hardness[a]
Beryllium I-70A	11.3 (6.3)	28.9 (42)		0.08	1.85 (0.067)	194 (112)	
Copper C10100 (OFHC)	16.9 (9.4)	11.7 (17)	6.9-36.5 (10-53)	0.35	8.94 (0.323)	391 (226)	10-60 Rockwell-B
Copper C17200 (BeCu)	17.8 (9.9)	12.7 (18.5)	107-134 (155-195)		8.25 (0.298)	107-130 (62-75)	27-42 Rockwell-C
Copper C2600 (brass)	20.5 (11.4)	9.65 (14.0)	12.4-35.9 (18-52)		8.50 (0.307)	116 (67)	62-80 Rockwell-B
Copper C2600 (cartridge brass)	20.0 (11.1)	11.0 (16)	7.6-44.8 (11-65)		8.52 (0.308)	121 (70)	55-93 Rockwell-B
Invar 36	1.26 (0.7)	14.7 (21.4)	27.6-41.4 (40-60)	0.29	8.05 (0.291)	11.1 (6.4)	160 Brinell

Table C12 - Mechanical properties of selected metals used for mechanical parts in optical instruments (Pg. 3 of 4)

Material	CTE (α) $\times 10^{-6}$/°C ($\times 10^{-6}$/°F)	Young's modulus* (E_M) $\times 10^{10}$ Pa ($\times 10^{6}$ lb/in.^{2})	Yield strength* $\times 10^{7}$ Pa (10^{3}lb/in.^{2})	Poisson's ratio (υ_M)	Density (ρ) g/cm^3 (lb/in.^{3})	Thermal Conductivity* (k) W/m-K (Btu/hr-ft-°F)	Hardness[a]
Super Invar	0.31 (0.17)	14.8 (21.5)	30.3 (44)	0.29	8.14 (0.294)	10.4 (6)	160 Brinell
Magnesium AZ-31B	25.2 (14)	4.48 (6.5)	14.5-25.5 (21-37)	0.35	1.77 (0.064)	97 (56)	73 Brinell
Magnesium M1A	25.2 (14)	4.48 (6.5)	12.4-17.9 (18-26)		1.77 (0.064)	138 (79.8)	42-54 Brinell
Steel 1015 (low carbon)	11.9 (6.6)	20.7 (30)	28.3-31.0 (41-45)	0.287	7.75 (0.28)		111-126 Brinell
Steel 304 (CRES)	17.3 (9.6)	19.3 (28)	51.7-103 (75-150)	0.29	8.03 (0.29)	16.2 (9.4)	83 Rockwell-B 42 Rockwell-C
Steel 416 (CRES)	9.9 (5.5)	20.0 (29)	27.6-103 (40-150)	0.3	7.75 (0.28)	24.9 (14.4)	82 Rockwell-B 42 Rockwell-C

Table C12 - Mechanical properties of selected metals used for mechanical parts in optical instruments (Pg. 4 of 4)

Material	CTE (α) $\times 10^{-6}$/°C ($\times 10^{-6}$/°F)	Young's modulus* (E_M) $\times 10^{10}$ Pa ($\times 10^6$ lb/in.^{2})	Yield strength* $\times 10^7$ Pa (10^3 lb/in.^{2})	Poisson's ratio (υ_M)	Density (ρ) g/cm^3 (lb/in.^{3})	Thermal Conductivity* (k) W/m-K (Btu/hr-ft-°F)	Hardness[a]
Titanium 6Al-4-V	8.8 (4.9)	11.4 (16.5)	82.7-106 (120-154)	0.34	4.43 (0.16)	6.6 (3.8)	36-39 Rockwell-C
SXA Metal Matrix (SiC & 2124 Al)	12.4 (6.9)	11.7 (17)			1.78 (0.064)	35 (20)	Variable within sample

* Range of values pertains to various tempers.

From: Yoder[8]

Table C13 - Typical physical characteristics of optical cements

Refractive index (n) after cure:	1.48 to 1.55
Thermal expansion coefficient (α):	
@ 27°C to 100°C	63×10^{-6}/°C (35×10^{-6}/°F)
@ 100°C to -200°C	56×10^{-6}/°C (31×10^{-6}/°F)
Young's modulus (E):	430 GPa (62×10^{6})
Shear strength:	360 GPa (5200 lb/in.^{2})
Specific heat (C_p):	837 J/kg K (0.2 - 0.4 BTU/lb °F)
Water absorption (bulk material)	0.3% after 24 hrs @ 25°C
Shrinkage during cure:	Approximately 6% (volumetric)
Viscosity:	200 to 320 cps
Density:	1.22 g/cm^{3} (0.044 lb/in.^{3})
Hardness (Shore D):	Approximately 90
Total mass loss in vacuum:	3% to 6%

From Paquin[50]

Table C14 - Typical characteristics of representative structural adhesives (Page 1 of 3)

Material	Mfr. code*	Recommended cure time @ °C	Uncured viscosity cps	Joint strength MPa (lb/in.^{2}) @ °C)	Temperature range of use °C (°F)	CTE (α) $\times 10^{-6}$ /°C ($\times 10^{-6}$ /°F)	Joint thickness mm (in.)
1-part epoxies:							
2214Hi-Temp	3M	40 min @121 @ 7000 Pa (10 lb/in.^{2}) load	aluminum filled paste	13.8 (2000) @-55 13.8 (2000) @ 24 20.7 (3000) @ 82 17.2 (2500) @ 121 6.2 (900) @ 177	-55 to 177 (-67 to 351)		
2-part epoxies:							
Milbond	SL	7 day @25		17.7 (2561) @-50 14.5 (2099) @ 25 6.8 (992) @ 70	-50 to 70 (-58 to 158)		0.381±0.025 (0.015±0.001)
EPO-TEK 314	ET	3 hr @120 1 hr @150 30 min @180	500	5.5 (800) @ 25	to 150 (302) continuous to 300 (572) intermittent	70 (39) @ <140 169 (94) @ >140	
EC2216B/A Gray	3M	Fix:30 min @93 Fix: 2 hr @65 Full: 7 day @75 all @7×10^4 Pa (10 lb/in.^{2}) load		13.8 (2000) @-55 17.2 (2500) @ 24 2.8 (400) @ 82	-55 to 150 (-67 to 302)	102 (57) @0-40 134 (74) @40-80	0.102±0.025 (0.004±0.001)

Table C14 - Typical characteristics of representative structural adhesives (Page 2 of 3)

Material	Mfr. code*	Recommended cure time @ °C	Uncured viscosity cps	Joint strength MPa (lb/in.^{2}) @ °C)	Temperature range of use °C (°F)	CTE (α) $\times 10^{-6}$ /°C ($\times 10^{-6}$ /°F)	Joint thickness mm (in.)
Trans-lucent		Fix: 1 hr @93 Fix: 6 hr @65 Full:30 day @75 all @ 7×10^4 Pa (10 lb/in.^{2}) load		20.7 (3000) @-55 8.3 (1200) @ 24 1.4 (200) @ 82	-55 to 150 (57 to 302	81 (45) @-50 to 0 207 (115) @ 60-150	0.102±0.025 (0.004±0.001)
27A/B	EC	24 hr @ 25 4 hr @ 45	400	13.8 (2000)	-65 to 105 (-85 to 157)		
285	EC	Variable with cure agent	thixotropic paste	14.5 (2100)	Variable with cure agent	27 (15)	
45	EC	24 hr @25 30 min @70 15 min @105	37,000	21.4 (3100) (-40 to 194)	-40 to 90		
Urethanes							
324	L	Fix: 3 min @20 Full: 24 hr @20 3 min @150	15,000	10.3 (1500) (-65 to 275)	-54 to 135	12.6 (7.0)	>1.0 (0.040)

Table C14 - Typical characteristics of representative structural adhesives (Page 3 of 3)

Material	Mfr. code*	Recommended cure time @ °C	Uncured viscosity cps	Joint strength MPa (lb/in.^{2}) @ °C)	Temperature range of use °C (°F)	CTE (α) $\times 10^{-6}$/°C ($\times 10^{-6}$/°F)	Joint thickness mm (in.)
3532B/A	3M	3 day @24 @ 3×10^4 Pa (5 lb/in.^{2}) load		17.2 (2500) @-55 13.8 (2000) @ 24 2.1 (300) @ 82 1.0 (150) @ 121			
UV curable							
UV-900	EC	UV Cure: 15 sec @ 200 W/in. Heat Cure: 10 min @120	900		-25 to 125 (-13 to 257)		<3.2 (0.125)
349	L	UV Cure @100 mW/cm Fix: 1 sec @ 0 gap 20 sec @ 0.25 gap Full: 36 sec @ 0.25 gap	7000	5.2 (750)	-54 to 130 (-65 to 266)		6.35 (0.25) max
Cyanoacrylates							
460	L	Fix: 1 min @22 Full: 7 days @22	50	12.1 (1750)	-54 to 71 (-65 to 160)	100 (56)	0.100 (0.004) max

* Mfr. Code: 3M = 3M, SL = Summers Laboratories; ET = Emerson & Cumings; L = Loctite
From Yoder[8] and manufacturer's data sheets.

Table C15a - Typical physical characteristics of representative elastomeric sealants (Page 1 of 3)

Material	Mfr. code*	Suggested cure time @ °C	Uncured viscosity**	Cured hardness (Shore A)	Temperature range of use °C (°F)	Shrinkage % after 3 days @ 25° C	Effluent or mass loss % after hrs @ °C	Manufacturer identified applicable MIL-spec
1-part silicone products:								
732	DC	24 hr @ 25	320 g/min	25	-55 to 200 (-67 to 392)	---	acetic acid	MIL-A-46106
RTV112	GE	3 day @ 25	200 p	25	<204 (400) continuous <260 (500) intermittant	1.0	acetic acid	---
NUVA-SIL84™	L	UV cure 1 min or 7 day @25° C	thixo-tropic	45	-70 to 260 (-65 to 500)	0.4	3.7 @ 96 hr @ 200	---
2-part silicone products:								
3112	DC	<12 hr @ 25	350 p	60	-55 to 250 (-67 to 482)	0.25	3.7 @ 24 day & 125 & 10^{-6} Torr	---

Table C15a - Typical physical characteristics of representative elastomeric sealants (Page 2 of 3)

Material	Mfr. code*	Suggested cure time @ °C	Uncured viscosity**	Cured hardness (Shore A)	Temperature range of use °C (°F)	Shrinkage % after 3 days @ 25° C	Effluent or mass loss % after hrs @ °C	Manufacturer identified applicable MIL-spec
93-500	DC	7 day @ 25 4 hr @ 65 1 hr @ 100 15 min @ 150	20 p	43	-65 to 200 (-85 to 392)		0.25 @ 24 hr & 125 & <10^{-6} Torr	---
RTV88	GE	<24 hr @ 25	8800 p	55	-54 to 260 (-65 to 500)	0.6	alcohol	---
RTV655	GE	45 min @ 125	4500 cps	50	-110 to 204 (-166 to 400)	0.2	---	---
RTV8111	GE	<24 hr @ 25	120 p	45	-54 to 204 (-65 to 400)	0.6	alcohol	MIL-S-23586E Type I, Class 1 Grade B-1
RTV8112	GE	<24 hr @ 25	120 p	45	-54 to 204 (-65 to 400)	0.6	alcohol	MIL-S-23586E Type I, Class 2 Grade A

Table C15a - Typical physical characteristics of representative elastomeric sealants (Page 3 of 3)

Material	Mfr. code*	Suggested cure time @ °C	Uncured viscosity**	Cured hardness (Shore A)	Temperature range of use °C (°F)	Shrinkage % after 3 days @ 25° C	Effluent or mass loss % after hrs @ °C	Manufacturer identified applicable MIL-spec
RTV8262	GE	<24 hr @ 25	500 p	55	-54 to 260 (-65 to 500)	0.6	alcohol	MIL-S-23586E Type II, Class 2 Grade A
Other products:								
EC801B/A	3M	tack free: <72 hr @ 25 full cure: 1 wk @ 77° F	heavy liquid	>35-60 (40 Rex)	-54 to 82 (-65 to 180)	---	---	MIL-S-7502 Class A

* Mfr. code: 3M = 3M; DC = Dow Corning; GE = General Electric; L = Loctite.
** Units: g/min = grams per minute extrusion rate; p = poise; cps = centipoise.

From Yoder[8] and manufacturer's data sheets.

Table C15b - Typical mechanical properties of representative elastomeric sealants (Page 1 of 2)

Material	Mfr. code*	Tensile Strength MPa (lb/in.^{2})	Density @ 25° C g/cm^3 (lb/in.^{3})	Poisson's ratio (assumed)	Young's modulus** MPa (lb/in.^{3})	CTE $\times 10^{-6}$/°C ($\times 10^{-6}$/°F)
1-part silicone products:						
732	DC	2.2 (325)	1.04 (0.037)	0.5		310 (172)
RTV112	GE	2.2 (325)	1.05 (0.038)	0.5		270 (150)
NUVA-SIL 84™	L	3.8 (550)	1.35 (0.049)	0.5		167 (93)
2-part silicone products:						
3112	DC	4.5 (650)	1.02 (0.037)	0.5		300 (167)
93-500	DC	69 (100)	1.02 (0.037)	0.5		300 (167)
RTV88	GE	5.9 (850)	1.48 (0.53)	0.5		210 (110)

Table C15b - Typical mechanical properties of representative elastomeric sealants (Page 2 of 2)

Material	Mfr. code*	Tensile Strength MPa ($lb/in.^2$)	Density @ 25° C g/cm^3 ($lb/in.^3$)	Poisson's ratio	Young's modulus MPa ($lb/in.^3$)	CTE $\times 10^{-6}/°C$ ($\times 10^{-6}/°F$)
RTV655	GE	6.2 (900)	1.03 (0.037)	0.5		330 (180)
RTV8111	GE	2.4 (350)	1.18 (0.043)	0.5		250 (140)
RTV8112	GE	2.4 (350)	1.18 (0.043)	0.5		250 (140)
RTV8262	GE	5.2 (750)	1.47 (0.053)	0.5		210 (110)
Other products:						
EC801	3M	~2.8 (400)	1.55 (0.056)			

* Mfr. code: 3M = 3M; DC = Dow Corning; GE = General Electric; L = Loctite.

From Yoder[8] and manufacturer's data sheets.

REFERENCES

1. Yoder, P.R., Jr., *Mounting Lenses in Optical Instruments, Tutorial Text Vol. TT21*, SPIE, Bellingham, WA, 1995.
2. Walles, S. and Hopkins, R.E., "The orientation of the image formed by a series of plane mirrors," *Appl. Opt. Vol. 3*, 1447, 1964.
3. Vukobratovich, D., Chapt. 2 in *Handbook of Optomechanical Engineering*, CRC Press, Boca Raton, FL, 1997.
4. Englehaupt, D., Chapt. 10 in *Handbook of Optomechanical Engineering*, CRC Press, Boca Raton, FL, 1997.
5. Genberg, V., Chapt. 8 in *Handbook of Optomechanical Engineering*, CRC Press, Boca Raton, FL, 1997.
6. Hatheway, A.E., "Review of finite element analysis techniques: capabilities and limitations," *SPIE Critical Review Vol. CR43*, 367, 1992.
7. Hatheway, A.E., "Unified thermal/elastic optical analysis of a lithographic lens," *SPIE Proceedings Vol. 3130*, 100, 1997.
8. Yoder, P. R., Jr., *Opto-Mechanical Systems Design,* 2nd Edition, Marcel Dekker, New York, 1993.
9. Marker, A.J. III, Hayden, J.S., and Speit, B., "Radiation resistant optical glasses," *SPIE Proceedings Vol. 1485*, 55, 1991.
10. *Schott Optical Glass Catalog*, Schott Glass Technologies, Inc., Duryea, PA.
11. Walker, B.H., "Select optical glasses," *The Photonics Design and Applications Handbook*, Lauren Publishing Co., Pittsfield, MA, H-356, 1993.
12. *Schott Product Information 2106/91, Lightweight Optical Glasses*, Schott Glass Technologies, Inc., Duryea, PA, 1991.
13. *Schott Product Information 10017e, Radiation Resistant Glasses*, Schott Glass Technologies, Inc., Duryea, PA, 1990.
14. Paquin, R., Chapt. 4 in *Handbook of Optomechanical Engineering*, CRC Press, Boca Raton, FL, 1997.
15. Smith, W.J., *Modern Optical Engineering,* 2nd Edition, McGraw-Hill, New York, 1990.
16. Kittel, D., "Precision mechanics," *SPIE Short Course Notes*, 1989.
17. Strong, J., *Procedures in Applied Optics*, Marcel Dekker, New York, 1988.
18. Smith, W.J., "Fundamentals of establishing an optical tolerance budget," *SPIE Proceedings Vol. 531*, 196, 1985.
19. Plummer, J.L., "Tolerancing for economies in mass production of optics," *SPIE Proceedings Vol. 181*, 90, 1979.
20. Adams, G., "Selection of tolerances," *SPIE Proceedings Vol. 892*, 173, 1988.
21. Willey, R.R. and Durham, M.E., "Maximizing production yield and performance in optical instruments through effective design and tolerancing," *SPIE Critical Review Vol. CR43*, 76, 1992.
22. Willey, R. and Parks, R., Chapt. 1 in *Handbook of Optomechanical Engineering*, CRC Press, Boca Raton, FL, 1997.
23. *MIL-HDBK-141, Optical Design*, U.S. Defense Supply Agency, Washington, DC, 1962.

24. Smith, W.J., Sect. 2 in *Handbook of Optics*, Optical Society of America, Washington, DC, 1978.
25. Yoder, P.R., Jr., "Two new lightweight military binoculars," *J. Opt. Soc. Am. Vol.50*, 491, 1960.
26. Durie, D.S.L., "A compact derotator design," *Opt. Eng. Vol. 13*, 19, 1974.
27. Yoder, P.R., Jr., "Study of light deviation errors in triple mirrors and tetrahedral prisms," *J. Opt. Soc. Am. Vol. 48*, 496, 1958.
28. *Sales Literature, Hard-Mounted Hollow Retroreflector*, PLX, Inc., Deer Park, NY.
29. Ulmes, J.J., "Design of a catadioptric lens for long-range oblique aerial reconnaissance," *SPIE Proceedings Vol. 1113*, 116, 1989.
30. Kingslake, R., *Optical System Design*, Academic Press, Orlando, FL, 1983.
31. Lohmann, A.W. and Stork, W., "Modified Brewster telescopes," *Appl. Opt. Vol. 28*, 1318, 1989.
32. Trebino, R., "Achromatic N-prism beam expanders: optimal configurations," *Appl.Opt. Vol. 24*, 1130, 1985.
33. Forkner, J.F., "Anamorphic prism for beam shaping," *U. S. Patent No. 4,623,225*, 1986.
34. Lipshutz, M.L., "Optomechanical considerations for optical beam splitters," *Appl. Opt. Vol. 7*, 2326, 1968.
35. Durie, D.S.L., "Stability of optical mounts," *Machine Des. Vol. 40*, 184, 1968.
36. Vukobratovich, D., Chapt. 3, "Optomechanical Systems Design," in *The Infrared & Electro-Optical Systems Handbook, Vol. 4*, ERIM, Ann Arbor, MI and SPIE, Bellingham, WA, 1993.
37. Yoder, P.R., Jr., "Non-image-forming optical components," *SPIE Proceedings Vol.531*, 206, 1985.
38. Delgado, R.F., "The multidiscipline demands of a high performance dual channel projector," *SPIE Proceedings Vol.389*, 75, 1983.
39. Yoder, P.R., Jr., "Design guidelines for bonding prisms to mounts," *SPIE Proceedings Vol. 1013*, 112, 1988.
40. Willey, R., *private communication*, 1991.
41. Beckmann, L.H.J.F., *private communication*, 1990.
42. Seil, K., "Progress in binocular design", *SPIE Proceedings Vol. 1533*, 48, 1991.
43. Seil, K., *private communication*, 1997.
44. Yoder, P.R., Jr., E.R. Schlessinger, and J.L. Chickvary, "Active annular-beam laser autocollimator system," *Appl. Opt. Vol. 14*, 1890, 1975.
45. Rourk, R.J., *Formulas for Stress and Strain*, 3rd Edition, McGraw Hill, New York, 1954.
46. *Engineering Design Guide*, Sorbothane, Inc., Kent, OH.
47. Jenkins, F.A. and White, H.E., *Fundamentals of Optics*, McGraw-Hill, New York, 1957.
48. Schubert, F., "Determining optical mirror size," *Machine Des. Vol. 51*, 128, 1979.
49. Rodkevich, G.V. and Robachevskaya, V.I., "Possibilities of reducing the mass of large precision mirrors," *Sov. J. Opt. Technol. Vol. 44*, 515, 1977.
50. Paquin, R., Chapt. 3 in *Handbook of Optomechanical Engineering*, CRC Press, Boca Raton, FL, 1997.
51. Seibert, G.E., "Design of Lightweight Mirrors", *SPIE Short Course Notes*, 1990.

52. Fitzsimmons, T.C. and Crowe, D.A., "Ultra-lightweight mirror manufacturing and radiation response study," *RADC-TR-81-226*, Rome Air Development Ctr., Rome, NY, 1981.
53. Lewis, W.C., "Space telescope mirror substrate," *OSA Optical Fabrication and Testing Workshop*, Tucson, Az, 1979.
54 Pepi, J.W. and Wollensak, R.J., "Ultra-lightweight fused silica mirrors for cryogenic space optical system," *SPIE Proceedings Vol. 183*, 131, 1979.
55 Cannon, J.E. and Wortley, R.W., "Gas fusion center-plane-mounted secondary mirror," *SPIE Proceedings Vol. 966*, 309, 1988.
56 Dahlgren, R. and Gerchman, M., "The use of aluminum alloy castings as diamond machining substrates for optical surfaces," *SPIE Proceedings Vol. 890*, 68, 1988.
57 Downey, C.H., Abbott, R.S., Arter, P.I., Hope, D.A., Payne, D.A., Roybal, E.A., Lester, D.F., and McClenahan, J.O., "The chopping secondary mirror for the Kuiper airborne observatory," *SPIE Proceedings Vol. 1167*, 329, 1989.
58. Vukobratovich, D., Gerzoff, A., and Cho, M.K., "Therm-optic analysis of bi-metallic mirrors," *SPIE Proceedings Vol. 3132*, 12, 1997.
59. Gould, G., "Method and means for making a beryllium mirror", *U. S. Patent No. 4,492,669*, 1985.
60. Paquin, R.A., Levenstein, H., Altadonna, L., and Gould, G., "Advanced lightweight beryllium optics," *Opt. Eng Vol. 23*, 157, 1984.
61. Paquin, R.A., "Hot isostatic pressed beryllium for large optics," *Opt. Eng. Vol. 25*, 1003, 1986.
62. Stern, A.K., *private communication*, 1998.
63. Kowalski, B.J., "A user's guide to designing and mounting lenses and mirrors", *Digest of Papers*, OSA Workshop on Optical Fabrication and Testing, North Falmouth, MA, 98, 1980.
64. Yoder, P.R., Jr., Chapt. 6 in *Handbook of Optomechanical Engineering*, CRC Press, Boca Raton, FL, 1997.
65. Strong, J., *Procedures in Applied Optics*, Marcel Dekker, New York, 1989.
66. Høg, E., "A kinematic mounting," *Astrom. Astrophys Vol. 4*, 107, 1975.
67. Mrus, G.J., Zukowski, W.S., Kokot, W., Yoder, P.R., Jr., and Wood, J.T., "An automatic theodolite for pre-launch azimuth alignment of the Saturn space vehicles", *Appl. Opt. Vol. 10*, 504, 1971.
68. Patrick, F.B., "Military optical instruments," Chapt. 7 in *Applied Optical and Optical Engineering Vol. V*, Academic Press, New York, NY, 1969.
69. Yoder, P.R., Jr., "High Precision 10-cm Aperture Penta and Roof-Penta Mirror Assemblies", *Appl. Opt. Vol. 10*, 2231, 1971.
70. Lipkins, J. and Bleier, Z., "Beryllium retroreflectors expand boresight uses," *Laser Focus World*, Nov. 1996.
71. Bacich, J., "Precision lens mounting," *U.S. Patent No. 4,733,945*, Mar. 1988.
72. Sarver, G., Maa, G., and Chang, L., "SIRTF primary mirror design, analysis, and testing," *SPIE Proceedings Vol. 1340*, 35, 1990.
73. Iraninejad, B., Vukobratovich, D., Richard, R., and Melugin, R., "A mirror mount or cryogenic environments," *SPIE Proceedings Vol. 450*, 34, 1983.
74. Zimmerman, J., "Strain-free mounting techniques for metal mirrors," *Opt. Eng. Vol. 20*, 187, 1981.

75. Addis, E.C., "Value engineering additives in optical sighting devices," *SPIE Proceedings Vol. 389*, 36, 1983.
76. Erickson, D.J., Johnston, R.A., and Hull, A.B., "Optimization of the opto-mechanical interface employing diamond machining in a concurrent engineering environment," *SPIE Critical Review Vol. CR43*, 329, 1992.
77. Malvick, A.J. and Pearson, E.T., "Theoretical elastic deformations of a 4-m diameter optical mirror using dynamic relaxation," *Appl. Opt. Vol. 7*, 1207, 1968.
78. Hindle, J.H., "Mechanical Flotation of Mirrors," *Amateur Telescope Making, Book One*, Scientific American, New York, NY, 1945.
79. Mehta, P.K., "Flat circular optical elements on a 9-point Hindle mount in a 1-g force field," *SPIE Proceedings Vol. 450*, 118, 1983.
80. Schwesinger, G., "Optical effect of flexure in vertically mounted precision mirrors,"", *J. Opt. Soc. Am. Vol. 44*, 417, 1954.
81. Malvick, A.J., "Theoretical elastic deformations of the Steward Observatory 230-cm and the Optical Sciences Center 154-cm mirrors," *Appl. Opt. Vol. 11*, 575, 1972.
82. Delgado, R.F. and Hallinan, M., "Mounting of optical elements," *Opt. Eng., Vol.14*, S-11, 1975. Reprinted in *SPIE Proceedings Vol. 770*, 173, 1988.
83. Bayar, M., "Lens barrel optomechanical design principles," *Opt. Eng., Vol. 20*, 181,1981.
84.Trsar, W. J., Benjamin, R. J., and Casper, J. F., "Production engineering and implementation of a modular military binocular," *Opt. Eng. Vol. 20*, 201, 1981.
85. Sheinis, A.I., Nelson, J.E., and Radovan, M.V., "Large prism mounting to minimize rotation in Cassegrain instruments," *SPIE Proceedings Vol. 3355*, 1998.
86. Epps, H.W. and Miller, J. S., "Echellette spectrograph and imager (ESI) for Keck Observatory," *SPIE Proceedings Vol. 3355*, 1998.
87. Sutin, B.M., "What an optical designer can do for you AFTER you get the design", *SPIE Proceedings Vol. 3355*, 1998.
88. Radovan, M.V., Nelson, J.E., Bigelow, B.C., and Sheinis, A.I., "Design of a collimator support to provide flexure control on Cassegrain instruments," *SPIE Proceedings Vol. 3355*, 1998.
89. Bigelow, B.C. and Nelson, J.E., "Determinate space-frame structure for the Keck II Echellete Spectrograph and Imager (ESI)," *SPIE Proceedings Vol. 3355*, 1998.
90. Vukobratovich, D., "Introduction to optomechanical design," *SPIE Short Course Notes*, 1993.
91. Iraninejad, B., Lubliner, J., Mast, T., and Nelson, J.E., "Mirror deformations due to thermal expansion of inserts bonded to glass," *Keck Observatory Report No. 160*, 1987 or *SPIE Proceedings Vol. 748*, 206, 1987.
92. Hookman, R., "Design of the GOES telescope secondary mirror mounting," *SPIE Proceedings Vol. 1167*, 368, 1989.
93. Shipley, A., Green, J.C., and Andrews, J.P., "The design and mounting of the gratings for the Far Ultraviolet Spectroscopic Explorer," *SPIE Proceedings Vol. 2542*, 185, 1995.
94. Shipley, A., Green, J., Andrews, J., Wilkinson, E., and Osterman, S., Final flight grating mount design for the Far Ultraviolet Spectroscopic Explorer," *SPIE Proceedings Vol. 3132*, 98, 1997.
95. *Tiodize Process Literature*, Tiodize Co., Inc., Huntington Beach, CA.

96. Schreibman, M. and Young, P., "Design of Infrared Astronomical Satellite (IRAS) primary mirror mounts," *SPIE Proceedings Vol. 250*, 50, 1980.
97. Young, P. and Schreibman, M., "Alignment design for a cryogenic telescope," *SPIE Proceedings Vol. 251*, 171, 1980.
98. Vukobratovich, D., Richard, R., Valente, T., Cho, M., *Final design report for NASA Ames/University of Arizona Cooperative Agreement No. NCC2-426 for period April 1, 1989-April 30, 1990*, Optical Sciences Center, Univ. of Arizona, Tucson, AZ.
99.*The Handbook of Plastic Optics*, 2nd Ed., U.S. Precision Lens, Inc., Cincinnati, OH, 1983.
100. Wolpert, H. D., Optical plastics: properties and tolerances in *The Photonic Design and Application Handbook*, Lauren Publishing Co., Pittsfield, H-321, 1989.
101. Tropf, W. J., Thomas, M. E. and Harris, T. J., Properties of crystals and glasses," Chapt. 33 in *OSA Handbook of Optics,* 2nd Edition, *Vol. II*, McGraw-Hill, New York, 1995.
102. Amirtharaj, P. M. and Seiler, D. G., Optical properties of semiconductors," Chapt. 36 in *OSA Handbook of Optics,* 2nd Edition, *Vol. II*, McGraw-Hill, New York, 1995.
103. Mohn, W.R. and Vukobratovich, D., "Recent applications of metal matrix composites in precision instruments and optical systems", *Opt. Eng. Vol. 27*, 90, 1988.

INDEX

Paul R. Yoder, Jr. currently is an independent consultant in optical engineering. For over 50 years, he has conducted theoretical and experimental research in optics; performed detailed design, analysis, and testing of optical instruments; and planned, organized and managed optical technology and electro-optical system projects ranging from conceptual studies and prototype developments to quantity hardware production. He formerly held various technical and engineering management positions with Taunton Technologies, Inc., The Perkin-Elmer Corporation and the U.S. Army's Frankford Arsenal. Yoder has authored over 50 technical papers in optical engineering as well as *Opto-Mechanical Systems Design* (Marcel Dekker, New York, 1986 and 2nd ed., 1993); Chap. 37, "Mounting Optical Components" in the OSA's *Handbook of Optics* (McGraw-Hill, New York, 1994); and Chap. 6 on "Optical Mounts" in the *Handbook of Optomechanical Engineering* (CRC Press, Boca Raton, 1997) and coauthored *BASIC-Programme fur die Optik* (Oldenbourg, Munich, 1986). He received his BS and MS degrees in physics from Juniata College and Pennsylvania State University, respectively; is a Fellow of the SPIE, a Fellow of the OSA, a member of Sigma Xi; and is listed in Who's Who in Science and Engineering. Yoder has served the SPIE as a member of the Board of Directors 1982-1984, 1990-1992 and 1994, was Chairman of the Publications Committee and member of the Executive Committee, 1991 and 1994 and has been a member of the Strategic Planning Committee since 1995. In 1996 he received the *Director's Award* from the SPIE and in 1997, the OSA's *Engineering Excellence Award.* He has served as Book Reviews Editor for *Optical Engineering* and Topical Editor for *Applied Optics*. A frequent organizer and chairman as well as active participant in SPIE and OSA symposia, he has taught numerous short courses on optical engineering, precision mounting of optical components, basic optomechanical design, and analysis of the optomechanical interface for the SPIE, industry, and U.S. government agencies. He also has taught two nationally-broadcast courses for the National Technological University Network and has lectured at the University of Arizona and the National University in Taiwan.